Lecture Notes in Electrical Engineering

Volume 579

The book series *Lecture Notes in Electrical Engineering* (LNEE) publishes the latest developments in Electrical Engineering - quickly, informally and in high quality. While original research reported in proceedings and monographs has traditionally formed the core of LNEE, we also encourage authors to submit books devoted to supporting student education and professional training in the various fields and applications areas of electrical engineering. The series cover classical and emerging topics concerning:

- Communication Engineering, Information Theory and Networks
- Electronics Engineering and Microelectronics
- Signal, Image and Speech Processing
- Wireless and Mobile Communication
- Circuits and Systems
- Energy Systems, Power Electronics and Electrical Machines
- Electro-optical Engineering
- Instrumentation Engineering
- Avionics Engineering
- Control Systems
- Internet-of-Things and Cybersecurity
- Biomedical Devices, MEMS and NEMS

For general information about this book series, comments or suggestions, please contact leontina.dicecco@springer.com.

To submit a proposal or request further information, please contact the Publishing Editor in your country:

China

Jasmine Dou, Associate Editor (jasmine.dou@springer.com)

India

Aninda Bose, Senior Editor (aninda.bose@springer.com)

Japan

Takeyuki Yonezawa, Editorial Director (takeyuki.yonezawa@springer.com)

South Korea

Smith (Ahram) Chae, Editor (smith.chae@springer.com)

Southeast Asia

Ramesh Nath Premnath, Editor (ramesh.premnath@springer.com)

USA, Canada:

Michael Luby, Senior Editor (michael.luby@springer.com)

All other Countries:

Leontina Di Cecco, Senior Editor (leontina.dicecco@springer.com)
Christoph Baumann, Executive Editor (christoph.baumann@springer.com)

**** Indexing: The books of this series are submitted to ISI Proceedings, EI-Compendex, SCOPUS, MetaPress, Web of Science and Springerlink ****

More information about this series at http://www.springer.com/series/7818

Luis Fernando D'Haro ·
Rafael E. Banchs · Haizhou Li
Editors

9th International Workshop on Spoken Dialogue System Technology

Springer

Editors
Luis Fernando D'Haro
Universidad Politécnica de Madrid
Madrid, Spain

Rafael E. Banchs
Nanyang Technological University
Singapore, Singapore

Haizhou Li
Department of Electrical and Computer Engineering
National University of Singapore
Singapore, Singapore

ISSN 1876-1100 ISSN 1876-1119 (electronic)
Lecture Notes in Electrical Engineering
ISBN 978-981-13-9445-4 ISBN 978-981-13-9443-0 (eBook)
https://doi.org/10.1007/978-981-13-9443-0

This Springer imprint is published by the registered company Springer Nature Singapore Pte Ltd.
The registered company address is: 152 Beach Road, #21-01/04 Gateway East, Singapore 189721, Singapore

Program Committee

Masahiro Araki, Kyoto Institute of Technology, Japan
Ron Artstein, University of Southern California, USA
Rafael E. Banchs, Nanyang Technological University, Singapore
Timo Baumann, Universität Hamburg, Germany
Jérôme Bellegarda, Apple, Inc., USA
Tim Bickmore, Northeastern University, USA
Jeffrey Bigham, Carnegie Mellon University, USA
Johan Boye, KTH Royal Institute of Technology, Sweden
Axel Buendia, SpirOps AI, France
Susanne Burger, Carnegie Mellon University, USA
Felix Burkhard, Institut für Sprache und Kommunikation, TU, Germany
Frédéric Béchet, Aix Marseille University, France
Zoraida Callejas, University of Granada, Spain
Nick Campbell, Trinity College Dublin, Ireland
Léonardo Campillos, LIMSI-CNRS, France
Luísa Coheur, IST/INESC-ID Lisboa
Marta Ruiz Costajussa, Universitat Politècnica de Catalunya, Spain
Luis Fernando D'Haro, Universidad Politécnica de Madrid, Spain
Justin Dauwels, Nanyang Technological University, Singapore
Yasuharu Den, Chiba University, Japan
Maxine Eskenazi, Carnegie Mellon University, USA
Anna Esposito, Università di Napoli, Italy
Morgan Fredriksson, Liquid Media/Nagoon, Sweden
Kotaro Funakoshi, Honda Research Institute Japan Co., Ltd.
Sadaoki Furui, Tokyo Institute of Technology, Japan
Milica Gasic, University of Cambridge, UK
Kallirroi Georgila, University of Southern California, USA
Emer Gilmartin, Speech Communications Lab, Trinity College Dublin, Ireland
Jonathan Ginzburg, Université Paris Diderot-Paris 7, France
David Griol, Universidad Carlos III de Madrid, Spain
Rainer Gruhn, Nuance Communications, Germany

Joakim Gustafson, KTH Royal Institute of Technology, Sweden
Sunao Hara, Okayama University, Japan
Martin Heckmann, Honda Research Institute Europe GmbH, Germany
Paul Heisterkamp, Daimler AG, Germany
Ryuichiro Higashinaka, NTT Corp, Japan
Julia Hirschberg, Columbia University, USA
Chiori Hori, Mitsubishi Electric Research Laboratories, USA
David House, KTH Royal Institute of Technology, Sweden
Kristiina Jokinen, University of Helsinki, Finland
Tatsuya Kawahara, Kyoto University, Japan
Harksoo Kim, Kangwon National University, Korea
Hong Kook Kim, Gwangju Institute of Science and Technology, Korea
Seokhwan Kim, Adobe Research, USA
Kazunori Komatani, Osaka University, Japan
Girish Kumar, Carousell, Singapore
Nio Lasguido, Rakuten Institute of Technology, Japan
Hung-Yi Lee, National Taiwan University, Taiwan
Kyusong Lee, Carnegie Mellon University, USA
Lin-Shan Lee, National Taiwan University, Taiwan
Fabrice Lefèvre, University of Avignon, LIA-CERI, France
Oliver Lemon, Heriot-Watt University, UK
Haizhou Li, National University of Singapore, Singapore
Pierre Lison, Norsk Regnesentral, Norway
Diane Litman, University of Pittsburgh, USA
José David Lopes, Heriot-Watt University, UK
Ramón López-Cozar Delgado, University of Granada, Spain
Joseph-Jean Mariani, LIMSI-CNRS, France
Yoichi Matsuyama, Carnegie Mellon University, USA
Michael McTear, Ulster University, UK
Etsuo Mizukami, National Institute of Information and Communications Technology (NICT), Japan
Samer Al Moubayed, KTH, Sweden
Satoshi Nakamura, Nara Institute of Science and Technology, Japan
Mikio Nakano, Honda Research Institute, Japan
Andreea Niculescu, Institute for Infocomm Research, Singapore
Takuichi Nishimura, National Institute of Advanced Industrial Science and Technology (AIST), Japan
Elmar Nöth, University of Erlangen-Nuremberg, Germany
Yoo Rhee Oh, Electronics and Telecommunications Research Institute, Korea
Kiyonori Otake, National Institute of Information and Communications Technology (NICT), Japan
Catherine Pelachaud, CNRS—ISIR, Sorbonne Université, France
Volha Petukhova, Saarland University, Germany
Roberto Pieraccini, Jibo Inc., USA
Oliver Pietquin, Google DeepMind, USA

Zahra Rahimi, University of Pittsburgh, USA
Norbert Reithinger, DFKI GmbH, Germany
Jiang Ridong, Institute for Infocomm Research, Singapore
Verena Rieser, Heriot-Watt University, UK
Sophie Rosset, LIMSI, CNRS, Université Paris-Saclay, France
Alexander Rudnicky, Carnegie Mellon University, USA
Sakriani Sakti, NAIST, Japan
Carlos Segura Perales, Telefonica I+D, Spain
Gabriel Skantze, KTH Royal Institute of Technology, Sweden
Svetlana Stoyanchev, Columbia University, USA
Sebastian Stüker, Karlsruhe Institute of Technology, Germany
Mariet Theune, University of Twente, Netherlands
María Inés Torres, Universidad del País Vasco, Spain
David Traum, USC Institute for Creative Technologies, USA
Stefan Ultes, University of Cambridge, UK
Hsin-Min Wang, Academia Sinica, Taiwan
Nigel Ward, University of Texas at El Paso, USA
Jason Williams, Microsoft Research, USA
Koichiro Yoshino, Kyoto University, Japan
Zhou Yu, University of California, Davis, USA
Tiancheng Zhao, Carnegie Mellon University, USA

Preface

The 9th International Workshop on Spoken Dialog Systems (IWSDS'18) was held on April 18–20, 2018, in Singapore; being the southernmost IWSDS ever, just one degree north of the Equator! The conference allowed participants to keep track of the state-of-the-art in spoken dialogue systems, while enjoying the year-round summer paradise island that is Singapore.

The IWSDS conference series brings together, on a yearly basis, international researchers working in the field of spoken dialogue systems and associated technologies. It provides an international forum for the presentation of current research, applications, technological challenges, and discussions among researchers and industrialists. The IWSDS'18 edition built over the success of the previous 8th editions:

- IWSDS'09 (Irsee, Germany),
- IWSDS'10 (Gotemba Kogen Resort, Japan),
- IWSDS'11 (Granada, Spain),
- IWSDS'12 (Paris, France),
- IWSDS'14 (Napa, USA),
- IWSDS'15 (Busan, Korea),
- IWSDS'16 (Saariselkä, Finland), and
- IWSDS'17 (Farmington, PA, USA).

IWSDS'18 conference theme was "Towards creating more human-like conversational agent technologies", inviting and receiving paper submissions on the following topics:

- Engagement and emotion in human–robot interactions.
- Digital resources for interactive applications.
- Multi-modal and machine learning methods.
- Companions, personal assistants, and dialogue systems.
- Proactive and anticipatory interactions.
- Educational and healthcare robot applications.
- Dialogue systems and reasoning.

- Big data and large-scale spoken dialogue systems.
- Multi-lingual dialogue systems.
- Spoken dialog systems for low-resource languages.
- Domain transfer and adaptation techniques for spoken dialog systems.

However, submissions were not limited to these topics, and submission of papers in all areas of spoken dialogue systems was encouraged. In particular, IWSDS'18 welcomed also papers that could be illustrated by a demonstrator, organizing the conference to best accommodate these papers whatever their category.

The program of IWSDS'18 included three keynotes by renowned international authorities in dialogue system research:

- Prof. Tatsuya Kawahara from Kyoto University in Japan,
- Prof. Alex Waibel from Carnegie Mellon University in USA and Karlsruhe Institute of Technology in Germany, and
- Prof. David Traum from University of Southern California in USA.

The keynote speech by Prof. Tatsuya Kawahara was entitled: "Spoken dialogue for a human-like conversational robot ERICA". He described a symbiotic human–robot interaction project, which aims at an autonomous android who behaves and interacts just like a human. This conversational android called ERICA is designed to conduct several social roles focused on spoken dialogue, such as attentive listening (similar to counseling) and job interview. Finally, he described the design principles, problems, and current solutions when developing the different spoken dialogue modules included in ERICA.

The keynote speech by Prof. Alex Waibel was entitled: "M3 Dialogs—Multimodal, Multilingual, Multiparty". He started describing that even though great progress has been made in building and deploying speech dialog systems, they are still rather siloed and limited in scope, domain, style, language, and participants. Most systems are strictly human–machine, one language, one request at a time, usually with a clear on–off signal and identification of who wants what from whom. Even though existing systems do this now rather well, they fall far short of the ease, breadth, and robustness with which humans can communicate. During his talk, Prof. Waibel claimed that a dialog is not only human–machine, but also human–human, human–machine–human, and machine–machine–human, and preferably all of the above in purposeful integration. Then, he outlined the flexibility we are missing in modern dialog systems, review several of efforts aimed at addressing them, and finished speculating on future directions for the research community.

The keynote speech by Prof. David Traum was entitled: "Beyond Dialogue System Dichotomies: Principles for Human-Like Dialogue". He started describing how many researchers have proposed related dichotomies contrasting two different kinds and aims of dialogue systems. One of the issues is whether human–system dialogue should even be human-like at all or humans should adapt themselves to the constraints given by the system. Then, he explored these dichotomies and presented "role-play dialogue" as a place where these dichotomies can find a commonality of purpose and where being human-like is important even simply for effective task performance. After that, he defined "Human-like Dialogue" (HLD) as distinct from purely human dialogue and also distinct from instrumental dialogue. Then, he finished giving some guideline principles on how we should create and evaluate the new generation of agents.

In addition, the IWSDS'18 included three special sessions:

- EMPATHIC: Empathic Dialog Systems for Elderly Assistance,
- HUMIC-DIAL: Designing Humor in HCI with Focus on Dialogue Technology,
- WOCHAT: Workshop on Chatbots and Conversational Agent Technologies.

The EMPATHIC session was organized by Prof. María Inés Torres, Universidad del País Vasco UPV/EHU (Spain), Prof. Kristiina Jokinen, AIRC-AIST (Japan), Prof. Gérard Chollet, Intelligent Voice (UK), and Prof. Marilyn Walker, University of California-Santa Cruz (USA). This session focused on the problem of generating Empathic Dialog Systems for Elderly Assistance. One of the more important applications of spoken dialog systems (SDS) is the development of personal assistants for elderly people. These kinds of systems are intended to provide personalized advice guidance through a spoken dialogue system to improve the quality of life and independency living status of the people as they aged. To this end, SDS has to deal not only with user goals but also implement health goals through negotiation strategies to convince the user to develop healthy habits. Such SDS should also include perceived user affective status to support the dialog manager decisions. This session also welcomed papers focused on affective computing in SDS, user-centered design, policies dealing with shared user-coach goals, management strategies to keep the user engagement, personalization and adaptation, ontologies, and knowledge representation.

The HUMIC-DIAL session was organized by Dr. Andreea I. Niculescu, Institute for Infocomm Research (I2R, Singapore), Dr. Rafael E. Banchs, Nanyang Technological University (Singapore), Dr. Bimlesh Wadhwa, National University of Singapore (NUS, Singapore), Prof. Dr. Anton Nijholt, University of Twente (The Netherlands), and Dr. Alessandro Valitutti, Università di Bari (Italy). After a successful first edition of HUMIC (HUMor in InteraCtion) at INTERACT 2017, for IWSDS'18, the organizers focused on humorous verbal dialogue interactions between humans and machines. Humor embracing various types of expression can be used to enhance the interaction outcome while being socially and culturally appropriate. Therefore, during this session the presented papers explored challenges in designing, implementing, and evaluating humorous interactions in spoken and written dialogues with artificial entities, as well as benefits and downsides of using humor in such interactive tasks.

The WOCHAT session was organized by Dr. Ryuichiro Higashinaka, Nippon Telegraph and Telephone Corporation (Japan), Prof. Ron Artstein, University of Southern California (USA), Prof. Rafael E. Banchs, Nanyang Technological University (Singapore), Prof. Wolfgang Minker, Ulm University (Germany), and Prof. Verena Rieser, Heriot-Watt University (UK). The session included a Shared

Task organized by Prof. Bayan Abu Shawar, Arab Open University (Jordan), Prof. Luis Fernando D'Haro, Universidad Politécnica de Madrid, Spain, and Prof. Zhou Yu, University of California, Davis (USA). This was the fifth event of a "Workshop and Special Session Series on Chatbots and Conversational Agents". WOCHAT aims at bringing together researchers working on problems related to chat-oriented dialogue with the objective of promoting discussion and knowledge sharing about the state-of-the-art and approaches in this field, as well as coordinating a collaborative effort to collect/generate data, resources, and evaluation protocols for future research in this area. The WOCHAT series also accommodated a Shared Task on Data Collection and Annotation for generating resources that can be made publicly available to the rest of the research community for further research and experimentation. In this shared task, human–machine dialogues are generated by using different online and offline chat engines, and annotations are generated following some basic provided guidelines.

IWSDS'18 received a total of 52 submissions, where each submission was reviewed by at least two program committee members. The committee decided to accept a total of 37 papers: 13 long papers, 6 short papers, 4 demo papers, 4 papers for the Empathic session, 7 papers for the WOCHAT session, 2 papers for the Humic session, and 1 invited paper.

Finally, we would like to take this opportunity to thank the IWSDS Steering Committee and the members of the IWSDS'18 Scientific Committee for their timely and efficient contributions and for completing the review process on time. In addition, we would like to express our gratitude to the members of the Local Committee who highly contributed to the success of the workshop, making it an unforgettable experience for all participants. Last, but not least, we want also to thank our sponsors: the Special Group on Discourse and Dialogue (SIGDial) and Chinese and Oriental Languages Information Processing Society (COLIPS) for their economical and logistic support; without it we and participants could not have such a remarkable conference.

With our highest appreciation,

Madrid, Spain	Luis Fernando D'Haro
Singapore	Rafael E. Banchs
Singapore	Haizhou Li
April 2019	

Contents

Part I
Language and Social Context Understanding

Attention Based Joint Model with Negative Sampling for New Slot Values Recognition

Mulan Hou, Xiaojie Wang, Caixia Yuan, Guohua Yang, Shuo Hu and Yuanyuan Shi

Abstract Natural Language Understanding (NLU) is an important component of a task oriented dialogue system, which obtains slot values in user utterances. NLU module is often required to return standard slot values and recognize new slot values at the same time in many real world dialogue such as restaurant booking. Neither previous sequence labeling models nor classifiers can satisfy both requirements by themselves. To address the problem, the paper proposes an attention based joint model with negative sampling. It combines a sequence tagger with a classifier by an attention mechanism. The tagger helps in identifying slot values in raw texts and the classifier simultaneously maps them into standard slot values or the symbol of new values. Negative sampling is used for constructing negative samples of existing values to train the model. Experimental results on two datasets show that our model outperforms the previous methods. The negative samples contribute to new slot values identification, and the attention mechanism discovers important information and boosts the performance.

M. Hou (✉) · X. Wang · C. Yuan · G. Yang
Center of Intelligence Science and Technology, Bejing University of Posts and Telecommunications, Bejing, China
e-mail: houmulan@bupt.edu.cn

X. Wang
e-mail: xjwang@bupt.edu.cn

C. Yuan
e-mail: yuancx@bupt.edu.cn

G. Yang
e-mail: yangguohua@bupt.edu.cn

S. Hu · Y. Shi
Beijing Samsung Telecom R&D Center, Bejing, China
e-mail: shuo.hu@samsung.com

Y. Shi
e-mail: yy.shi@samsung.com

L. F. D'Haro et al. (eds.), *9th International Workshop on Spoken Dialogue System Technology*, Lecture Notes in Electrical Engineering 579,
https://doi.org/10.1007/978-981-13-9443-0_1

1 Introduction

Task oriented dialogue system, which has been widely used in a variety of different applications, is designed to accomplish a specific task through natural language interactions. One of its most important components is Natural Language Understanding (NLU). NLU aims at collecting information related to the task.

Semantic frames are commonly applied in NLU [11], each of which contains different slots. One of the goals of NLU is to fill in the slots with values extracted from the user utterances. In previous work, sequence labeling models are usually used for slot values recognition. For example, Tur et al. [10] used Conditional Random Field (CRF) with domain-specific features for the task. With the success of deep neural networks, Yao et al. [14] proposed a RNN model with Named Entities (NER) as features. They also used Long Short-Term Memory (LSTM) [13] and some other deeper models. Ma et al. [4] combined Convolutional Neural Network (CNN), LSTM and CRF in a hierarchical way, where features extracted by a CNN are fed to a LSTM, a CRF in top level is used to label slot values.

Nevertheless, only the labeling of the slot values is not enough in some applications. The slot values labeled in utterances should be normalized to some standard values for database search. For example, in a restaurant booking system, there are standard values of slot 'food' like 'Asian oriented'. If a user wondered a restaurant which serves 'pan Asian' food, the system should normalize the 'pan Asian' in utterance into the standard value of 'Asian oriented' in database. There were two different ways for addressing this problem. One is two-stage methods. Lefvévre [3] proposed a 2+1 model. It used a generative model consisted of two parts, namely semantic prior model and lexicalization model, to determine the best semantic structure and then treated the normalized slot values as hidden variables to figure it out. Yeh [15] employed fuzzy matching in Apache Solr system for the normalization. Two-stage methods are either prone to accumulating errors or too complicated to compute. The other way is directly mapping an utterance to one of the standard values instead of identifying the values in raw texts. A lot of classifiers were used for building the mappings. Bhagat et al. [1] tried several different models including Vote model, Maximum Entropy, Support Vector Machine (SVM). Mairesse et al. [5] proposed a two-step method: a binary classifiers was first used to determine if a slot appears in the utterance or not, and then a series classifiers were used to map the utterance to standard values of that slot. Mota et al. [7] built different classifiers for different slots respectively.

There is an important problem in above classification based methods however. These models failed in dealing with the situation where a slot value out of the standard value set is mentioned in an utterance. This value should not be classified into any existing standard values and should be recognized as a new value. To our knowledge, there is no research on this problem in classification based NLU.

The problem might be thought as one type of zero-shot problems in word sense or text classification and others. But there is a significant difference between new slot values and other zero-shot problems. The sense of a new word might be very different from that of other known words. But a new slot value is still a value of the same slot. It should share some important similarities with other known slot values. That is the starting point for us to construct training samples for unknown new values. We first distinguish two different types of samples of the standard values of a specific slot S. Utterances including any known standard value or its variants of the slot S are positive samples, and the others are negative ones. We further divide the negative samples into two types, the first is negative samples of S, i.e. samples including values of other slots or including no value of any slot, and the second is negative samples of any known standard values of S. The latter is therefore can be used to build a classifier (together with positive samples of the standard values of S) for identifying if an utterance includes a known standard value or a new value of S. The paper proposes a negative sampling based method to construct samples of the latter.

Meanwhile, sequence labeling is able to locate slot values in original utterances even if they are unseen in standard value set. The slot values themselves are also important information for classification. The paper proposes a joint model of sequence labeling and classification by attention mechanism, which focuses on important information automatically and takes advantage of the raw texts at the same time. Sequence labeling here aims at slot-value detection and classification is used to obtain the standard values directly.

Overall, we propose an attention based joint model with negative sampling. Our contributions in this work are two-fold: (1) negative sampling for existing values for a certain slot S enables our model to effectively recognize new slot values; (2) joint model collaborated by attention mechanism promotes the performance. We evaluate our work on a public dataset DSTC and a dataset Service from an enterprise. All the results demonstrate that our model achieves impressive improvements on new slot values with less damage on other sub-datasets. The F1 score evaluated on new slot values raises up to 0.8621 in DSTC and 0.7759 in Service respectively.

This paper is organized as follows: Sect. 2 details on our attention based joint model with negative sampling. We explain experiment settings in Sect. 3, then evaluate and analyse our model in Sect. 4. In Sect. 5 we will conclude our work.

2 Attention Based Joint Model with Negative Sampling (AJM_NS)

We assume that slots are independent of each other so they can be handled separately. A vocabulary of values for slot S is defined as $R^S = \{S_{old}\} \bigcup \{NEW, NULL\}$, where $S_{old} = \{s_0, s_1, ...s_k\}$ refers to the set of standard values for which there is some labeled data in training set. NEW refers to a new slot value. It will be assigned to an utterance providing a new slot value for slot S which is outside S_{old}, and $NULL$ refers to no value in an utterance. For a user input x_i, the aim of the model is to map the x_i into one of values in R^S. Since there is no training data for a new slot value (if we have some training samples for a value, it belongs to S_{old}), classification based models on the dataset are unable to address the problem, while sequence taggers need another step to normalize the labels.

We describe our attention based joint model, followed by the negative sampling methods.

2.1 Attention Based Joint Model

A sequence tagger and a classifier complement each other. A sequence tagger recognizes units of a slot value in an utterance, while a classifier map an utterance as a whole into a slot value. In order to benefit from both of them, we combine them into a joint model.

Specifically, we adopt the bi-directional LSTM [2] as a basic structure. The output of each timestep is used to output a slot tag by a softmax operation on a linear layer as shown in Eq. 1:

$$\hat{s}_t = \text{softmax}(\mathbf{W}_s \boldsymbol{h}_t + \boldsymbol{b}_s) \tag{1}$$

$\boldsymbol{h}_t = (\overrightarrow{\boldsymbol{h}_t}, \overleftarrow{\boldsymbol{h}_t})$ refers to the hidden state of time t by concatenating the hidden state in forward and backward direction. In each direction of LSTM, like in forward LSTM, hidden state $\overrightarrow{\boldsymbol{h}_t}$ is a function of the current input and the inner memory, as defined in Eq. 2

$$\overrightarrow{\boldsymbol{h}_t} = f(\overrightarrow{\boldsymbol{h}_{t-1}}, w_t, \overrightarrow{\boldsymbol{C}_{t-1}}) \tag{2}$$

where w_t denotes the input word at time t and $\overrightarrow{\boldsymbol{C}_{t-1}}$ is the previous cell state. We compute function f using the LSTM cell architecture in [16]. So as on backward direction.

The hidden state of the last timestep T is used to output the class label according to Eq. 3:

$$\hat{y} = \text{softmax}(\mathbf{W}_c \boldsymbol{h}_T + \boldsymbol{b}_c) \tag{3}$$

We further combine them by attention mechanism [13]. Figure 1 illustrates the procedure (Fig. 1).

Upon attention mechanism, the model automatically focuses on locations of important information and constructs a context vector $\boldsymbol{H}$ which is defined in Eq. 4.

$$\boldsymbol{H} = \sum_{t}^{T} \alpha_t \boldsymbol{v}_t \tag{4}$$

where $\boldsymbol{v}_t = (\boldsymbol{e}_t, \boldsymbol{h}_t)$ concatenates word embeddings and hidden states of LSTM and α_t is defined in Eq. 5.

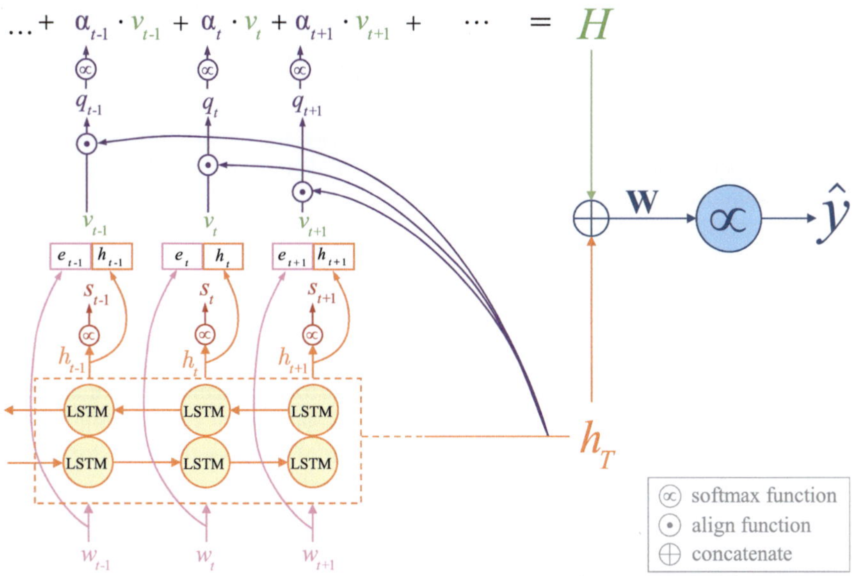

Fig. 1 In this figure, attention based joint model combines sequence tagging and classifying and adopts attention mechanism for further improvements. Legend in the right corner shows the meaning of operations

$$\alpha_t = \frac{\exp(q_t)}{\sum_k^T \exp(q_k)} \tag{5}$$

Our model computes q_t by an align function in Eq. 6 which is the same way as [9]:

$$q_t = (\tanh(\mathbf{W}\boldsymbol{v}_t))^\top \boldsymbol{h_T} \tag{6}$$

It is regarded as a similarity score of the utterance representation $\boldsymbol{h_T}$ and the information $\boldsymbol{v_t}$ of each timestep.

Finally we concatenate context vector H and the sentence embedding $\boldsymbol{h_T}$, and feed it into a softmax layer as shown in Eq. 7 to predict the class label of standard slot values.

$$\hat{\boldsymbol{y}} = \text{softmax}(\mathbf{W}(\boldsymbol{H}, \boldsymbol{h_T}) + \boldsymbol{b}) \tag{7}$$

All parameters are learned simultaneously to minimize a joint loss function shown in Eq. 8, i.e. the weighted sum of two losses for sequence tagging and classification respectively.

$$L = \gamma L_{tagging} + (1 - \gamma) L_{classification} \tag{8}$$

$$L_{tagging} = \frac{1}{N}\sum_i^N \frac{1}{T_i}\sum_t^{T_i} L(\hat{\boldsymbol{s}}_t^i, \boldsymbol{s}_t^i) \tag{9}$$

$$L_{classification} = \frac{1}{N}\sum_{i}^{N} L(\hat{y}_i, y_i) \tag{10}$$

γ is a hyperparameter to balance the sequence tagging and classifying module. N in Eq. 9 refers to the size of training data and T_i is the length of the i-th input. $L(\cdot)$ is cross-entropy loss function.

2.2 *Negative Sampling*

Model fails in recognizing new slot values without training data for them as mentioned before. If we regard all the samples for new slot values of a specific slot as the negative samples of existing ones, construction of samples for new slot values can then convert to the construction of negative samples of old ones.

As mentioned in Sect. 1, a new slot value is still a value of the same slot. It should share some important similarities with other known slot values. Here we think the similarities are hidden in contexts of the value, i.e. the contexts are shared among different values of a same slot. It is therefore a possible way to construct a negative sample by just replacing the slot values in a positive sample with a non-value. But there are so many choices for non-value, how to choose a proper one?

Mikolov et al. [6] have already used negative sampling in CBOW and Skip-gram models. They investigated a number of choices for distribution of negative samples and found that the unigram distribution $U(word)$ raised to the 3/4rd power (i.e., $U(word)^{3/4}/Z$) outperformed significantly the unigram and the uniform distributions. Z is the normalization constant and $U(word)$ is the word frequency in another word, which is calculated by $U(word) = count(word)/\,|Data|$. We use the same method but leave the word frequency alone. In our work a negative sample is a complete slot value that sometimes consists of several words, different from the negative samples of a single word in [6]. That results in repeating sampling until a segment of the same length as the existing value is formed. Figure 2 shows the construction of a negative example for Service dataset.

	我	的	手	机	可	以	边	打	电	话	边	视	频	吗	?
	O	O	O	O	O	O	B-func	I-func	I-func	I-func	I-func	I-func	I-func	O	O
negative sampling:	手	速	录	支	无	用	间								
	我	的	手	机	可	以	手	速	录	支	无	用	间	吗	?
	O	O	O	O	O	O	B-func	I-func	I-func	I-func	I-func	I-func	I-func	O	O

	Can	I	have	a	video	chat	on	my	phone	?
	O	O	B-func	I-func	I-func	I-func	O	O	O	O
negative sampling:	**You**	**use**	**battery**	**let**						
	Can	I	**You**	**use**	**battery**	**let**	on	my	phone	?
	O	O	B-func	I-func	I-func	I-func	O	O	O	O

Fig. 2 Negative sampling for service dataset. Lower part is a translation of the example

3 Experiments Setting

3.1 Dataset

We evaluate our model on two dataset: Dialogue State Tracking Challenge (DSTC) and a dataset from an after-sale service dialogue system of an enterprise (Service).

DSTC is an English dataset from a public contest [12] and we use DSTC2 and DSTC3 together. It collects 5510 dialogues about hotels and restaurants booking. Each of the utterance in dialogues gives the standard slot values, according to which slot tags can be assigned to word sequence. Based on the independency assumption, we build datasets for each slot: keep all B- or I- tags of the slot labels and reset the rest to 'O'. However we find out that not all slots are suitable for the task, since there are too few value types of the slot. At last we choose the dataset for slot 'food' only in our experiments.

Service is a Chinese dialogue dataset which is mainly about consultation for cell phones and contains a single slot named 'function'. It has both sequence tags and slot values on each utterance.

We divide two datasets into training, dev and test set respectively, and then construct some negative samples into training set for both of them. All of the utterances corresponding to infrequent slot values in training set are put into test set to form corpus of new slot values. These values thus have no samples in training data. Table 1 shows the statistics of the final experimental data and Table 2 tells about the diversity of slot values.

Table 1 Statistics of two dataset

Corpus		DSTC			Service		
		Train	Dev	Test	Train	Dev	Test
Original data	Old	2805	937	917	3682	514	1063
	New	0	113	275	0	15	64
	Null	2244	840	953	427	64	109
Negative samples		561	0	0	736	0	0
Overall size		5610	1890	2145	4845	593	1236

Table 2 Value types

Corpus	DSTC			Service		
	Train	Dev	Test	Train	Dev	Test
Old	66	64	65	80	55	67
New	0	21	21	0	13	44

3.2 Evaluation Measurements

We take weighted $F1$ score as the evaluation criterion in our experiments. It is defined as in Eqs. 11 and 12.

$$F1 = \sum_{i}^{N} \omega_i F1_{s_i} \tag{11}$$

with

$$\omega_i = \frac{n_{s_i}}{n}, \quad F1_{s_i} = 2\frac{P_{s_i} \times R_{s_i}}{P_{s_i} + R_{s_i}} \tag{12}$$

where n refers to the size of the test set and n_{s_i} denotes the size of class s_i. P and R is precision score and recall score defined in [8].

We also evaluate on the sub-dataset of old values by Eq. 13.

$$F1_{old} = \sum_{i=0}^{k} \omega_i^{old} F1_{s_i} \tag{13}$$

where $\omega_i^{old} = \frac{n_{s_i}}{n_{old}}$.

For sequence tagging we still consider F1 score as criterion which can be calculated by running the official script conlleval.pl[1] of CoNLL conference.

3.3 Baseline

There are no previous models and experimental results reported especially on new slot values recognition. We compare our model to existing two types of NLU methods for the task.

(1) The pipeline method:labeling the words with slot value tags first and then normalizing them into standard values. Here, a bi-directional LSTM as same as that in our model is used for tagging, and the fuzzy matching[2] is then used to normalize extracted tags like that in [15]. The model is denoted by LSTM_FM.

(2) The classification: classifying the utterance to standard values directly. A bi-directional LSTM is used to encode user input, a full-connected layer is then used for the classification. The model is denoted by LSTM_C.

---

[1] https://www.clips.uantwerpen.be/conll2000/chunking/output.html.

[2] http://chairnerd.seatgeek.com/fuzzywuzzy-fuzzy-string-matching-in-python.

3.4 Hyperparameters

We adopt bi-directional LSTM as the basic structure. Hyperparameter γ is 0.1. The longest input is 30, size of LSTM cell is 64, and dimension of word embedding is 100. We use minibatch stochastic gradient descent algorithm with Adam to update parameters. Learning rate is initialized as 0.005. Each batch keeps 512 pieces of training data. We choose the model performs best in dev set as the test one.

4 Result and Analyses

4.1 Comparisons Among Different Models

We evaluate our model on two dataset described in Sect. 3.1. Our model can output both classification results of a utterance and the labeled tags in a utterance. Tables 3 and 4 shows the results of classification and labeling respectively.

As we can see in Table 3, our model outperforms both baseline models significantly in classification task. It achieves 13.44 and 16.17% improvements compared to LSTM_FM and LSTM_C model on DSTC dataset, and achieves 8.55 and 5.85% improvements on Service dataset. Especially, it shows big advantage on new slot values recognition, where the $F1$ scores achieve at least 20% raises on both DSTC and Service data.

Similar to the performance in the classification, our model also achieves best results in slot value tagging as we can see in Table 4. It performs significant better than the pipeline method, especially for the new value. We also give the tagging results of LSTM_FM trained by adding negative samples used in our model (denoted by LSTM_FM_NS in Table 4). We find negative samples are helpful to NEW slot values significantly, but they hurt the performance of old values. We give more details of negative samples and attention mechanism in our model and baseline models in next subsection.

Table 3 Classification results of different models

	DSTC				Service			
	All	NEW	S_{old}	NULL	All	NEW	S_{old}	NULL
LSTM_FM	0.8491	0.3063	0.9670	0.8923	0.8981	0.5693	0.9320	0.5693
LSTM_C	0.8290	0.0000	0.9249	0.9761	0.9210	0.0000	0.9720	**0.9643**
AJM_NS (ours)	**0.9632**	**0.8621**	**0.9738**	**0.9822**	**0.9749**	**0.7759**	**0.9881**	0.9633

Table 4 Tagging results of different models

	DSTC			Service		
	All	NEW	S_{old}	All	NEW	S_{old}
LSTM_FM	0.8546	0.2363	0.9837	0.8850	0.2615	0.9269
LSTM_FM_NS	0.8289	0.2844	0.9709	0.8386	**0.4853**	0.8655
AJM_NS (ours)	**0.9024**	**0.5684**	**0.9946**	**0.9132**	0.3399	**0.9573**

Table 5 Comparison inside the model with F1 scores for classification

	DSTC				Service			
	All	NEW	S_{old}	NULL	All	NEW	S_{old}	NULL
Full (AJM_NS)	**0.9632**	**0.8621**	0.9738	**0.9822**	**0.9749**	**0.7759**	**0.9881**	**0.9633**
Attention only (JM_NS)	0.9515	0.8129	**0.9739**	0.9699	0.9700	0.7207	0.9862	0.9585
NS only (AJM)	0.8247	0.0000	0.9426	0.9492	0.9234	0.0000	0.9761	0.9511

Table 6 Confusion matrix of DSTC

DSTC

	NEW	S_{old}	NULL			NEW	S_{old}	NULL
NEW	0	184	91		NEW	225	33	17
S_{old}	0	916	1	$\Rightarrow$	S_{old}	9	908	0
NULL	0	9	944		NULL	13	4	936

4.2 Analyses

In order to analyze our model, we compare it to the model dropping out attention mechanism only and the other dropping negative samples only. We refer to the former model as JM_NS and the latter as AJM.

From Table 5 we can find out that the one dropping out negative samples (AJM) failed in dealing with new slot values recognition. It shows that the negative sampling is the key for the success in new slot values recognition. The negative samples actually enables the model to distinguish old and new slot values. For more details, the changes of confusion matrices are shown in Tables 6 and 7. The left part of '$\Rightarrow$' in the table is the confusion matrix of the model without negative samples(AJM), and the right part is from the original full model (AJM_NS). With the training of negative samples, classification results related to NEW value change better significantly, while change little on other classes, i.e. negative samples bring less damage to other classes.

We also add same negative samples for training other models. The result in Table 8 shows that LSTM_C_NS(LSTM_C with negative samples) now achieve good performance of recognizing new slot values. As for LSTM_FM_NS, the $F1$ score drops a lot for old values while for new slot values it raises up on the contrary. It shows that, although negative samples still work, they damage other classes significantly

Table 7 Confusion matrix of Service

Service								
	NEW	S_{old}	NULL			NEW	S_{old}	NULL
NEW	0	55	9		NEW	45	17	2
S_{old}	0	1063	0	⇒	S_{old}	4	1057	2
NULL	0	2	107		NULL	3	1	105

Table 8 Classification results based on negative samples

	DSTC				Service			
	All	NEW	S_{old}	NULL	All	NEW	S_{old}	NULL
LSTM_FM_NS	0.8572	0.3536	0.9286	0.9241	0.8642	0.6203	0.9009	0.6488
LSTM_C_NS	0.9543	0.8261	0.9637	**0.9822**	0.9684	0.7103	0.9825	**0.9815**
JM_NS	0.9515	0.8129	**0.9739**	0.9699	0.9700	0.7207	0.9862	0.9585
AJM_NS	**0.9632**	**0.8621**	0.9738	**0.9822**	**0.9749**	**0.7759**	**0.9881**	0.9633

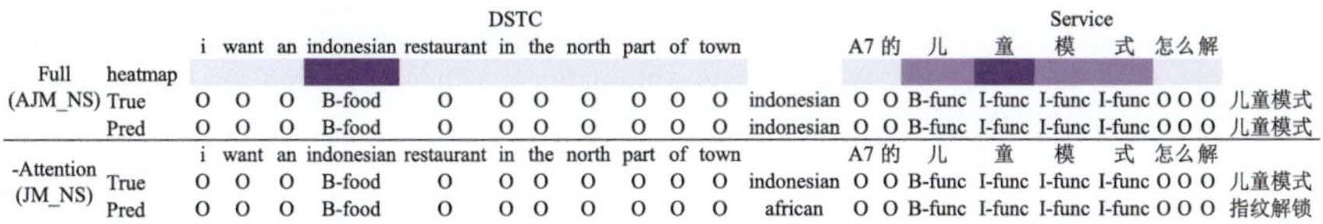

Fig. 3 Comparison between the full model (AJM_NS) and the one dropping out attention mechanism (JM_NS). The heatmap in full model is the visualization of weights for different words. The deeper color means a larger weight

in pipeline model. We can also find out that our model AJM_NS still beats the rest models on the whole dataset even if all of them use negative samples.

When we abandon attention mechanism (JM_NS), the model is slightly inferior to the full one (AJM_NS), i.e. the attention mechanism can further improve the performance by focusing on the important subsequences. Since it introduces the original word embeddings at the same time, it corrects some mistakes in the model dropping out attention mechanism (JM_NS) in which the final label is wrongly classified even with correct sequence tags. We visualize a sample of attention in Fig. 3.

5 Conclusion

In lots of industrial or commercial applications, it is necessary for a NLU module to not only fill the slot with predefined standard values but also recognize new slot values due to the diversity of users linguistic habits and business update.

The paper proposes an attention based joint model with negative sampling to satisfy the requirement. The model combines a sequence tagger with a classifier by an attention mechanism. Negative sampling is used for constructing negative samples for training the model. Experimental results on two datasets show that our model outperforms the previous methods. The negative samples contributes to new slot values identification, and the attention mechanism improves the performance.

We may try different methods of negative sampling to further improve the performance in following works, such as introducing prior knowledge. At the same time, scenario of multiple slot in an utterance will also be explored as it happens a lot in daily life.

References

1. Bhagat R, Leuski A, Hovy E (2005) Statistical shallow semantic parsing despite little training data. In: Proceedings of the Ninth international workshop on parsing technology, pp 186–187. Association for Computational Linguistics
2. Graves A, Jaitly N, Mohamed AR (2013) Hybrid speech recognition with deep bidirectional LSTM. In: 2013 IEEE Workshop on Automatic Speech Recognition and Understanding (ASRU), pp 273–278. IEEE
3. Lefévre, F (2007) Dynamic Bayesian networks and discriminative classifiers for multi-stage semantic interpretation. In: IEEE International Conference on Acoustics, Speech and Signal Processing, ICASSP 2007, vol 4, pp IV–13. IEEE
4. Ma X, Hovy E (2016) End-to-end sequence labeling via bi-directional LSTM-CNNS-CRF. arXiv:1603.01354
5. Mairesse F, Gasic M, Jurcícek F, Keizer S, Thomson B, Yu K, Young S (2009) Spoken language understanding from unaligned data using discriminative classification models. In: IEEE International Conference on Acoustics, Speech and Signal Processing, 2009. ICASSP 2009, pp 4749–4752. IEEE
6. Mikolov T, Sutskever I, Chen K, Corrado GS, Dean J (2013) Distributed representations of words and phrases and their compositionality. Advanc Neural Informat Process Syst 3111–3119
7. Pedro Mota Luísa Coheur AM (2012) Natural language understanding as a classification process: report of initial experiments and results. In: INForum
8. Perry JW, Kent A, Berry MM (1955) Machine literature searching x. machine language; factors underlying its design and development. J Associat Informat Sci Technol 6(4):242–254
9. Seo M, Kembhavi A, Farhadi A, Hajishirzi H (2016) Bidirectional attention flow for machine comprehension. arXiv:1611.01603
10. Tur G, Hakkani-Tür D, Heck L, Parthasarathy S (2011) Sentence simplification for spoken language understanding. In: 2011 IEEE International Conference on Acoustics, Speech and Signal Processing (ICASSP), pp 5628–5631. IEEE
11. Wang YY, Deng L, Acero A (2011) Semantic frame-based spoken language understanding. Spoken language understanding: systems for extracting semantic information from speech, pp 41–91
12. Williams J, Raux A, Ramachandran D, Black A (2013) The dialog state tracking challenge. In: Proceedings of the SIGDIAL 2013 conference, pp 404–413
13. Yao K, Peng B, Zhang Y, Yu D, Zweig G, Shi Y (2014) Spoken language understanding using long short-term memory neural networks. In: 2014 IEEE Spoken Language Technology Workshop (SLT), pp 189–194. IEEE
14. Yao K, Zweig G, Hwang MY, Shi Y, Yu D (2013) Recurrent neural networks for language understanding. In: Interspeech, pp 2524–2528

15. Yeh PZ, Douglas B, Jarrold W, Ratnaparkhi A, Ramachandran D, Patel-Schneider PF, Laverty S, Tikku N, Brown S, Mendel J (2014) A speech-driven second screen application for tv program discovery. In: AAAI, pp 3010–3016
16. Zaremba W, Sutskever I, Vinyals O (2014) Recurrent neural network regularization. arXiv:1409.2329

Dialogue Act Classification in Reference Interview Using Convolutional Neural Network with Byte Pair Encoding

Seiya Kawano, Koichiro Yoshino, Yu Suzuki and Satoshi Nakamura

Abstract Dialogue act classification is an important component of dialogue management, which captures the user's intention and chooses the appropriate response action. In this paper, we focus on the dialogue act classification in reference interviews to model the behaviors of librarians in the information seeking dialogues. Reference interviews sometimes include rare words and phrases. Therefore, the existing approaches that use words as units of input often do not work well here. We used the byte pair encoding compression algorithm to build a new vocabulary for the inputs of the classifier. By using this new unit as a feature of the convolutional neural network-based classifier, we improved the accuracy of the dialogue act classification while suppressing the size of vocabulary.

1 Introduction

Requests from users for information retrieval systems are often ambiguous, so this property makes it difficult to provide the exact information related to the real demand of a user in the information seeking process [5, 16]. It is known that clarifications such as "confirmation" or "asking for background information" help to find the requested information. A reference interview, a chat-style information-seeking dialogue at the reference service in a library, is an example of information seeking dialogue with these clarification action. Conducting this kind of reference interview in advance improves the accuracy of information provision in the reference service [6, 7]. We focused on reference interview for modeling the librarians' behavior to create a system that can provide information through interactions, even if the intention of the user at the first utterance is ambiguous.

To model the response strategy, we focused on the task of dialogue act classification in the reference interview by using Inoue's dialogue act tag set [3]. We

S. Kawano (✉) · K. Yoshino · Y. Suzuki · S. Nakamura
Graduate School of Information Science, Nara Institute of Science and Technology,
Takayama-cho, Ikoma, Nara 6300192, Japan
e-mail: kawano.seiya.kj0@is.naist.jp

L. F. D'Haro et al. (eds.), *9th International Workshop on Spoken Dialogue System Technology*, Lecture Notes in Electrical Engineering 579,
https://doi.org/10.1007/978-981-13-9443-0_2

constructed classifiers with convolutional neural network (CNN), known as the state-of-the-art classifier using statistical methods [11]. To model the classifier with neural networks, we need enough training data with labels, which is difficult to obtain because the number of labeled dialogue data is limited. A trained model sometimes does not work well due to the lack of training data for rare and unusual words.

Subword approach is known as for reducing this problem [4]. On the other hand, information-seeking dialogues include phrases such as “May I help you?” and “Hold on, please”. Using word units for this kind of expressions wastes the feature space and decrease the accuracy of classification. We implemented byte pair encoding (BPE) compression algorithm for effective use of the CNN feature space. Our investigation confirmed that the BPE-based features improve the accuracy of the dialogue act classification while suppressing the size of the vocabulary to be used.

2 Reference Interview in Libraries

Librarians in libraries provide documents that may contain answers to the user’s questions. They try to clarify the information requirements of the user through a reference interview by asking about the subject, background, purpose, and motivation [8]. This kind of dialogue attracts high attention in the field of dialogue system research and is known as “information navigation” [16].

2.1 Corpus

We use chat logs of a virtual reference service QuestionPoint as the corpus [6, 7]. This corpus consists of 700 sessions of 12,634 utterances. Meta-data labels of the participants, dates, and times are given for each utterance. Personal information (user’s name, email address, etc.) are anonymized.

Table 1 shows an example dialogue of the virtual reference interview. In this example, the librarian clarifies users background information (examination at school the next day, the grade of the user, search histories, etc.).

2.2 Dialogue Act in Reference Interview

In dialog systems, it is impractical to define comprehensive behaviors of the system by rules. Recent works tackle this problem with data-driven approaches, which learn behaviors of the system from dialogue corpora with statistical methods such as reinforcement learning [15, 17]. However, a data-driven approach requires very large-scale datasets [16]. Using dialogue act is known to avoid this problem.

Table 1 Example of virtual reference interview

ID	Utterance
P1	Here is a current in a metal wire due to the motion of electrons. sketch a possible path for the motion of a single electron in this wire, the direction of the electric field vector, and the direction of conventional current
P2	You can just describe what they would look like
L3	Just a moment, please....
P4	Thanks
L5	Is this for a school assignment and if so what is your grade level?
P6	Im a junior in high school... its for a physics class... i have a test tomorrow and this stuff and Im still shakey on it
L7	What part of your physics books this question comes from: electricity?
P8	Ya
L9	Let me check
L10	Hold on please
P11	Ok
L12	I am still checking
L13	Hold on please
L14	http://www.swansontec.com/set.htm
L15	The source that I just sent has good graphics that shows the electic currents
L16	And the graphic is animated so you can see the movement
L17	Can you see the page?
P18	Yes
L19	Let me check for more hold on please

Dialogue acts are defined as tags that indicate the intention of each utterance in dialogues [13].

In the dialogue acts of reference interviews defined by Inoue [3], librarians and users have two dialogue act categories to process the interview: (1) information transfer to request or provide information and (2) task management to assign or commit to tasks.

They also have two other dialogue act categories for smooth communication: (3) social relationship management to manage socio-emotional aspects of communication, and (4) communication management to manage physical aspects of communication. These four fundamental categories of dialogue acts are called dialogue act functions (DAF). They have detailed tags to model the behavior of participants, which is called dialogue act domain (DAD). The detail of the dialogue act definition are shown in Table 2.

Table 2 Dialogue act tags in reference interview

Dialogue act function (5 classes)	Dialogue act domain (19 classes)
Information transfer – Infomration provision – Infomration request	Information problem Search process Information object Feedback Other
Task management	Librarian's task User's task Other
Social relationship management	Greeting Valediction Exclamation Apology Gratitude Downplay Closing ritual Rapport building
Communication management	Channel checking Pausing Feedback

3 Dialogue Act Classification for Reference Interview

A reference interview is an open-domain task, so the dialogues often contain out of vocabulary (OOV) words and low-frequency words. It is very difficult to train a good statistical model to classify utterances into dialogue act classes if there are many OOVs. Furthermore, the feature space will be wasted on some typical and frequent expressions if we use conventional word-based features. Such expressions can be compressed into one dimension of the feature vector. Therefore, we trained a domain-dependent tokenizer based on BPE, which is optimized with entropy, from the reference interview corpus to make efficient inputs for the dialogue act classifier.

3.1 Byte Pair Encoding

Byte pair encoding (BPE) is a simple form of data compression that recursively concatenates frequent consecutive symbols into one symbol to reduce the entropy [2, 12]. Symbols (vocabularies) defined in BPE for texts start from a set of characters. Thus, it can reduce the number of low-frequency words that often are OOVs in the test-set. On the other hand, BPE can create a long symbol if the set of characters is frequent. Although BPE was originally proposed in the field of data compression, Sennrich et al. [9] applied BPE to create a vocabulary for neural machine translation in order to reduce the number of OOVs. They also reported that reducing of the number of OOVs improved the bilingual evaluation understudy (BLEU) score of machine translation.

In this paper, BPE is regarded as a domain-dependent feature extractor and trained as a tokenizer to create a new unit. We need to give a size of vocabulary before the BPE training, which is decided by the number of initial symbols (=characters) and the number of merge operations of BPE. While Sennrich et al. [9] considered subwords in each word, we considered spaces as one token and trained the tokenizer to extract not only subwords but also set phrases. We used SentencePiece[1] as the implementation of byte pair encoding.

3.2 Dialogue Act Classification Using Convolutional Neural Network

We used convolutional neural networks (CNN) [10] for dialog act classification. We tokenized each utterance into sequences of BPE units and made a matrix for each utterance as shown in Fig. 1. Each unit was converted into a fixed length embedding vector. These vectors were placed as columns according to the sequence of the original units. We used 0-padding because the number of columns was set as the maximum number for units of one utterance. Our CNN consists of one convolution and global max-pooling layer, four hidden layers and a softmax output layer. Batch normalization was set for each layer, and RMSProp was used as the optimizer. The initial value of the word embedding vector was set randomly.

4 Experimental Evaluation

4.1 Experimental Settings

In the Online Computer Library Center (OCLC) virtual reference interview dataset, 200 sessions are annotated with dialogue acts (200 labeled sessions) by Inoue [3] and the other 500 sessions do not have any annotations of dialogue acts (500 unlabeled sessions). In the dialogue act classification, we use 200 labeled session of 5,327 utterances with 10-fold cross-validation. The remaining 500 session of 7,307 utterances are used to learn the tokenizer and embed of the trained BPE units.

We examined the CNN-based classifier with the BPE units under several vocabulary settings. We prepared CNN based classifiers with word units and character

[1]http://github.com/google/sentencepiece.

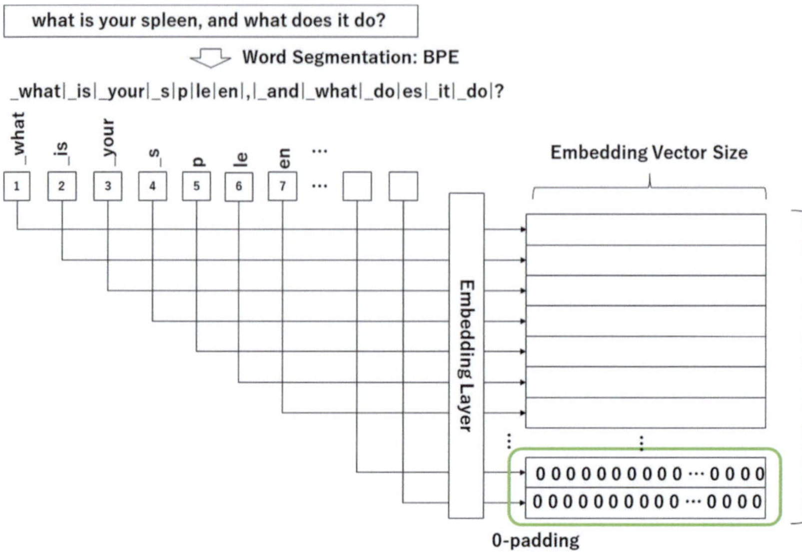

Fig. 1 The input generation to CNN

Table 3 Parameter settings for CNN

Arguments	Hyper-parameter
Dimension of word embedding	50, 100
Max length of segmentation	300, 100, 50
Number of filters	128 256
Kernel size of convolution	2 3 5
Stride length of convolution	1

units for comparison. The optimal parameters of CNN adopted for the experiment are listed in Table 3.

We also prepared classifiers based on multi-layer perceptron (MLP) and random forests (RF) algorithm by using features in previous works [14, 18]: (1) Bag-of-words (BoW), (2) Bag-of-bigrams (BoB), (3) Text segmentation length, (4) Speaker type (librarian of user), and (5) Message position in the dialogue.

4.2 *Experimental Results and Discussions*

Table 4 summarizes the results of the dialogue act classification in each setting. We targeted from 100 to 1000 as the vocabulary size of each BPE. However, the vocabulary is trained on difference dataset (500 unlabeled sessions), and some words

Table 4 Accuracies different dialogue act classifications

Methods	Vocab	DAF	DAD	OOV	Maximum word length	Average word length	Average text length
BPE-unit-level CNN	97	0.8601*	0.7175	0.0	4	1.5	60.6
BPE-unit-level CNN	**197**	**0.8684*****	**0.7256***	0.0	10	2.5	46.35
BPE-unit-level CNN	295	0.8622***	0.7209 *	0.0	10	2.8	41.1
BPE-unit-level CNN	395	0.8620**	0.7130	0.0	12	3.2	37.8
BPE-unit-level CNN	494	0.8570**	0.7122	0.0	13	3.4	35.8
BPE-unit-level CNN	592	0.8585**	0.7141	0.2	13	3.5	34.3
BPE-unit-level CNN	686	0.8585	0.7092	0.4	13	3.7	33.2
BPE-unit-level CNN	784	0.8556	0.7091	0.7	13	3.8	32.3
BPE-unit-level CNN	881	0.8536	0.7040	1.2	13	3.8	31.5
BPE-unit-level CNN	977	0.8541	0.7046	1.6	13	4.0	30.9
Character-level CNN	67	0.8538	0.7124	0.0	1	1.0	75.8
Word-level CNN	6091	0.8438	0.6937	333.9	80	6.9	16.5
Word-level LSTM	6091	0.8286	0.6745	333.9	80	6.9	16.5
MLP (BoW + BoB)	6091	0.8498	0.7119	333.9	80	6.9	16.5
MLP (All features)	6091	0.8515	0.7145	333.9	80	6.9	16.5
RF (BoW + BoB)	6091	0.8292	0.6790	333.9	80	6.9	16.5
RF (All features)	6091	0.8367	0.7008	333.9	80	6.9	16.5

Paired t-test with MLP (All feature): $*p < 0.05$, $**p < 0.01$, $***p < 0.001$

were unseen in the training data (200 labeled sessions) of the classifier. Vocab is the size of the vocabulary, DAF and DAD are the accuracies of labeling for each category, OOV is the average number of OOVs in cross-validation. Maximum and Average Word Length means the maximum and the average of the units used in the classifier. As seen in Table 4, the BPE reduced the number of OOVs compared to the word-based methods, although there are fewer BPE units than words. In the following text segmentation example of BPE, if the vocabulary size is 100, tokens are similar to characters and only frequent words are tokenized. Frequently words (such as "how", "help", "information", etc.) are tokenized according to the increasing vocabulary size.

- **Original utterance**: do you want information on pilot mountain or rock climbing? how can i help you?
- **Vocab size = 100**: _d o _you _w an t _i n f or m at i on _ on _p i l o t _m ou n t a in _ or _ r o c k _c l i m b ing ? _h o w _c an _i _h e l p _you ?
- **Vocab size = 500**: _do _you _w ant _information _on _p il ot _m ount ain _or _ ro c k _c li m b ing ? _how _can _i _help _you ?
- **Vocab size = 1000**: _do _you _want _information _on _p il ot _m ount ain _or _ ro ck _c li m b ing ? _how _can _i _help _you ?

In the dialogue act classification results, the accuracy of dialogue act classification was improved by the proposed BPE-unit-level CNN on each vocabulary size from 100 to 1000 compared to other models, even if they do not use additional information such as the role of the speaker or appearance position in dialogue. The word and Character-level CNN did not show better performance compared to the conventional method based on MLP.

In respects to DAF, the proposed methods estimated the dialogue acts with high accuracy. However, accuracies of DAD were not high enough, being only 0.7256 in the best condition. This can be improved by considering some additional information such as dialogue history. We also analyzed misclassified examples and found that there were some ambiguities caused by the annotation. Some utterances of participants had several roles, but the original annotation scheme did not allow to annotate multiple dialogue acts to one utterance. Such ambiguity of annotations should be eliminated to improve annotation. Below are some examples of utterances that should have multiple dialogue acts.

- thank you so much. this looks great. can you find any reasons why tea would do this? (Social:Gratitude, Info:Problem)
- "Name", welcome to maryland askusnow! i'm looking at your question right now; it will be just a moment. (Social:Greeting, Task:Librarian, Comm:Pausing)

As a solution to the problem, we can introduce the ISO24617-2 dialogue act annotation scheme [1]. The scheme has general-purpose functions (GPF) for utterances that control the contents of dialogues, and domain-specific functions (DSF) that process the dialogues. This scheme allows to annotate multiple DSF tags for one utterance.

5 Conclusion

In this paper, we proposed a dialogue act classification model based on BPE tokenizer and CNN-based classifier in the reference interview. Experimental results show that the classification accuracy of the proposed model was significantly higher than that of any baseline model. Our proposed model efficiently built the input the classifier with BPE-based tokenizer. It performed better than the classifiers that use words and characters as input units. Our model performed well for the DAF category. However, improvement in the DAD category remains as a future challenge. In addition, it is necessary to investigate the effectiveness of our method in other dialogue domains, and compare the other approaches like lemmatization or word-CNN with pre-trained embedding model.

In the error analysis, we found that some problems were caused by the annotation scheme. The lack of data was also a problem, therefore, as future work, we plan to improve the number and quality of the data. By using the classifier that we proposed, we will develop a dialogue manager that models the strategy of librarians in reference interviews to help find the exact information for the user, even if the requests of the users are ambiguous.

Acknowledgements This research and development work was supported by the JST PREST (JPMJPR165B) and JST CREST(JPMJCR1513).

References

1. Bunt H, Alexandersson J, Choe JW, Fang AC, Hasida K, Petukhova V, Popescu-Belis A, Traum DR (2012) Iso 24617-2: a semantically-based standard for dialogue annotation. In: LREC, pp 430–437
2. Gage P (1994) A new algorithm for data compression. C Users J 12(2):23–38
3. Inoue K (2013) An investigation of digital reference interviews: a dialogue act approach. PhD thesis, Syracuse University
4. Mikolov T, Sutskever I, Deoras A, Le HS, Kombrink S, Cernocky J (2012) Subword language modeling with neural networks. http://www.fit.vutbr.cz/imikolov/rnnlm/char.pdf
5. Misu T, Kawahara T (2006) Dialogue strategy to clarify users queries for document retrieval system with speech interface. Speech Commun 48(9):1137–1150
6. Radford ML, Connaway LS (2005) Seeking synchronicity: evaluating virtual reference services from user, non-user, and librarian perspectives. In: Proposal for a research project, submitted February, vol 1
7. Radford ML, Connaway LS, Confer PA, Sabolcsi-Boros S, Kwon H (2011) Are we getting warmer? Query clarification in live chat virtual reference. Reference & User Services Quarterly, pp 259–279
8. Ross CS, Radford ML, Nilsen K (2009) Conducting the reference interview: a how-to-do-it-manual for librarians. Neal-Schuman Publishers, Inc.
9. Sennrich R, Haddow B, Birch A (2015) Neural machine translation of rare words with subword units. arxiv:abs/1508.07909
10. Severyn A, Moschitti A (2015) Unitn: training deep convolutional neural network for twitter sentiment classification. In: SemEval@ NAACL-HLT, pp 464–469
11. Shi H, Ushio T, Endo M, Yamagami K, Horii N (2016) A multichannel convolutional neural network for cross-language dialog state tracking. In: 2016 IEEE Spoken Language Technology Workshop (SLT), pp 559–564. IEEE
12. Shibata Y, Kida T, Fukamachi S, Takeda M, Shinohara A, Shinohara T, Arikawa S (2000) Speeding up pattern matching by text compression. In: Algorithms and complexity, pp 06–315
13. Stolcke A, Ries K, Coccaro N, Shriberg E, Bates R, Jurafsky D, Taylor P, Martin R, Van Ess-Dykema C, Meteer M (2000) Dialogue act modeling for automatic tagging and recognition of conversational speech. Computat Linguist 26(3):339–373
14. Webb N, Hepple M, Wilks Y (2005) Dialogue act classification based on intra-utterance features. In: Proceedings of the AAAI workshop on spoken language understanding, vol 4, p 5
15. Williams JD, Young S (2007) Partially observable markov decision processes for spoken dialog systems. Comput Speech Lang 21(2):393–422
16. Yoshino K, Kawahara T (2015) Conversational system for information navigation based on pomdp with user focus tracking. Comput Speech Lang 34(1):275–291
17. Young S, Gašić M, Thomson B, Williams JD (2013) Pomdp-based statistical spoken dialog systems: A review. Proceedings of the IEEE 101(5):1160–1179
18. Yu B, Inoue K (2012) An investigation of digital reference interviews: dialogue act annotation with the hidden Markov support vector machine. OCLC/ALISE research grant report published electronically by OCLC research

“I Think It Might Help If We Multiply, and Not Add”: Detecting Indirectness in Conversation

Pranav Goel, Yoichi Matsuyama, Michael Madaio and Justine Cassell

Abstract The strategic use of indirect language is crucial in business negotiations, doctor-patient interactions, instructional discourse, and multiple other contextss. Being indirect allows interlocutors to diminish the potential threats to their interlocutors' desired self-image—or, face threat—that may arise by being overtly direct. Handling indirectness well is important for spoken dialogue systems, as being either too indirect or too direct at the wrong time could harm the agent-user relationship. We take the first step towards handling users' indirection by exploring different supervised machine learning approaches for the task of automatically detecting *indirectness* in conversations. Accurate automated *detection* of indirectness may help conversational agents better understand their users' intents, gauge the current relationship with the user in order to appropriately plan a response, and inform the strategic use of indirectness to manage the task goals and social goals of the interaction. To our knowledge we are the first to use a multi-modal approach to detecting indirect language: we rely on both verbal and nonverbal features of the interaction. Our best model acheives a 62% F1 score on our dataset, outperforming non-neural baselines including approaches used by past work for related tasks such as uncertainty detection.

P. Goel (✉)
Computer Science, University of Maryland (Work done as an Intern at Carnegie Mellon University), College Park, Baltimore, USA
e-mail: pgoel1@cs.umd.edu; pranav.goel.cse14@iitbhu.ac.in

Y. Matsuyama · M. Madaio · J. Cassell
School of Computer Science, Carnegie Mellon University, Pittsburgh, USA
e-mail: yoichim@cs.cmu.edu

M. Madaio
e-mail: mmadaio@cs.cmu.edu

J. Cassell
e-mail: justine@cs.cmu.edu

L. F. D'Haro et al. (eds.), *9th International Workshop on Spoken Dialogue System Technology*, Lecture Notes in Electrical Engineering 579,
https://doi.org/10.1007/978-981-13-9443-0_3

1 Introduction

Indirect delivery, or indirectness, is the linguistic phenomenon where the speaker intentionally does not communicate their intention straightforwardly. This is done either by distancing themselves from their intended meaning or the proposition they are communicating, by introducing vagueness, reducing certainty or intensity, or making their statements appear more subjective, among others [13, 29, 38, 43, 44]. It is a conversational strategy that is seen as intrinsic to conversational discourse [31]. In a variety of linguistic contexts, such as negotiation, counseling, health care [51], and education [29, 41], it is often important to speak indirectly to communicate most effectively. The stakes in such conversations are quite high, ranging from profitable business deals to life-saving medical advice. In many cases, delivering information (e.g. a dire medical diagnosis) in a direct manner, while perhaps successful in communicating information in the short-term, can have serious consequences for the long-term relationship and even the success of the interaction. The strategic use of *indirectness* is thus critical to softening the blow of such direct delivery, resulting in more effective communication [6]. Another example is business negotiations, where being indirect is used to present tentative views, weaken one's commitment to a particular bid, and build trust between negotiators [52].

To use indirectness strategically, interlocutors should be able to *detect* it in each others' utterances and are in fact often expected to do that. Consider bosses mitigating orders in office meetings to sound less threatening and more friendly [9]. The employee is expected to parse the request despite its indirectness and understand that it is actually imperative to get the requested work done. A study by [39] concluded that humans interact or behave towards computers as they do towards other humans including in terms of their social response, and another study found that humans increase or decrease the use of their indirection according to the messaging style deployed by computer agents [5]. These studies indicate that users will, in fact, be indirect when conversing with agents. Combined with the rapid increase in the use of computer-mediated discourse and virtual assistants in many domains (including the domains above) [20, 26, 40], the issue of *detecting indirectness* becomes particularly important. Knowing when a user is being indirect can help such systems better understand, respond to, and build the user-agent relationship.

The prevalence and importance of indirectness in educational interactions [37] coupled with the rise of tutoring dialogue systems such as "teachable agents" [20] make education a productive domain for us to target in this work. Students use such indirectness when proposing answers to their teachers [41], and students peer tutoring one another [34] use indirectness to either communicate uncertainty or to reduce the threat to their partner's self-image and self-esteem that might result from overly direct feedback [29, 37]. Without such indirectness, peer tutors' requests and feedback may threaten what Brown and Levinson (following Goffman) call one's "positive face", or desire to be seen in a positive light [6]. Madaio et al. [29] found that peer tutors who had *greater self-efficacy* (i.e. confidence in their tutoring ability) used *more indirectness*, suggesting that it plays a strategic face-saving role, and thus a relationship-building role. Therefore, *intelligent tutoring systems* would benefit

Table 1 Examples showing how politeness, uncertainty and indirectness are different phenomena. The first two examples are from our reciprocal peer tutoring corpus (see Sect. 3), example 3 from the corpus of requests annotated for politeness by [10] and example 4 from the Wikipedia corpus annotated for uncertainty detection by [13]

#	Example	Indirect	Polite	Uncertain
1	Can you please ***just*** stay with me and not doodle	✓	✗	✗
2	***I think*** I solved it, *would you mind* checking it?	✓	✓	✗
3	***Nice work*** so far on your rewrite.	✗	✓	✗
4	The club enjoyed ***most of its success*** in its early years.	✗	✗	✓

from detecting indirectness used by the students. This is even more important if the system (or agent) aims to build a social motivational relationship with the students. It is also notable that [29] found that tutors with a *stronger relationship* or "rapport" with their partners were *more direct*. This suggests that having a spoken dialogue system be continuously polite (as done in [21]) may in fact be perceived as distancing and harm the rapport between the agent and user [37]. Note that being polite is one function of being indirect, but politeness and indirectness are different phenomena (Sect. 2). Automatically detecting indirectness in user utterances can thus help all kinds of dialogue systems better estimate the state of the social relationship—an aid in designing an appropriate response—and more effectively communicate information.

To detect indirectness in conversation, we use various machine learning approaches such as Support Vector Machines and Logistic Regression with various feature combinations (Sect. 4). This is motivated in part by studies carried out by [38], where they gave empirical proof for the lack of effectiveness of a simple keyword search based approach for uncertainty or hedge detection. We also use several neural network based methods, which have been shown to give improved performance over non-neural baselines for many NLP applications including detection of uncertainty in texts [1] and politeness classification in requests [2] (these phenomena are related to, but distinct from, indirectness (Sect. 2)). To the best of our knowledge, this is the first work to leverage machine learning methods for automated detection of indirectness in dialogue. We take the first step towards multi-modal detection of indirect language in conversations by leveraging verbal (text from dialogue transcript) as well as visual (features extracted from video recordings) modalities. Our approach and results can help a variety of spoken dialogue systems such as tutoring dialogue systems and socially-aware conversational agents use automatic indirectness detection as the first step to handle indirectness in their interaction with users.

2 Related Work

Prior work has either focused on automated detection of a specific function of indirectness (such as politeness [4]), or targeted some specific manifestation of the phenomenon (such as uncertainty [14, 33]) using only a single modality (i.e. text) [10,

13, 38]. This differentiates our work which attempts to detect indirectness, and not just a specific function or manifestation, and leverages multiple modalities together. For example, [10] explored a politeness classifier based on syntactic and lexical features by incorporating various components of politeness theory. Aubakirova and Bansal [2] then used neural networks for the same task on the same corpus. Their annotation for politeness includes being indirect as *one of the ways* of being polite in requests. However, indirectness is not always interpreted as politeness and can even be associated with lack of politeness [4]). Indirectness has many more functions in addition to marking politeness [9, 33], while one can be polite in ways other than being indirect [10]. Examples 1 and 3 in Table 1 help illustrate the difference. Indirectness is often produced through the use of hedges, which are "single- or multi-word expressions used to indicate uncertainty about the propositional content of an utterance or to diminish its impact" [14, 33]. Thus, uncertainty is just one of the ways in which indirectness can manifest in conversations or it can be seen as an indirectness strategy [3]. This can be seen in examples 1 and 2 in Table 1). A statement may also be uncertain (due to lack of exact percentages or numeric data when trying to quantify something) without being indirect (consider example 4 in Table 1 which was annotated as 'uncertain' in [13]).

Many studies in NLP have explored the detection of such hedges, focusing only on the uncertainty aspect [13, 38], especially for text. The ConLL 2010 shared task on hedge or uncertainty detection [13] facilitated automated separation of 'uncertain' and 'factual' statements by providing two annotated datasets—a BioScope corpus (abstracts and articles from biomedical literature) and a Wikipedia corpus. Recently, [1] carried out deep neural network based experiments on the ConLL 2010 shared task datasets for uncertainty detection to explore different kinds of attention mechanisms using the task setting. However, text-based corpora don't allow for use of the rich data communicated by nonverbal behavior.

Even in the spoken dialogue setting, past work has again focused on detecting uncertainty but not the broader phenomenon of being indirect. Liscombe et al. [28] used prosody to automatically detect student 'certainness' in spoken tutorial dialogue. Dral et al. [11] also investigated automatic detection of uncertainty using predefined prosodic markers. If the targeted prosodic markers could not be identified for a certain utterance, they fell back on a defined list of lexical markers to classify an utterance as certain or uncertain. The phenomenon of indirectness we study in our corpus relates more to the general face threat mitigation needs in dialogue [29]. While both uncertainty and indirectness can serve the goal of saving face in conversations, indirectness can do so in multiple ways without needing to be uncertain.

Prior work on spoken dialogue and text-based corpora used one modality at a time when classifying an instance. While verbal (text) and vocal (speech) modalities have been explored, no past work has leveraged the 'visual' modality to the best of our knowledge. The use of nonverbal behaviors (including visual features such eye contact and smiling) has been motivated by [11] for uncertainty detection (based on the experiments carried out by [25]), and by [47] as crucial for face threat mitigation. We try to combine the use of text and such visual behaviours in our experiments.

3 Corpus Collection and Annotation

Our dialogue corpus was collected from 12 American-English speaking pairs (or dyads) of teenagers (mean age = 13.5) tutoring each other in basic linear algebra. They worked together for 5 weekly hour-long sessions for a total of about 60 hours. Each session included some social interaction as well as one of the members of the dyad tutoring the other (the roles are reversed later in the session). Indirectness was annotated only for the 'tutoring' periods. Audio and video data were recorded, transcribed, and segmented for clause-level dialogue annotation of various conversational strategies including *indirectness* or indirect delivery. The corpus was coded for four types of indirectness—apologizing, hedging language (e.g. use of qualifiers), the use of vague category extenders, and "subjectivizing" [14, 33, 51]. These are detailed in Table 2. For all the four codes, the Krippendorff's alpha for five trained annotators was at least 0.7. Once the annotators reached sufficient inter-rater reliability, the corpus was divided amongst the annotators, each labeling one-fifth of the corpus. An utterance was classified as indirect or not based on its inclusion in any of these four categories. After some data cleaning and simple pre-processing steps (not detailed here for brevity), we got a total of 23437 utterances, with 1113 out of them labeled as 'indirect'.

This corpus was also annotated for certain visual behaviors:

- Eye Gaze—Three types of gaze were annotated—Gaze at the Partner (gP), Gaze at the Worksheet (gW), and Gaze Elsewhere (gE).
- Smile—A smile is defined by the elongation of the participants lips and rising of their cheeks. Smiles were annotated from the beginning of the rise to the end of the decay (as per the parameters explained in [19]). Laughter (including smiling) has shown to be an effective method of face threat mitigation [48], and therefore might be used in conjunction with indirect language.
- Head Nod—Temporal intervals of head nod were encoded (beginning of the head moving up and down until the moment the head came to rest).

Inter-rater reliability for visual behavior was 0.89 for eye gaze, 0.75 for smile count (how many smiles occur), 0.64 for smile duration, and 0.99 for head nod. Further

Table 2 Annotation of codes under the 'indirect' label. Distribution = % of all indirect utterances

Code	Definition	Example	Distribution (%)
Apology	Apologies used to soften direct speech acts	Sorry, its negative 2	7.7
Qualifiers	Qualifying words for reducing intensity or certainty	You just add 5 to both sides	66.1
Extenders	Indicating uncertainty by referring to vague categories	You have to multiply and stuff	3.6
Subjectivizer	Making an utterance seem more subjective to reduce intensity	I think you divide by 3 here	22.6

details of extraction including ground truth definitions for each behavior can be found in [53], who found these behaviors to be important for automatic detection of conversational strategies such as social norm violation, self-disclosure, praise, and reference to shared experience. In particular, they found gaze behaviors, head nods, and smiling helpful. In addition, [22] showed that head tilt was one of the strongest nonverbal cues of interpersonal intimacy.

4 Approaches

Our (supervised) ML methods include non-neural network based approaches relying on various sets of features and different neural network based architectures (due to their use in related tasks like politeness and uncertainty detection, as mentioned in Sect. 1). We mention the variations but focus on the best performing models.

4.1 Non-neural Network Based

These methods involve explicit feature representation and then using a learning algorithm for training.

Feature Representation: We tried various ways to represent the utterances. Table 3 summarizes the features we considered. We briefly discuss these features below:

- Unigram or bag-of-words: We set a rare threshold of 10 for our experiments, which means that a word (or the target n-gram) will be considered only if it occurs at least 10 times in the training set.
- Pair-based features: To capture some context beyond just the words, we use a feature representation consisting of bigrams, Part-of-Speech (POS) bigrams, and word-POS pairs (also used by [53] for conversational strategy classification). The rare threshold is again set to 10.
- Pre-trained word vectors: Word2Vec [32] and GloVe [36] are useful ways to represent text in a vector space of manageable dimensionality. This methodology has been successfully applied to many NLP tasks [12, 32]. We tried various available pre-trained models like *Twitter word2vec* [16] trained on 400 million Twitter tweets, GloVe representations [36] trained on Wikipedia articles (called *GloVe wiki*) and web content crawled via Common Crawl (called *GloVe Common Crawl*), and word vectors trained on Wikipedia articles using Word2Vec by [1] which we call *Wikipedia Word2Vec*.
- Word2Vec trained on our dataset: We learn word vector representations on our own training data. We refer to this model as *RPT Word2Vec* (for Reciprocal Peer Tutoring). We tune the various training parameters on the validation dataset (see Training Detail) resulting in window size = 9, dimensionality = 300, and training algorithm = continuous bag-of-words (CBOW).

Table 3 Summary of the feature sets (dimensionality) used to represent utterances

Extracted from text		Extracted from video
Bag-of-words/n-gram	Word embeddings	
Unigram ($\sim$900)	Twitter Word2Vec (400)	Visual (10)
Pair-based [Bigrams, POS bigrams, Word-POS pairs] ($\sim$3700)	GloVe wiki (300)	
	GloVe Common Crawl (300)	
	Wikipedia Word2Vec (400)	
	RPT Word2Vec (300)	

Note that for word embeddings, we get the representation of each word of a sentence. To get the representation for the overall sentence we take the *mean* of the individual word embeddings.

- Visual features: Visual behaviors annotated for our corpus were explained in Sect. 3. Three types of eye gaze, smile, and head nod were annotated for both the tutor and tutee at each turn, giving a set of 10 visual features (Table 3).

Training Detail: Our corpus contains 60 sessions of peer tutoring interaction. Out of these, we take 48 sessions as the training dataset, 6 as validation, and 6 as the test set. This is repeated 5 times to get five train-validation-test splits. We use the same splits across every experiment. The splits were done on the basis of dialogue sessions across all speaker dyads. The validation set is used to decide the best performing approaches, tune hyperparameters and to choose the training settings for Word2Vec. For each feature representation, we tried the following supervised machine learning algorithms—Logistic Regression, Naive Bayes, Random Forest and Support Vector Machine (SVM), with Logistic Regression and SVM performing best on the validation set.

4.2 Neural Network Based

We applied various neural architectures to see if they could perform better than the non-neural models for indirectness detection. Since this task has not been explored directly in the past (Sect. 2), we tried many different architectures that have worked well in the past classification based NLP work. These include fully connected or feedforward neural networks, Convolutional Neural Networks (CNNs) and Recurrent Neural Networks (RNNs) (including variants like Long Short-Term Memory (LSTM) [18] network and Gated Recurrent Units [7]) which have all been applied successfully to various NLP tasks [8, 17, 27] including politeness classification and hedge detection [1, 2]. CNNs and RNNs expect sequential input, and hence we concatenate the word vector representations to form the sentence representation.

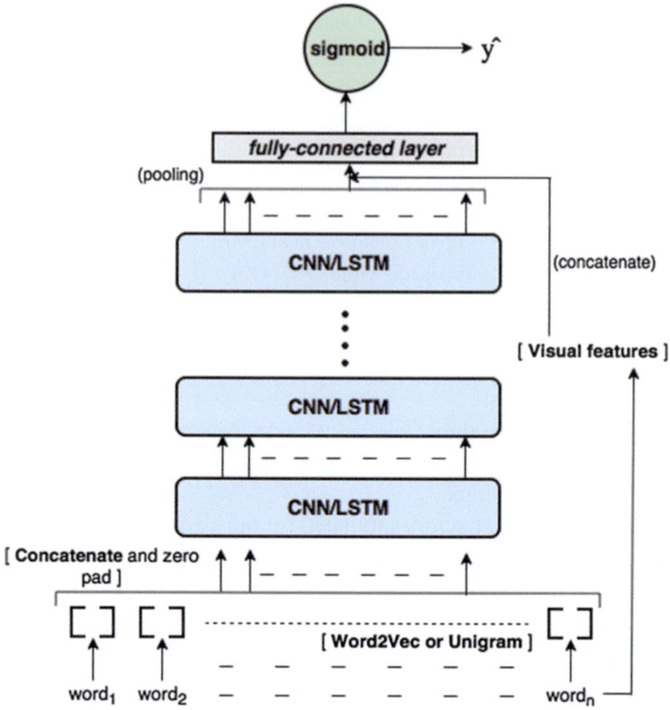

Fig. 1 A general representation of the different neural architectures (and combinations) tried

Combining LSTMs and CNNs in sequence, and having one or more fully connected layers after convolutional/recurrent layers (Fig. 1) has also proven to be effective in NLP [27, 54]. For pooling the sequential output of CNN/LSTM, we tried taking the maximum, mean, or only the final vector in the sequence.

In our experiments, we incorporated visual features by concatenating the feature vector (of dimensionality 10, see Table 3) with the input to the first feedforward or fully connected layer (which could also be the output layer) of the deep neural network (Fig. 1) for all the various architectures we tried. This method is inspired by other similar ways previously used to incorporate external features in neural architectures [35]. Recently, the attention mechanism has been successfully applied to augment CNNs and LSTMs [49, 50]. Certain portions of a sequence are more predictive of the output than others, and the selective mechanism of attention allows the network to focus on the most relevant parts of an input or hidden layer , which is useful for long input sequences. We tried all these various networks and compared their performances on the validation set. The two best performing architectures for indirectness detection are discussed below.

Table 4 Hyperparameter settings for the Stacked LSTMs approach which resulted in the best validation set performance

	Initial embedding layer	#Stacked LSTM layers (dimensionality)	Dropout rate	Sigmoid threshold
RPT dataset	Twitter Word2Vec	4 (400, 300, 200, 100)	0.5	0.3
BioScope	Wikipedia Word2Vec	3 (400, 200, 100)	0.2	0.4
Wikipedia	GloVe wiki	4 (400, 300, 200, 100)	0.5	0.3

4.2.1 Stacked LSTMs

Stacking multiple LSTM layers one after the other has been an effective method for various NLP tasks such as dialogue act classification [23] (which is close to our task since we are classifying a conversational strategy in dialogue). Some of the architectural decisions include the number of layers to be stacked, having a fully connected layer after the stacked LSTMs or not, the activation function for the output layer, etc. All such hyperparameters were tuned on the validation set.

We experimented with two common choices for the final output layer—softmax, which gives a probability distribution over the number of classes or a single neuron and sigmoid activation, which gives a real-valued score between 0 and 1 ($\hat{y}$ in Fig. 1). Using a sigmoid function requires choosing a threshold value, such that an utterance assigned score above this value gets labeled '1' (presence of indirectness). We tuned this threshold and also found that sigmoid worked better for our task. The sequential output from the final LSTM was pooled using last pooling. This vector can then be concatenated with the vector representing the visual features, before going to either some fully connected layers (Fig. 1) or to the output layer itself. We applied Dropout [45] at each LSTM layer, tuning the dropout rate on the validation set. After fine-tuning these hyperparameters, we get the best performing setting reported in Table 4. The network parameters for the neural model were learned by minimizing the binary cross-entropy loss [42] between the actual and predicted labels. We optimized this function by back-propagating through layers via Mini-Batch Gradient Descent using a batch size of 512, 25 training epochs and Adam optimization algorithm [24] with the parameters set as $\alpha = 0.001, \beta_1 = 0.9, \beta_2 = 0.999\ and\ \varepsilon = 10^{-9}$.

4.2.2 Attention-Based CNN

Adel and Schütze [1] applied attention mechanism in different ways to the task of uncertainty detection on the ConLL 2010 shared tasks datasets [13]. They outperformed the best shared task system on the Wikipedia dataset while matching the state-of-the-art on the Biomedical dataset (Table 6). We tried using their methodology on our peer-tutoring dataset for indirectness detection. As per [1], the attention layer a for input x is given as: $\alpha_i = \dfrac{exp(f(x_i))}{\sum_j exp(f(x_j))}$; $a_i = \alpha_i \cdot x_i$ where f is a scoring function, the α_i are attention weights and each input x_i gets re-weighted (selectively focused upon) by its corresponding attention weight α_i The most basic definition for f is the linear scoring function (on the input x): $f(x_i) = W^T x_i$. W are parameters learned in training. We applied attention on the input sequence itself (Att_Inp CNN) and on the hidden layer of the convolution (Att_Conv CNN) (for details, refer [1]). We also tried using attention on LSTM which was not as effective as using CNN.

5 Results and Discussion

We use the F1 score as the evaluation metric to measure the performance of our various models. Since our dataset is unbalanced (skewed towards the '0' class representing an absence of indirectness), accuracy would not be a good choice. F1 score was also used in the ConLL 2010 shared task on uncertainly detection. The best performing systems for that shared task [13] used essentially SVM on bag-of-words based features, for both the Wikipedia and BioScope datasets [15, 46]. Attention-based CNN model by [1] gave state-of-the-art results on the shared task datasets. We try these approaches for our task of indirectness detection, along with various other neural architectures (Sect. 4.2). To further establish the effectiveness of our stacked LSTMs approach (Sect. 4.2.1), we see its performance on the uncertainty detection shared task datasets as well (with the tuned hyperparameter settings reported in Table 4).

We show results on the test set for those variants of the neural and non-neural models which performed the best on validation set in Tables 5 and 6, and observe:

- SVM outperforms Logistic Regression for almost every feature representation by a significant margin (Table 5). Adding visual features does not seem to offer much improvement in terms of results, pointing to the need of looking at different nonverbal behaviors, or fusing them with verbal features in a different way.
- Using bag-of-words or n-grams obtained from our peer tutoring dataset gives better performance than using word2vec models pre-trained on other, much larger datasets (Table 5). This may indicate that the indirectness detection task relies on the specific domain of the dataset, something hinted for hedging by [38]. Among the pre-trained word2vec models, Twitter Word2Vec gave the best performance, and many utterances in our corpus (of American teenagers tutoring and talking to one another) do share the short length and informal nature of Twitter tweets.
- Using Pre-trained Word2Vec + Stacked LSTMs as well as Attention-based CNN performs better on our dataset. These neural models outperform SVM by roughly 3–4% F1 score (second column of Table 6). The observation holds true for the uncertainty detection on the Wikipedia dataset as well (a performance gain of about 6%). This indicates that neural models constitute a viable approach for indirectness detection as well as uncertainty detection. For BioScope corpus, however, SVM on bag-of-words based features perform the best. One potential reason is that the annotation for uncertainty in the BioScope corpus primarily relies on certain keywords alone.

Table 5 F1 score (%) on test set for various features in Table 3 and combinations fed to non-neural ML algorithms (Sect. 4.1) for indirectness detection on reciprocal peer-tutoring dataset

	Logistic Reg.	SVM
Unigram	57.71	59.1
Unigram + Visual	57.74	**59.3**
Pair-based	57.09	58.28
Pair-based + Visual	55.89	58.41
Twitter Word2Vec	44.83	53.86
GloVe Wiki	37.91	45.25
GloVe Common Crawl	38.94	45.06
Wikipedia Word2Vec	44.56	49.54
RPT word2vec	44.95	39.36

Table 6 F1 score (%) for the various neural models compared with SVM approach for two different tasks on different datasets. Results marked with * have been taken from previous literature as explained in Sect. 5

	Reciprocal peer-tutoring corpus (indirectness detection)	Wikipedia (uncertainty detection)	BioScope (Uncertainty Detection)
Att_Inp CNN	**62.03**	65.13*	84.99*
Att_Conv CNN	61.4	**66.49***	84.69*
Pre-trained W2V + Stacked LSTM	61.15	66.07	82.62
Pre-trained W2V + Stacked LSTM + Visual	61.35	–	–
Unigram + Stacked LSTM	56.5	43.71	73.03
Unigram + Stacked LSTM + Visual	57.11	–	–
SVM on Bag-of-Words	58.28	60.2*	**85.2***

- The best results we obtained on automatic detection of indirectness in peer-tutoring is 62.03% F1 score (Table 6), and the neural methods performed well for uncertainty detection in other domains as well.

6 Conclusion and Future Work

Indirectness is often used to mitigate face threat in conversations held in various settings like business negotiations, doctor-patient discourse, counseling, conference talks, and tutoring. Detecting indirectness can help virtual conversational agents and spoken dialogue systems respond to the user in a more appropriate manner. This may entail being more effective in their task goals (business deals, medical advice, or tutoring instructions) or in managing their interpersonal relationship with the user (i.e. mitigating face threat and building trust). To achieve this, we train our models on a corpus of peer tutoring dialogues, using nonverbal behaviors like eye gaze, head nod, and smiling in conjunction with the text of the transcripts. Insights from our experiments can help inform the design of future spoken dialogue systems that can automatically detect indirect delivery in user utterances. For example, the Twitter-like nature of collaborative educational conversations between teenagers can be exploited like we did via a word2vec model pre-trained on tweets. We also observe that neural approaches (like stacked LSTMs and attention-based CNNs) outperform non-neural approaches (like SVM). This indicates that capturing indirectness requires powerful pattern recognition methods like deep neural networks. Such observations can help spoken dialogue systems, regardless of the domain.

We intend for this work to be the first step towards automatic detection of indirect language using multimodal data. For future work, we plan to leverage more visual behaviors by studying how various nonverbal behaviors inform the use of indirectness via dedicated experiments and annotate more behaviors such as head tilts [22] and laughter [48] which can be relevant to mitigating face-threat. We also aim to use acoustic or paralinguistic features to create a fully multimodal system for indirectness detection. Another line of work we hope to explore is properly incorporating our findings into an Intelligent Tutoring Agent or a general-purpose socially-aware spoken dialogue system [30] that can detect and use indirectness strategically. As indirectness is ubiquitous in interpersonal communication, incorporating its detection in spoken dialogue systems may ultimately lead to more natural, human-like interactions with users.

References

1. Adel H, Schütze H (2017) Exploring different dimensions of attention for uncertainty detection. In: EACL 2017, European chapter of the association for computational linguistics, Valencia, Spain, 3–7 April 2017
2. Aubakirova M, Bansal M (2016) Interpreting neural networks to improve politeness comprehension. arXiv preprint arXiv:1610.02683
3. Biber D (1991) Variation across speech and writing. Cambridge University Press
4. Blum-Kulka S (1987) Indirectness and politeness in requests: same or different? J Pragmat 11(2):131–146
5. Brennan SE, Ohaeri JO (1994) Effects of message style on users' attributions toward agents. In: Conference on human factors in computing systems: conference companion on human factors in computing systems, vol 24, pp 281–282
6. Brown P, Levinson SC (1987) Politeness: some universals in language usage, vol 4. Cambridge University Press
7. Cho K, Van Merriënboer B, Bahdanau D, Bengio Y (2014) On the properties of neural machine translation: encoder-decoder approaches. arXiv preprint arXiv:1409.1259
8. Collobert R, Weston J, Bottou L, Karlen M, Kavukcuoglu K, Kuksa P (2011) Natural language processing (almost) from scratch. J Mach Learn Res 12(Aug), 2493–2537
9. Cutting J (2007) Vague language explored. Springer
10. Danescu-Niculescu-Mizil C, Sudhof M, Jurafsky D, Leskovec J, Potts C (2013) A computational approach to politeness with application to social factors. arXiv preprint arXiv:1306.6078

11. Dral J, Heylen D et al (2008) Detecting uncertainty in spoken dialogues: an explorative research to the automatic detection of a speakers' uncertainty by using prosodic markers
12. Enríquez F, Troyano JA, López-Solaz T (2016) An approach to the use of word embeddings in an opinion classification task. Expert Syst Appl 66:1–6
13. Farkas R, Vincze V, Móra G, Csirik J, Szarvas G (2010) The conll-2010 shared task: learning to detect hedges and their scope in natural language text. In: Proceedings of the fourteenth conference on computational natural language learning—shared task. Association for Computational Linguistics, pp. 1–12
14. Fraser B (2010) Pragmatic competence: the case of hedging. In: New approaches to hedging, pp 15–34
15. Georgescul M (2010) A hedgehop over a max-margin framework using hedge cues. In: Proceedings of the fourteenth conference on computational natural language learning—shared task. Association for Computational Linguistics, pp 26–31
16. Godin F, Vandersmissen B, De Neve W, Van de Walle R (2015) Multimedia lab@ acl w-nut ner shared task: named entity recognition for twitter microposts using distributed word representations. ACL-IJCNLP 2015:146–153
17. Graves A, Jaitly N (2014) Towards end-to-end speech recognition with recurrent neural networks. ICML 14:1764–1772
18. Hochreiter S, Schmidhuber J (1997) Long short-term memory. Neural Comput 9(8):1735–1780
19. Hoque M, Morency LP, Picard RW (2011) Are you friendly or just polite?—analysis of smiles in spontaneous face-to-face interactions. In: International conference on affective computing and intelligent interaction. Springer, pp. 135–144
20. Jin J, Bridges SM (2014) Educational technologies in problem-based learning in health sciences education: a systematic review. J Med Internet Res 16(12)
21. Johnson, W.L., Rizzo, P.: Politeness in tutoring dialogs:run the factory, thats what id do. In: Intelligent Tutoring Systems, pp. 206–243. Springer (2004)
22. Kang SH, Gratch J, Sidner C, Artstein R, Huang L, Morency LP (2012) Towards building a virtual counselor: modeling nonverbal behavior during intimate self-disclosure. In: Proceedings of the 11th international conference on autonomous agents and multiagent systems, International foundation for autonomous agents and multiagent systems, vol 1, pp 63–70
23. Khanpour H, Guntakandla N, Nielsen R (2016) Dialogue act classification in domain-independent conversations using a deep recurrent neural network. In: COLING, pp 2012–2021
24. Kingma D, Ba J (2014) Adam: a method for stochastic optimization. arXiv preprint arXiv:1412.6980
25. Krahmer E, Swerts M (2005) How children and adults produce and perceive uncertainty in audiovisual speech. Lang Speech 48(1):29–53
26. Kulik JA, Fletcher J (2016) Effectiveness of intelligent tutoring systems: a meta-analytic review. Rev Educ Res 86(1):42–78
27. Lee JY, Dernoncourt F (2016) Sequential short-text classification with recurrent and convolutional neural networks. arXiv preprint arXiv:1603.03827
28. Liscombe J, Hirschberg J, Venditti JJ (2005) Detecting certainness in spoken tutorial dialogues. In: INTERSPEECH, pp 1837–1840
29. Madaio M, Cassell J, Ogan A (2017) The impact of peer tutors use of indirect feedback and instructions. International society of the learning sciences, Philadelphia
30. Matsuyama Y, Bhardwaj A, Zhao R, Romeo O, Akoju S, Cassell J (2016) Socially-aware animated intelligent personal assistant agent. In: SIGDIAL conference, pp 224–227
31. McQuiddy IW (1986) Some conventional aspects of indirectness in conversation. Ph.D. thesis, University of Texas at Austin
32. Mikolov T, Chen K, Corrado G, Dean J (2013) Efficient estimation of word representations in vector space. arXiv preprint arXiv:1301.3781
33. Neary-Sundquist C (2013) The use of hedges in the speech of ESL learners. Elia (13):149
34. Palinscar AS, Brown AL (1984) Reciprocal teaching of comprehension-fostering and comprehension-monitoring activities. Cogn Instr 1(2):117–175

35. Park E, Han X, Berg TL, Berg AC (2016) Combining multiple sources of knowledge in deep CNNS for action recognition. In: 2016 IEEE Winter Conference on Applications of Computer Vision (WACV). IEEE, pp 1–8
36. Pennington J, Socher R, Manning C (2014) Glove: global vectors for word representation. In: Proceedings of the 2014 conference on empirical methods in natural language processing (EMNLP), pp 1532–1543
37. Person NK, Kreuz RJ, Zwaan RA, Graesser AC (1995) Pragmatics and pedagogy: conversational rules and politeness strategies may inhibit effective tutoring. Cogn Instr 13(2):161–188
38. Prokofieva A, Hirschberg J (2014) Hedging and speaker commitment. In: 5th International workshop on emotion, social signals, sentiment & linked open data, Reykjavik, Iceland
39. Reeves B, Nass CI (1996) The media equation: how people treat computers, television, and new media like real people and places. Cambridge University Press
40. Reynolds M (2017) Chatbots learn how to drive a hard bargain
41. Rowland T (2007) well maybe not exactly, but its around fifty basically?: vague language in mathematics classrooms. In: Vague language explored. Springer, pp 79–96
42. Rubinstein R (1999) The cross-entropy method for combinatorial and continuous optimization. Methodol Comput Appl Probab 1(2):127–190
43. Rundquist S (1990) Indirectness in conversation: flouting grices maxims at dinner. Annu Meet Berkeley Linguist Soc 16:509–518
44. Searle JR (1975) Indirect speech acts
45. Srivastava N, Hinton G, Krizhevsky A, Sutskever I, Salakhutdinov R (2014) Dropout: a simple way to prevent neural networks from overfitting. J Mach Learn Res 15(1):1929–1958
46. Täckström O, Velupillai S, Hassel M, Eriksson G, Dalianis H, Karlgren J (2010) Uncertainty detection as approximate max-margin sequence labelling. In: Proceedings of the fourteenth conference on computational natural language learning—shared task. Association for Computational Linguistics, pp 84–91
47. Trees AR, Manusov V (1998) Managing face concerns in criticism integrating nonverbal behaviors as a dimension of politeness in female friendship dyads. Hum. Commun. Res. 24(4):564–583
48. Warner-Garcia S (2014) Laughing when nothings funny: the pragmatic use of coping laughter in the negotiation of conversational disagreement. Pragmatics 24(1):157–180
49. Yin W, Schütze H, Xiang B, Zhou B (2015) Abcnn: attention-based convolutional neural network for modeling sentence pairs. arXiv preprint arXiv:1512.05193
50. Yu H, Gui L, Madaio M, Ogan A, Cassell J, Morency LP (2017) Temporally selective attention model for social and affective state recognition in multimedia content
51. Zhang G (2013) The impact of touchy topics on vague language use. J Asian Pac Commun 23(1):87–118
52. Zhao D, Nie J (2015) Vague language in business negotiation-from a pragmatics perspective. Theory Pract Lang Stud 5(6):1257
53. Zhao R, Sinha T, Black AW, Cassell J (2016) Automatic recognition of conversational strategies in the service of a socially-aware dialog system. In: 17th Annual meeting of the special interest group on discourse and dialogue, p 381
54. Zhou C, Sun C, Liu Z, Lau F (2015) A c-lstm neural network for text classification. arXiv preprint arXiv:1511.08630

Automated Classification of Classroom Climate by Audio Analysis

Anusha James, Yi Han Victoria Chua, Tomasz Maszczyk, Ana Moreno Núñez, Rebecca Bull, Kerry Lee and Justin Dauwels

Abstract While in training, teachers are often given feedback about their teaching style by experts who observe the classroom. Trained observer coding of classroom such as the Classroom Assessment Scoring System (CLASS) provides valuable feedback to teachers, but the turnover time for observing and coding makes it hard to generate instant feedback. We aim to design technological platforms that analyze real-life data in learning environments and generate automatic objective assessments in real-time. To this end, we adopted state-of-the-art speech processing technologies and conducted trials in real-life teaching environments. Although much attention has been devoted to speech processing for numerous applications, few researchers have attempted to apply speech processing for analyzing activities in classrooms. To address this shortcoming, we developed speech processing algorithms that detect speakers and social behavior from audio recordings in classrooms. Specifically, we aim to infer the climate in the classroom from non-verbal speech cues. We extract non-verbal speech cues and low-level audio features from speech segments and train classifiers based on those cues. We were able to distinguish between positive and negative CLASS climate scores with 70–80% accuracy (estimated by leave-one-out crossvalidation). The results indicate the potential of predicting classroom climate automatically from audio recordings.

1 Introduction

CLASS [1] assesses the interactional quality between students and teachers in preschool classrooms and two of its ten subscales are related to the climate in the classrooms. Classroom climate is the overall emotional tone of the class observed

A. James · Y. H. V. Chua · T. Maszczyk · J. Dauwels (✉)
School of Electrical and Electronic Engineering (EEE), Nanyang Technological University, Singapore, Singapore
e-mail: jdauwels@ntu.edu.sg

A. M. Núñez · R. Bull · K. Lee
National Institute of Education, Singapore, Singapore

L. F. D'Haro et al. (eds.), *9th International Workshop on Spoken Dialogue System Technology*, Lecture Notes in Electrical Engineering 579,
https://doi.org/10.1007/978-981-13-9443-0_4

from the warmth and respect in interactions. Students in a positive climate class are enthusiastic and happy, whereas negative climate reflects anger or aggression. Establishing a positive climate in class is important for effective teaching [1]. It is tedious and time-consuming to manually observe and code CLASS scores. There is a need to explore indicators that are easily and quickly extracted and correlate with CLASS climate dimensions. As classroom discourse is an important indicator of teacher instruction and classroom climate, we explore the idea here to analyze teaching practices automatically by means of speech processing technologies. Specifically, we have designed a pipeline to infer the classroom climate automatically. The processing contains the following steps: speaker diarization, sociometric analysis and machine learning. Audio recordings of preschool classrooms used posed a few technical challenges. In this paper, we describe each step of the processing pipeline, and show numerical results both for speaker diarization and estimation of the classroom climate. Overall, we achieved a classification accuracy for classroom climate between 70 and 80%. These results are promising, since preschool classrooms tend to be noisy, and the audio was only recorded by a microphone worn by the teacher.

The rest of the paper is structured as follows. In Sect. 2, we briefly summarize similar work, while in Sect. 3, we discuss about the data and challenges that motivated this framework. In Sect. 4, we elaborate on the algorithms developed, and in Sect. 5 we discuss the results and findings, followed by conclusions in Sect. 6.

2 Related Work

In a related study, Donnelly et al. [2] developed a pipeline that automates Nystraand's CLASS coding scheme for dialogic instruction [3]. Natural language processing, utterance timing, and acoustic features were exploited to automatically detect teacher's questions ($F_1 = 0.69$) [2] and classify instructional segments ($F_1 = 0.60$), such as question-and-answer [4] in middle-school classroom audio. Most of Blanchard's studies on classroom discourse classification are centered around automated analysis of the teacher speech signals only, whereas we attempt to analyze both the teacher and children speech. Similarly, Wang et al. developed tools to automatically segment recordings from the Learning Environment Analysis system (LENA) [5, 6] into teacher speech, student speech, overlap, silence and discussion. Next, they classified the teaching behaviors by leveraging on conversational features (i.e., length of time student/teacher spoke/discussion). When given feedback on these features, teachers promoted desirable teaching behaviors by giving more time to students to speak, and by making more time for discussions [5]. Lastly, a large-scale study, spanning 1720 h of audio from college science courses [7], applied decibel analysis to automatically classify between lecture and non-lecture activities. The aforementioned studies all rely on speech features to automatically predict types of classroom activities, while in our study we aim to predict classroom assessment scores; specifically, as a first step, we attempt inferring the classroom climate.

3 Data and Challenges

The data is collected from 92 classrooms in reschools in Singapore by researchers of the National Institute of Education (NIE). The recordings typically last 20 min, with 10–15 students, and feature different classroom activities, such as small team activities (children sitting at tables with teacher walking around), and teacher-student discussions (students sitting around teacher). Each video was captured by one stationary camera, and the audio is recorded by a microphone worn by the teacher.

For the climate labels, two independent annotators scored the overall classroom climate in each video according to the rubrics outlined in the CLASS manual [1]. Annotators looked at dimensions of positive affect, relationships, positive communication, and respect during teacher-student and student-student interactions when coding for positive climate. Similarly, when coding for negative climate, they assessed dimensions of negativity, punitive control and disrespect.

In this paper, we analyzed only the microphone audio recordings. Out of 92 videos, we selected 12 videos to label audio ground truth (GT). Annotators were trained to label the audio w.r.t speakers, speaking time, and non-speech activities. The nomenclature is as follows: Teacher (S0), Children (S1), Overlap—when teacher and children speak together (S2), Silence (S3), and Noise (S4). The two annotators achieved an agreement of 80–95%, while labelling independently. We analyzed 80 audio recordings, which is about 27 h of audio (20 min $\times$ 80 audio recordings).

Building a robust speaker diarization system in this setting is a challenging task due to varying classroom settings. The audio is only recorded by a single microphone worn by the teacher, capturing not only teacher and children speech, but also background noise (e.g., crying and feet stamping). The speech of the children was not always captured with adequate fidelity and intelligibility, since the teacher is often walking around and the children might be far away from the microphone and might be speaking softly. As a result, the children speech is occasionally misclassified as noise. Most, if not all, state-of-the-art diarization algorithms were designed for single non-moving microphone. Therefore, the research is novel in the following ways. First, unlike studies which investigate structured settings, an ecological data set is considered here. Learning environments are more dynamic and less controlled than structured dialogs. Second, while the ground truth CLASS scores are coded w.r.t visual and audio information, we only used audio information to predict CLASS climate scores.

4 Proposed Approach

In this section, we present the proposed speech processing pipeline. First, the system classifies speech and non-speech events from the recordings (speech detection), it identifies "who spoke when" (speaker group diarization). Next, it extracts con-

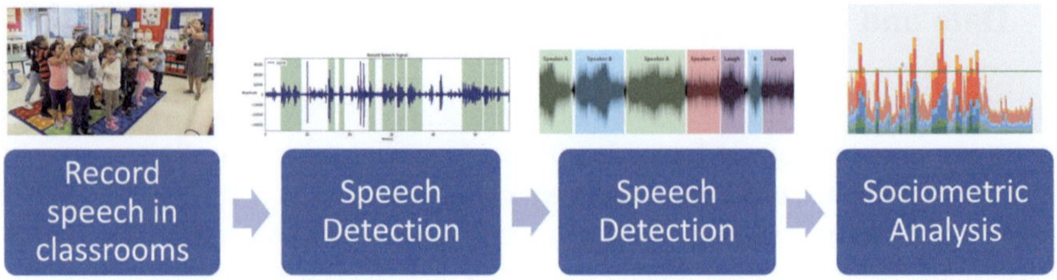

Fig. 1 Block diagram of the proposed approach

versational and low-level audio features of teacher, students, overlap, silence and noise segments and at last, it infers from those features the classroom climate. In the following we explain each step in detail (Fig. 1).

4.1 Speech Detection

It is advantageous to first separate speech from non-speech acoustic activity like music, background noise, noise made using toys, etc. [8]. We implemented the following approach: Silence is removed from audio by means of thresholding, and next the resulting audio is passed to a non-speech event detector. Some of the input features for this detector include energy and harmonic ratio to detect non-speech [9]. As a result, only the speech segments are retained.

4.2 Speaker Group Diarization

In this step, we identify different group of speakers present, and attribute each speaker with their spoken speech segments. We applied the LIUM toolkit [10], which performs diarization by extracting acoustic features, segmenting, and clustering. LIUM's default settings yielded poor results as it was originally designed for broadcast news recordings, and had difficulties parsing overlap speech of children and teacher.

4.3 Automated Speaker Labelling

LIUM provides general cluster names such as S100, S120, but for our purposes, these segments need to be labelled specifically as teacher (S0), children (S1), overlap (S2), silence (S3) and noise (S4). To this end, we developed an automated classifier to label each cluster. We computed the Euclidean distance between low-level audio features (e.g., MFCCs, pitch) extracted from each LIUM cluster and the manually labeled GT

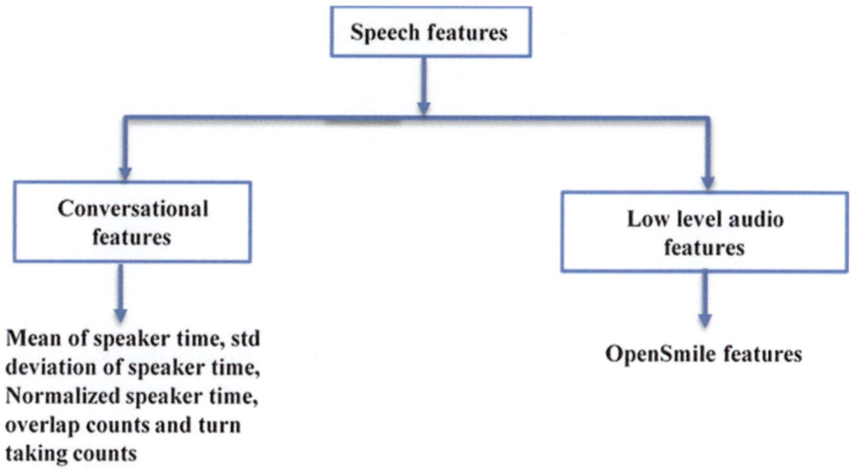

Fig. 2 Features for climate prediction

segments (of the 12 audio recordings used to train the speaker diarization system). We then assign the label (S0–S4) with the smallest Euclidean distance to each LIUM cluster.

4.4 Feature Extraction and Selection

We computed conversational features from speech sequences, e.g., normalized speaker time, speaker duration, and in addition, we extracted 988 low-level audio features for each speech segment by openSMILE [11]. A total of 5940 features for teacher, children and overlap segments were obtained (see Fig. 2). As the number of features grows, the computational complexity will increase, and performance of the classifiers may degrade. Therefore, we applied Kruskal-Wallis test, and correlation-based feature selection algorithms to select the most relevant features, which we then feed into classifiers for prediction of classroom climate. We assess these classifiers by 10-fold cross validation (CV).

5 Results and Discussion

5.1 Speaker Group Diarization

We trained and tested the speaker diarization system on 8 audio recordings (4 positive, 4 negative climate). To obtain reliable evaluation results, we applied leave-one-out

Table 1 Confusion matrices showing the results of speaker group diarization for 8 classroom conversations

(a) **Average of normalized confusion matrices (%)**			
True\estimate	Teacher	Children	Overlap
Teacher	77.3	21	1.7
Children	23.7	71.6	4.7
Overlap	75.4	21.6	3.1
(b) **Absolute confusion matrices (in seconds)**			
Teacher	2773.4	741.2	61.4
Children	618.4	1885.6	142.6
Overlap	483	169.2	29.9
(c) **Confusion matrix normalized by total time (%)**			
Teacher	40.2	10.7	0.9
Children	9.0	27.3	2.1
Overlap	7.0	2.5	0.4

CV. We trained the speaker diarization system on 7 audio recordings, tested it on the 8th audio recording, and repeated this procedure for all 8 recordings, and averaged the results. The proposed system yields an average accuracy of 77% for teacher and 72% for children (Table 1a). This level of accuracy is reasonable as the students' speech was not captured with high fidelity. The accuracy for overlap is low, as LIUM is not designed to detect overlap. The overlap segments are often misclassified as teacher speech, since the voice of the teacher is typically louder as the microphone is attached to the teachers. However, overlap is relatively infrequent in our recordings and not critical in our analysis (see Table 1c). Silence rarely occurs as the classes are quite active (Fig. 3).

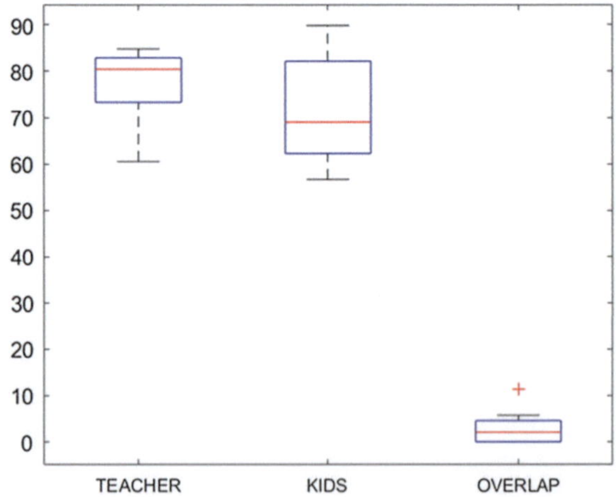

Fig. 3 Boxplot of accuracy of speaker diarization of 8 classroom conversations

5.2 Climate Prediction

Conversational features and low-level audio features are extracted from the individual speaker segments. The Kruskal-Wallis test is applied on all 5940 features to refine the discriminative features (with p-values < 0.005) for climate prediction (Fig. 4a). A total of 116 features (2%) satisfy this condition, of which 37 (32%), 36 (31%), and 43 (37%) features are associated with teacher, children and overlap respectively (Fig. 4b). Spectral features like MFCC, pitch and LSP have the smallest p-values. Such features carry emotional information because of its dependency on the tension of vocal folds [12]. The emotional content of speech is related to the acoustic characteristics of voice which can indicate emotions like joy, surprise, anger, or disgust [12]. The importance of these primary spectral features in emotion recognition has been established in [12, 13]. Typically, low-level audio features are better capable of climate detection than conversational features.

We reduced the feature set further by means of the random forest classifier approach [14] and by a correlation-based approach [15]. We trained 9 classifiers with this reduced feature set and applied 10-fold CV on 80 recordings. These 80 recordings are different from the 12 recordings used to train the speaker diarization system. The training labels are "Positive climate" (+1) and "Negative climate" (−1). The accuracy of the classifier for the 10 folds is shown in Fig. 5 and the corresponding mean accuracy is shown in Table 2. Overall, the classifiers yielded accuracies from

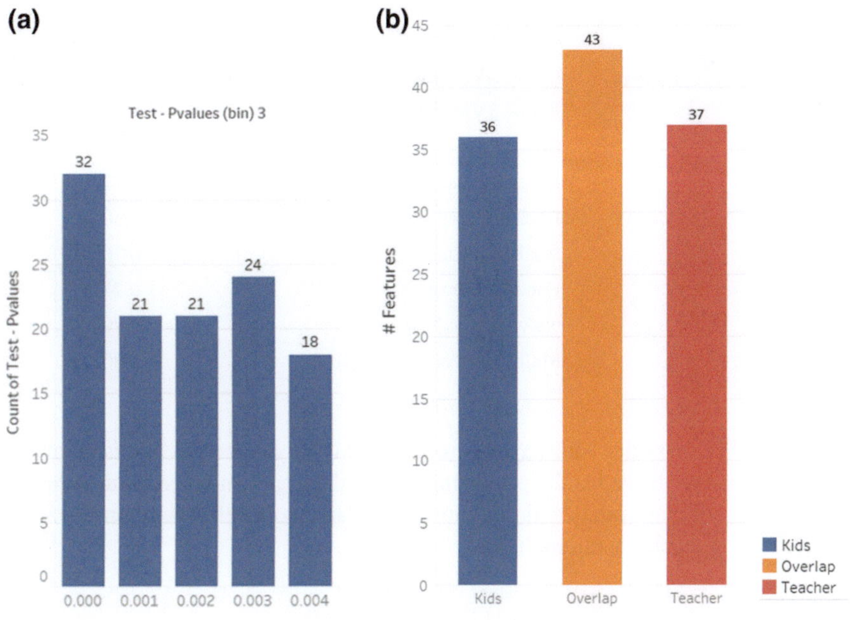

Fig. 4 Statistics of salient features. **a** Distribution of p-values; **b** Number of salient features associated with teacher and children speech, and overlap segments

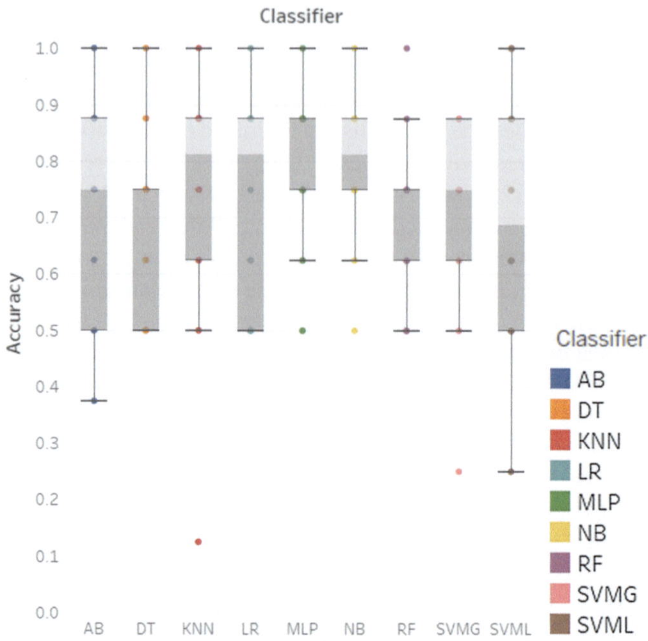

Fig. 5 Boxplot of accuracy of classifiers in 10 folds

Table 2 Mean of classification accuracy of 10 folds for different classifiers

Classifier	Accuracy
Linear SVM (Support Vector Machine)	0.71
Radial SVM (SVMG)	0.73
Logistic Regression(LR)	0.78
K Nearest Neighbors (kNN)	0.73
Decision Tree (DT)	0.60
AdaBoost Classifier (AB)	0.70
Random Forest Classifier(RF)	0.74
Naïve Bayes(NB)	0.79
Multilayer Perceptron (MLP)	0.80

70 to 80%. The classifier Multilayer Perceptron (MLP) yielded the best accuracy of 80%. We consider these results promising due to the noisy and unstructured nature of the preschool classrooms and that our entire framework relies only on the audio captured using teacher's microphone.

6 Conclusion

The experimental results indicate that prediction of classroom climate might be possible from non-speech verbal and prosodic cues, even from recordings captured by a single microphone worn by the teacher in noisy classrooms. The ability to predict climate scores from low-level indicators could be useful in the development of automated feedback tools for aiding professional teacher development. More testing is warranted to investigate how such a system can be adapted to suit the needs of professional development and classroom assessment. In future work, we will investigate video features and further validate and evaluate our system on a larger and more diverse dataset across more teachers, classroom sessions and class activities.

Acknowledgements This project is supported by a grant from Centre for Research and Development in Learning (CRADLE@NTU).

References

1. Pianta RC, La Paro KM, Hamre BK (2008) Classroom Assessment Scoring System™: manual K–3. Paul H Brookes Publishing
2. Donnelly PJ, Blanchard N, Olney AM, Kelly S, Nystrand M, D'Mello SK (2017) Words matter: automatic detection of teacher questions in live classroom discourse using linguistics, acoustics, and context. In: Proceedings of the 7th International Learning Analytics & Knowledge Conference, pp 218–227. ACM
3. Nystrand M (2004) CLASS 4.0 user's manual
4. Donnelly PJ, Blanchard N, Samei B, Olney AM, Sun X, Ward B, Kelly S, Nystrand M, D'Mello SK (2016) Multi-sensor modeling of teacher instructional segments in live classrooms. In: Proceedings of the 18th ACM international conference on multimodal interaction, pp 177–184. ACM
5. Wang Z, Miller K, Cortina K (2013) Using the LENA in teacher training: promoting student involvement through automated feedback. Unterrichtswissenschaft 4
6. Wang Z, Pan X, Miller KF, Cortina KS (2014) Automatic classification of activities in classroom discourse. Comput. & Educ 78:115–123
7. Owens MT et al (2017) Classroom sound can be used to classify teaching practices in college science courses. Proc Natl Acad Sci 114(12):3085–3090
8. Anguera X, Bozonnet S, Evans N, Fredouille C, Friedland G, Vinyals O (2012) Speaker diarization: a review of recent research. IEEE Trans Audio, Speech, Lang Proc 20(2):356–370
9. Giannakopoulos T, Pikrakis A (2014) Introduction to audio analysis: a MATLAB approach. Academic Press, Oxford
10. Meignier S, Merlin T (2009) Lium Spkdiarization: an Open Source Toolkit for Diarization. In: CMU SPUD Workshop
11. Eyben F, Wollmer M, Schuller B (2010) Opensmile: the munich versatileand fast open-source audio feature extractor. In: Proceedings of the 18th ACM international conference on multimedia. ACM, pp 1459–1462
12. Sezgin MC, Gunsel B, Kurt GK (2012) Perceptual audio features for emotion detection. EURASIP J. Audio, Speech, Music Proc 1:16
13. Gunes H, Schuller B, Pantic M, Cowie R (2011) Emotion representation, analysis and synthesis in continuous space: a survey. In: Face and gesture, pp 827–834
14. Breiman L (2011) Random forests. Mach Learn 4(1):5–32
15. Hall MA (1999) Correlation-based feature selection for machine learning

Automatic Turn-Level Language Identification for Code-Switched Spanish–English Dialog

Vikram Ramanarayanan, Robert Pugh, Yao Qian and David Suendermann-Oeft

Abstract We examine the efficacy of text and speech-based features for language identification in code-switched human-human dialog interactions at the turn level. We extract a variety of character- and word-based text features and pass them into multiple learners, including conditional random fields, logistic regressors and deep neural networks. We observe that our best-performing text system significantly outperforms a majority vote baseline. We further leverage the popular i-Vector approach in extracting features from the speech signal and show that this outperforms a traditional spectral feature-based front-end as well as the majority vote baseline.

1 Introduction

Code-switching refers to multilingual speakers' alternating use of two or more languages or language varieties within the context of a single conversation or discourse in a manner consistent with the syntax and phonology of each variety [1–4]. Increasing globalization and the continued rise of multilingual societies around the world makes research and development of automated tools for the processing of code-switched speech a very relevant and interesting problem for the scientific community since it has applications in multiple domains, including consumer/home electronics and business applications. In our case, an important additional motivating factor

V. Ramanarayanan (✉) · R. Pugh · Y. Qian · D. Suendermann-Oeft
Educational Testing Service R&D, 90 New Montgomery St, #1500,
San Francisco, CA, USA
e-mail: vramanarayanan@ets.org

R. Pugh
e-mail: rpugh@ets.org

Y. Qian
e-mail: yqian@ets.org

D. Suendermann-Oeft
e-mail: suendermann-oeft@ets.org

L. F. D'Haro et al. (eds.), *9th International Workshop on Spoken Dialogue System Technology*, Lecture Notes in Electrical Engineering 579,
https://doi.org/10.1007/978-981-13-9443-0_5

for studying and developing tools to elicit and process code-switched or crutched[1] language comes from the education domain, specifically language learning. Recent findings in the literature suggest that strategic use of code-switching of bilinguals L1 and L2 in instruction serves multiple pedagogic functions across lexical, cultural and cross-linguistic dimensions, and could enhance students' bilingual development and maximize their learning efficacy [6, 7]. This seems to be a particularly effective strategy especially when instructing language learners with low proficiency [8]. Therefore, the understanding of code-switched dialog and development of computational tools for automatically processing such code-switched conversations would provide an important pedagogic aid for teachers and learners in classrooms, and potentially even enhance learning at scale and personalized learning.

Automated processing of code-switched speech and dialog poses an interesting, albeit challenging problem for the scientific community. This is because the hurdles observed during traditional dialog processing tasks such as automatic speech recognition (ASR), spoken language understanding (SLU), natural language generation (NLG) and dialog management (DM) are exacerbated in the case of code-switched speech where the language the speaker is using at any given instant is not known apriori. Integrating an explicit *language identification* (or LID) step into the ASR module can alleviate these issues and improve user experience greatly. Take for example a use case of designing conversational applications for non-native English language learners (ELLs) from multiple native language (or L1) backgrounds. Many such learners tend to "crutch" on their L1 while speaking in the target language (or L2) that they are learning, especially if they are low proficiency learners [9], resulting in mixed-language speech. In such a case, LID becomes important not only for ASR, but also for DM, where the dialog designer/language expert may want the conversational agent to perform different dialog actions depending on whether the speaker used his/her L1 alone, the L2 alone, or a mixture of both during the previous turn.

Researchers have made significant progress in the automated processing of code-switched text in recent years [10–12]. Particularly relevant to our work is prior art on predicting code-switch points [13] and language identification [14, 15]. Researchers have made much progress on LID in code-switched text (tweets, in particular) thanks to recent workshops dedicated to the topic [12]. One of the top-performing systems used character n-gram, prefix and suffix features, letter case and special character features and explored logistic regression and conditional random field (CRF) learners to achieve the best performance for Spanish-English codeswitched text [16]. Yet another successful system leveraged bi-directional long short term memory networks (BLSTMs) and CRFs (along with word and character embedding features) on both Spanish-English and Standard Arabic-Egyptian language pairs [17].

While there is comparatively less work in the literature on automated analysis of code-switched speech and dialog, the number of corpora and studies is steadily growing in several language pairs—for instance, Mandarin–English [18, 19], Cantonese–English [20] and Hindi–English [21]. As far as dialog is concerned, the Bangor

[1]Crutching refers to language learners relying on one language to fill in gaps in vocabulary or knowledge in the other [5].

Corpus consists of human-human dialog conversations in Spanish–English, Welsh–English and Spanish–Welsh [22]. More recently, Ramanarayanan and Suendermann-Oeft (2017) also proposed a multimodal dialog corpus of human-machine Hindi–English and Spanish–English code-switched data [23]. To our knowledge, there is limited research on LID in code-switched speech and dialog—while certain works do use an LID system [24, 25] to improve the performance of code-mixed ASR, the LID component is baked into and uses the ASR setup. While this is perhaps the optimal way to proceed if one is only concerned with one or two language pairs, as we scale up code-switched dialog systems to multiple language pairs, building ASRs for each of the languages involved becomes difficult, especially keeping in mind SLU and DM. Hence, this paper explores an ASR-free approach to turn-level LID in code-switched dialog, exploring the efficacy of both text-based and speech-based features on a single corpus of code-switched data. To our knowledge, this is the first such exploration of both text and speech features for turn-level LID in human-human dialog data.

The rest of this paper is organized as follows: Sect. 2 describes the Bangor Miami corpus used for our turn-level LID experiments. We then elucidate the various text and speech features used in our experiments in Sect. 3, followed by the experimental setup in Sect. 4. Section 5 presents the results of our LID experiments as well as analyses on the different factors affecting classification accuracy. Finally, we conclude with a discussion of current observations and an outlook for future work in Sect. 6.

2 Data

We used the Bangor Miami corpus[2] of code-switched human–human dialog in English and Spanish for our turn-level LID experiments. The corpus consists of 56 audio recordings and their corresponding transcripts of informal conversations between two or more speakers, involving a total of 84 speakers living in Miami, Florida (USA). In total, the corpus consists of 242,475 words of text from 35 h of recorded conversation. 63% of the transcribed words are English, 34% Spanish, and 3% are undetermined (Table 1). The manual transcripts include beginning and end times of utterances and per word language identification. For our experiments, we excluded turns containing tokens with ambiguous or undetermined language.

The audio was split into turns as specified by the provided timestamps. Each turn was downsampled to 8 kHz and converted to a single channel. The transcriptions were processed by performing whitespace tokenization on each turn, and removing event descriptions (such as “&=laugh”) and unintelligible tokens.

[2]http://bangortalk.org.uk/.

Table 1 Corpus statistics

Item	Bangor Miami
Number of turns collected	35428
Utterance-level language use or codeswitching percentage	English: 65%
	Spanish: 29%
	Both: 6%

3 Feature Extraction

We performed turn-level LID experiments using speech-only features as well as text features (which serve as a benchmark since they directly contain linguistic information), and compared them to a majority vote (or chance) baseline. In this section, we will first describe the various text and speech features explored, followed by the machine learning setup in the subsequent section.

3.1 Text Features

Following earlier work [16, 17], we experimented with the following low-level binary text features that capture the presence or absence of the following:

- **Word n-grams**: We used a bag-of-words representation, trying uni- and bi-grams.
- **Character n-grams**: The set of unique character n-grams ($1 \leq n \leq 4$), without crossing word-boundaries. For example, the word sequence "la sal" would produce the following character n-grams {'l', 'a', 's', 'al', 'la', 'sa', 'sal'}.
- **Character Prefixes/Suffixes**: All affixes with length ≤3. For example, the word "intricate" would have prefixes {'i', 'in', 'int'}, and suffixes {'ate', 'te', and 'e'}.

Additionally, for one experiment (LSTM), we used randomly initialized word embeddings as input features, which were trained as part of the network.

3.2 Speech Features

We explored two featuresets for our speech experiments—OpenSMILE features and i-Vector features. We used the OpenSMILE toolkit [26] to extract features from the audio signal—specifically, the standard emobase2010 feature set containing 1582 features that is tuned for recognition of paralinguistic information in speech. These consist of multiple low-level descriptors—intensity, loudness, MFCCs, pitch, voicing probability, F0 envelope, Line Spectral Frequencies (LSFs) and zero crossing rate, among others—as well as their functionals (such as standard moments).

We also trained a GMM-based i-Vector system (see [27]) using the Kaldi toolkit [28]. Initially introduced for speaker recognition [29], i-Vectors have also been shown to be particularly useful features for language recognition (see for example [30]). The i-Vector extraction procedure can be viewed as a probabilistic compression process that maps input speech features into a reduced dimensionality space using a linear Gaussian model—for more details, see [29]. The front-end for the extracts were 20-dimensional MFCCs including C0, using a 20 ms Hamming window with 10 ms time shift along with their first and second derivatives. We deleted non-speech segments within utterances through an energy-based voice active detection (VAD) method, and performed utterance-based cepstral mean normalization on the acoustic feature vectors. We trained a GMM and a full covariance matrix as the Universal Background Model (UBM) by using the entire Fisher English [31] and Spanish corpora, in addition to the Bangor corpus data (we did not include code-switched turns). We then used the Bangor Corpus to train an i-Vector extractor T-matrix. The number of Gaussian components and the i-Vector dimensions were set to 1024 and 800, respectively.

4 Experiments

We randomly partitioned the Bangor corpus data into train and test sets using an 80–20% split. For experiments with text-based featuresets, we first extracted the word and character level features described in Sect. 3. We then tried two approaches to predicting one of 3 classes—English, Spanish or Code-switched—at the turn-level: (i) Use a CRF to make word-level predictions, and aggregate them to form a turn-level prediction, and (ii) aggregate the features at the turn level and try a variety of learners, including logistic regression and deep neural networks to make language predictions at the turn level. Additionally, we tried passing sequences of word-embeddings (randomly initialized and trained on the train partition of the Bangor corpus) to an LSTM and making an LID prediction for each turn. We experimented with different learner configurations and parameter settings and summarize the best performing featureset and learner combination in the Results section.

We set up the speech experiments using the following steps:

1. Partition the full audio files into train and test sets. We hold out any code-switched turns from the training partition at this step in order to train a two-class UBM model for i-Vector extraction.
2. Segment each turn into a sequence of two-second segments, and extract an 800-dimensional i-Vector for each. For this step, 80% of the code-switched i-Vectors are randomly moved back to the training partition while the remaining 20% were moved to the test partition. (Note that during this process we ensure that we use the *same train-test partitions as we are using for the text-based systems* to enable a fair comparison of systems).

3. For each turn, we generate a three-dimensional vector consisting of (a) the Euclidean distance of the segments from the average English segment, (b) the distance from the average Spanish segment, and (c) the length of the turn in seconds. We also experimented with using Cosine distance, as well as each segment's PLDA score for the respective classes, instead of the Euclidean distance.
4. Optionally, use SMOTE oversampling [32] to overcome class imbalance, ensuring that there Spanish and English classes have the same size (code-switched turns were not oversampled[3]).
5. Fit an appropriate learner (such as a Linear Discriminant Analysis classifier) in order to predict the turn-level language of each turn in the test set.

5 Observations and Analysis

Table 2 lists the best performing text and speech systems, including the feature sets and model details. We observe that the best text system significantly outperforms the majority vote baseline by a huge margin, with an overall weighted average F1 score of 0.97 and an F1 score of 0.91 for the code-switched category. We also observe that character- and word-level features combined with a conditional random field (CRF) classifier perform slightly better than word embedding features fed into a long-short term memory neural network (LSTM).

On the other hand, LID performance dips for speech-based systems relative to their text-based counterparts for the systems we investigated, with the best performing speech system (using i-Vectors and turn length) yielding an overall weighted average F1 score of 0.67. In this case, the F1 score for the code-switched class is pretty low—0.19. A closer examination of the confusion matrices for the best performing text and speech systems (Tables 3 and 4 respectively) provides more insight into the performance gap between the two—the text systems have a lot less confusability between classes, especially for the code-switched class. While this is not entirely surprising given that we are not directly incorporating linguistic information into our speech feature front-ends, and the large amount of far-field noise and background chatter in the speech data, this result is still in much need of improvement. Having said that, all speech systems still perform well over the majority vote baseline F1 score of 0.51.

In order to investigate to what extent the smaller sample size of the code-switched category is responsible for bringing down the LID performance of the speech-based systems, we looked at the performance of our speech systems on a 2-class (English–Spanish) classification problem at both the segment and the turn level by removing all code-switched turns from our training and test data. For the 2-class segment classification, we experimented with using the segment i-Vectors, using the Euclidean

[3]We did experiment with oversampling the code-switched class as well, but this resulted in a degradation in performance. This could probably be due to the relatively few samples in the code-switched class to begin with.

Table 2 Performance of speech and text systems (Eng: English, Spa: Spanish, CS: Code-switched). The weighted average F1 score of the best-performing text and speech systems are in bold

System	Featureset	Machine learner	F1 per class			Weighted ave. F1
			Eng	Spa	CS	
Text	Word, character n-grams (1–4), character affixes (1–3) for current, previous, and next word	CRF, aggregated for turn-level predictions	0.98	0.95	0.91	**0.97**
	Word embeddings	LSTM	0.98	0.95	0.9	0.96
	Word n-grams (1–2), character n-grams (1–4), character affixes (1–3), turn-length in tokens	Logistic regression	0.97	0.93	0.67	0.94
Speech	Euclidean distance of segments from mean English and mean Spanish i-Vectors, length of turn in seconds	Linear Discriminant Analysis classifier (with SMOTE oversampling)	0.68	0.71	0.19	**0.67**
	OpenSMILE features	Linear Discriminant Analysis classifier	0.75	0.20	0.16	0.55
	Sequence of i-Vectors	LSTM[a]	0.77	0.36	0.15	0.61
Majority baseline	N/A	N/A	0.79	0.0	0.0	0.51
Random baseline	N/A	N/A	0.44	0.32	0.09	0.39

[a]We used an LSTM implementation with 200 units and a *tanh* activation function. We optimized on a categorical cross-entropy loss function using the Adam optimizer

Table 3 Confusion matrix for best-performing text system

	English	Codeswitched	Spanish
English	4513	8	18
Codeswitched	20	352	37
Spanish	122	7	1936

Table 4 Confusion matrix for best-performing speech system

	English	Codeswitched	Spanish
English	3259	504	776
Codeswitched	217	97	95
Spanish	1585	1	2953

and Cosine distance between each segment and the mean vector for each class, as well as the segment-level PLDA scores for English and Spanish as features, with a the same set of learners tested on other experiments, and report the weighted average F1 for the best performing system among these. We further examined the effect of class imbalance in the Bangor corpus—recall that the corpus contains 64% English turns, 30% Spanish turns, and 6% code-switched turns—on system performance. Table 5 lists the results of these experiments. We observe that in general, the English–Spanish "monolingual" classifiers do much better in terms of class-specific F1 scores

Table 5 Speech system varieties

	Segment or turn	SMOTE?	N (train)	N (test)	English	Spanish	CS	Avg (weighted)
Monolingual	Segment	No	29,583	7,073	0.85	0.55	N/A	0.76
		Yes	39,488	10,016	0.72	0.81	N/A	0.77
	Turn	No	26,708	6,604	0.85	0.50	N/A	0.74
		Yes	36,062	9,078	0.75	0.71	N/A	0.73
3-Class	Turn	No	28,415	7,013	0.77	0.50	0.21	0.66
		Yes	37,769	9,487	0.68	0.71	0.19	0.67

than the "3-class" classifiers. Using the synthetic minority oversampling technique (SMOTE) to overcome class imbalance also helped boost performance, suggesting that class imbalance and the small amounts of code-switched data contributed to a performance drop. Finally, we also hypothesize that audio quality issues in the noisy Bangor corpus speech data might have contributed to the lower numbers of the speech system. Systematically investigating this hypothesis is a subject for future research.[4]

6 Discussion and Outlook

We have presented an experimental evaluation of different text and speech-based featuresets in performing language identification at the turn level in human-human dialog interactions. While the best text-based system performed excellently and at par with the state of the art in the field, the best speech-based i-Vector system did not perform as well, but still significantly outperformed the majority vote chance baseline. We observed that one of the reasons for the relatively poor performance of the speech-based system could be the relatively noisy audio that contains significant amounts of far-field and background noise. This, along with the greater percentage of English than Spanish or code-switched turns in the database, might have contributed to the performance drop. However, note that as in the case of the text-based systems, we are not directly using any linguistic or syntactic information for the speech-based systems, which undoubtedly impacts the efficacy of the latter, since such information is extremely useful for the task of language identification.

[4]In order to roughly test this hypothesis, we ran experiments wherein we used the relatively cleaner Fisher corpora (of both Spanish and English speech) for both training and testing. In this case, the F1 score obtained was 0.96, highlighting both the mismatch between the Fisher and Bangor corpora as well as the effect of noise in the Bangor corpus. Of course, there is the possibility that the 2-class classification of English and Spanish turns from monolingual turns in code-switched speech might pose more challenges than LID in *non*-code-switched speech. Nevertheless, while this test was not a systematic one (and hence reported only as a footnote), this clearly points toward the effect of dataset quality on system performance.

That having been said, we will explore a number of potential avenues for improving the performance of the speech-based LID systems. Chief among these, as mentioned earlier, will be to investigate techniques for noise-robustness in order to improve the LID performance of speech systems and bring them at par with their text counterparts. In addition, we would like to explore the performance of more feature–learner systems, including a more comprehensive study of deep neural network-based learners. Finally, it will be important to see how such LID systems perform on different code-switched datasets, both within and across language pairs, in order to truly test the robustness of systems across languages and dataset bias.

Going forward, understanding and processing code-switched speech has many implications for building code-switching dialog systems. For instance, integrating an explicit language identification step into the automatic speech recognition (ASR) module could help enhance the recognition performance. However, such solutions still require one to develop an ASR for each of the languages being analyzed. This starts becoming increasingly impractical if one wants to scale applications to multiple language pairs—an example use case is in the case of designing dialog solutions for non-native English language learners (ELLs) from multiple native language (or L1) backgrounds. In such cases, research into end-to-end spoken language understanding or SLU (where we directly go from the speech input to the SLU hypothesis) becomes very useful; and language identification would be a key component of such modules. Over and above SLU applications, such an LID module might also help inform pragmatic considerations during dialog management and the language generation module for the generation of appropriate mixed-language output. We therefore believe that in many ways this study has just scratched the surface of interesting and relevant research directions in the automated processing and modeling of code-switched dialog.

References

1. Milroy L, Muysken P (1995) One speaker, two languages: cross-disciplinary perspectives on code-switching. Cambridge University Press
2. Wei L (2000) The bilingualism reader. Psychology Press
3. MacSwan J (2004) Code switching and grammatical theory. In: The handbook of bilingualism, vol 46, p 283
4. Myers-Scotton C (2006) Codeswitching with English: types of switching, types of communities. World Engl Crit Concepts Linguist 4(3):214
5. OConnor BH, Crawford LJ (2015) An art of being in between: the promise of hybrid language practices. In: Research on preparing inservice teachers to work effectively with emergent bilinguals. Emerald Group Publishing Limited, pp 149–173
6. Wheeler RS (2008) Code-switching. In: Educational leadership
7. Jiang Y-LB, García GE, Willis AI (2014) Code-mixing as a bilingual instructional strategy. Biling Res J 37(3):311–326
8. Ahmad BH, Jusoff K (2009) Teachers code-switching in classroom instructions for low English proficient learners. Engl Lang Teach 2(2):49
9. Littlewood W, Yu B (2011) First language and target language in the foreign language classroom. Lang Teach 44(1):64–77

10. Solorio T, Blair E, Maharjan S, Bethard S, Diab M, Gohneim M, Hawwari A, AlGhamdi F, Hirschberg J, Chang A et al (2014) Overview for the first shared task on language identification in code-switched data. In: Proceedings of the first workshop on computational approaches to code switching. Citeseer, pp 62–72
11. Bali K, Vyas Y, Sharma J, Choudhury M (2014) I am borrowing ya mixing? An analysis of English-Hindi code mixing in facebook. In: Proceedings of the first workshop on computational approaches to code switching, EMNLP 2014, p 116
12. Molina G, Rey-Villamizar N, Solorio T, AlGhamdi F, Ghoneim M, Hawwari A, Diab M (2016) Overview for the second shared task on language identification in code-switched data. In: EMNLP 2016, p 40
13. Solorio T, Liu Y (2008) Learning to predict code-switching points. In: Proceedings of the conference on empirical methods in natural language processing. Association for Computational Linguistics, pp 973–981
14. Barman U, Das A, Wagner J, Foster J (2014) Code mixing: a challenge for language identification in the language of social media. In: EMNLP 2014, vol 13
15. King B, Abney SP (2013) Labeling the languages of words in mixed-language documents using weakly supervised methods. In: HLT-NAACL, pp 1110–1119
16. Shirvani R, Piergallini M, Gautam GS, Chouikha M (2016) The Howard University system submission for the shared task in language identification in Spanish-English codeswitching. In: Proceedings of the second workshop on computational approaches to code switching, pp 116–120
17. Samih Y, Maharjan S, Attia M, Kallmeyer L, Solorio T (2016) Multilingual code-switching identification via LSTM recurrent neural networks. In: EMNLP 2016, p 50
18. Li Y, Yu Y, Fung P (2012) A Mandarin-English code-switching corpus. In: LREC, pp 2515–2519
19. Lyu D-C, Tan T-P, Chng E-S, Li H (2015) Mandarin-English code-switching speech corpus in South-East Asia: SEAME. Lang Resour Eval 49(3):581–600
20. Chan JY, Ching P, Lee T (2005) Development of a Cantonese-English code-mixing speech corpus. In: INTERSPEECH, pp 1533–1536
21. Dey A, Fung P (2014) A Hindi-English code-switching corpus. In: LREC, pp 2410–2413
22. Donnelly K, Deuchar M (2011) The bangor autoglosser: a multilingual tagger for conversational text. ITA11. Wrexham, Wales
23. Ramanarayanan V, Suendermann-Oeft D (2017) Jee haan, I'd like both, por favor: elicitation of a code-switched corpus of Hindi-English and Spanish-English human-machine dialog. Proc Interspeech 2017:47–51
24. Vu NT, Lyu DC, Weiner J, Telaar D, Schlippe T, Blaicher F, Chng E-S, Schultz T, Li H (2012) A first speech recognition system for Mandarin-English code-switch conversational speech. In: 2012 IEEE international conference on acoustics, speech and signal processing (ICASSP). IEEE, pp 4889–4892
25. Yeh C-F, Sun L-C, Huang C-Y, Lee L-S (2011) Bilingual acoustic modeling with state mapping and three-stage adaptation for transcribing unbalanced code-mixed lectures. In: 2011 IEEE international conference on acoustics, speech and signal processing (ICASSP). IEEE, pp 5020–5023
26. Eyben F, Weninger F, Gross F, Schuller B (2013) Recent developments in openSMILE, the Munich open-source multimedia feature extractor. In: Proceedings of the 21st ACM international conference on multimedia. ACM, pp 835–838
27. Qian Y, Evanini K, Wang X, Suendermann-Oeft D, Pugh RA, Lange PL, Molloy HR, Soong FK (2017) Improving sub-phone modeling for better native language identification with non-native English speech. Proc Interspeech 2017:2586–2590
28. Povey D, Ghoshal A, Boulianne G, Burget L, Glembek O, Goel N, Hannemann M, Motlicek P, Qian Y, Schwarz P et al (2011) The Kaldi speech recognition toolkit. In: IEEE 2011 workshop on automatic speech recognition and understanding, no. EPFL-CONF-192584. IEEE Signal Processing Society

29. Dehak N, Kenny PJ, Dehak R, Dumouchel P, Ouellet P (2011) Front-end factor analysis for speaker verification. IEEE Trans Audio Speech Lang Process 19(4):788–798
30. Martinez D, Plchot O, Burget L, Glembek O, Matějka P (2011) Language recognition in ivectors space. In: Twelfth annual conference of the international speech communication association
31. Cieri C, Miller D, Walker K (2004) The fisher corpus: a resource for the next generations of speech-to-text. In: LREC, vol 4, pp 69–71
32. Chawla NV, Bowyer KW, Hall LO, Kegelmeyer WP (2002) Smote: synthetic minority over-sampling technique. J Artif Intell Res 16:321–357

Part II
Dialogue Management and Pragmatic Models

Spoken Dialogue System for a Human-like Conversational Robot ERICA

Tatsuya Kawahara

Abstract This article gives an overview of our symbiotic human-robot interaction project, which aims at an autonomous android who behaves and interacts just like a human. A conversational android ERICA is designed to conduct several social roles focused on spoken dialogue, such as attentive listening (similar to counseling) and job interview. Design principles in developing these spoken dialogue systems are described, in particular focused on the attentive listening system. Generation of backchannels, fillers and laughter is also addressed to make human-like conversation behaviors.

1 Introduction

In the past decade, spoken dialogue systems (SDSs) have become prevailing in smartphones, car navigation systems, and smart speakers. They provide information services on many domains such as weathers and public transportation as well as some chatting function. They are useful, but the dialogue itself is very constrained in that users need to think over what the system can, and then utter one simple sentence with clear articulation before getting a response. Apparently, there is a big gap from human dialogue such as those provided by a tourist guide and a hotel concierge.

A majority of the current SDSs assume one sentence per one turn, and they respond only when the users ask. In human-human dialogue, on the other hand, we utter many sentences per one turn while the listeners make backchannels occasionally. This kind of human-like dialogue manner is necessary for humanoid robots which will be engaged in social or household roles, because humans naturally expect the humanoid robots to show human-like behaviors. We also expect the humanoid robots with the human-like conversation capability will be used in a variety of domains.

Above all, most current SDSs regard dialogue as a means to conduct some tasks by machines such as operations and information retrieval. In these tasks, an objective

---

T. Kawahara (✉)
School of Informatics, Kyoto University, Kyoto, Japan
e-mail: kawahara@i.kyoto-u.ac.jp

L. F. D'Haro et al. (eds.), *9th International Workshop on Spoken Dialogue System Technology*, Lecture Notes in Electrical Engineering 579,
https://doi.org/10.1007/978-981-13-9443-0_6

goal is defined and should be completed as soon as possible. Thus, their design principle and evaluation criteria are to make the dialogue as efficient as possible. For humans, on the other hand, a dialogue itself can be a task, for example, explanation, persuasion and consulting. Note that the goals of these kinds of dialogue are not necessarily definite, and thus interaction and redundancy are essential. We expect humanoid robots will be engaged in some of these real communication services in the future symbiotic society.

This is the motivation of our Symbiotic Human-Robot Interaction (HRI) project sponsored by the JST ERATO program, which started in 2016. The major focus is placed on the development of an autonomous android ERICA with human-like verbal and non-verbal interaction capability [1–3]. It is the major distinction of this project that places a priority on the human-like interaction rather than pragmatic applications. Another distinction of the project is the real collaboration between the (author's) spoken dialogue research laboratory and (Prof. Ishiguro's) robotics research laboratories.

2 ERICA Project

2.1 Research Goal

The goal of the project is an autonomous android ERICA, who looks, behaves and interacts exactly like a human. This involves not only spoken dialogue but also eye gaze, head movement, and gestures. A snapshot of ERICA is shown in Fig. 1.

Our ultimate goal is to pass a "Total Turing Test", convincing people it is comparable to a human. It would not be easy to make it entirely indistinguishable from a human in the foreseeable future, but our goal is to make the interaction with ERICA

Fig. 1 A snapshot of ERICA

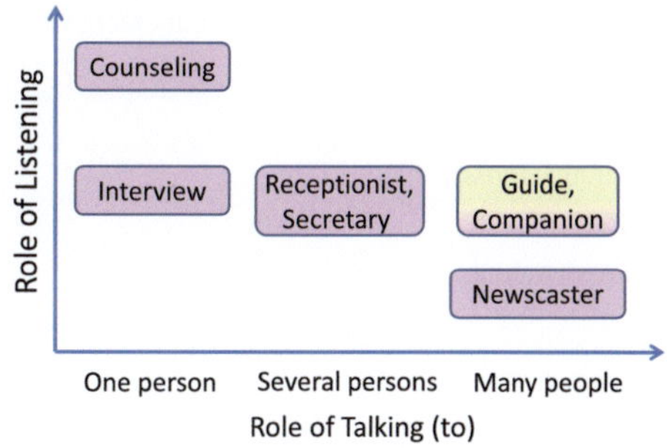

Fig. 2 Social roles of ERICA

is as engaging as that with a human. In terms of spoken dialogue, a Turing Test can be conducted by the comparison against the remote-operated android or WoZ setting.

As this is still challenging in an open domain, we set up several social tasks designated for ERICA, as listed in Fig. 2. We believe these are promising applications of humanoid robots. On the other hand, we hope the challenge of the Total Turing Test will reveal what is missing in the current interaction and critical for natural interaction.

2.2 *Social Interaction Tasks of ERICA*

We have explored a number of social roles suited to the android ERICA that can take advantage of human-like presence and would realize human-like dialogue. A majority of tasks conducted by the current SDSs such as information services are not adequate; they are better suited to smartphones and smart speakers. While most conventional robots are engaged in physical tasks such as moving objects, recently many kinds of communication robots are designed and introduced in public spaces. They serve as a companion [4] or a receptionist [5]. They are effective for attracting people, but the resulting interaction is usually very shallow and short such as greeting and short chatting. In recent years, chatting systems are also developed intensively, but most of the generated responses are boring or non-sense.

In contrast with these tasks, we design "social interaction" tasks in which human-like presence matters and deep and long interaction is exchanged. Here, dialogue itself is a task, and the goal of dialogue may be mutual understanding or appealing. We assign a realistic social role to ERICA, so matched users will be seriously engaged beyond chatting. Specifically, we set up the following three tasks. Note that these are all dyadic style, and face-to-face interaction including non-verbal communication is important.

1. Attentive listening
   In this task, ERICA mostly listens to senior people talking about topics such as memorable travels and recent activities [6]. Attentive listening is being recog-

Table 1 Comparison of three social interaction tasks

	Attentive listening	Job interview	Speed dating
Dialogue initiative	User	System	Both (mixed)
Utterance mostly by	User	User	Both
Backchannel by	System	System	Both
Turn-switching	Rare	Clear	Complex

nized as effective for maintaining the communication ability of senior people, and many communication robots are designed for this task. The role of ERICA is to encourage users to speak for long. In this sense, attentive listening is similar to counseling [7].

2. Job interview (practice)
   While dialogue systems have been investigated for casual interviews [8], a job interview is very important for both applicants, typically students, and companies hiring them. Each side makes a lot of preparations including rehearsal. In this setting, ERICA plays the role of interviewer by asking questions. She provides a realistic simulation, and is expected to replace a human interviewer in the future.
3. Speed dating (practice)
   Speed dating is widely held for giving an opportunity for people to find a partner. In Japan, there was a convention of arranged meeting for marriage, named *Omiai*, which is set up for one couple. In these settings, two persons meet for the first time and talk freely to introduce themselves, and see if the counterpart can be a good match. There was a study [9] that analyzed a corpus of speed dating. In our setting, ERICA plays the role of the female participant by talking about topics such as hobbies and favorite foods. She provides a realistic simulation and gives proper feedbacks according to the dialogue.

While these three tasks share the key characteristics that face-to-face interaction is important, they are different in the nature of dialogue as listed in Table 1.

2.3 Dialogue Data Collection

We have set up an environment for multi-modal recording of dialogue sessions of ERICA and subjects in the Wizard of Oz (WoZ) setting. We recruited four female actresses as an operator of ERICA. They not only respond to the user utterances but also operate ERICA's motion and eye gaze using haptic devices.

A subject sits in front of ERICA over a round table, and is engaged in dialogue for about 10 min. Microphone arrays, cameras, and Kinect sensors are set up on and around the table.

As of April 2018, we have collected 19 sessions of attentive listening, 30 sessions of job interview, and 33 sessions of speed dating.

2.4 Research Issues

There are many research topics involved in this project.

1. Robust automatic speech recognition (ASR)
   Humanoid robots need to deal with distant and conversational speech. This calls for integration of front-end microphone-array processing with back-end acoustic and language models.
   We note that when people speak without a microphone, the speaking style becomes so casual that it is not easy to detect utterance units. This is a serious problem in human-robot interaction, but circumvented in smartphones by using the push-to-talk interface and in smart speakers by forcing a magic word.
2. Flexible dialogue
   Humanoid robots need to deal with conversations without definite goals. This requires language understanding and generation without well-defined semantic slots. Moreover, natural turn-taking and backchanneling capability are essential for human-like interactions.
3. Natural speech synthesis
   Speech synthesis should be designed for the conversational style rather than text-reading applications, which are conventional targets of text-to-speech (TTS). Moreover, a variety of non-lexical utterances such as backchannels, fillers and laughter are needed with a variety of prosody.

Moreover, latency is critical for human-like conversation. In our corpus analysis of the WoZ dialogue, we find the average turn-switch interval is approximately 500, and 700 ms would be too late for smooth conversation. Cloud-based ASR and TTS services can hardly meet this requirement. The ASR system has been based on Julius,[1] but recently we have developed an acoustic-to-word end-to-end ASR system, which realizes a real-time factor of 0.03 [10]. The speech synthesis is based on VoiceText,[2] but enhanced with non-lexical tokens mentioned above. All downstream natural language processing (NLP) modules and motion generation modules are tuned to run within 200 ms.

Hereafter, this article focuses on the flexible dialogue designed for the social interaction tasks mentioned in the previous section.

3 Spoken Dialogue System Design

There are several different approaches to spoken dialogue design, which are described as follows.

[1] http://julius.osdn.jp.

[2] http://voicetext.jp.

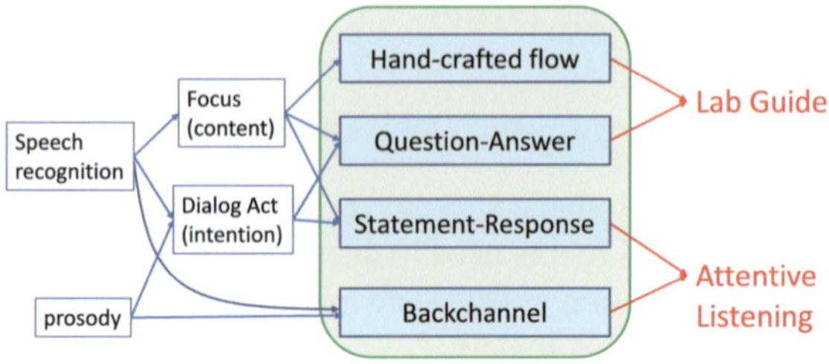

Fig. 3 Hybrid architecture of spoken dialog systems

- State-transition flow
  A flow of dialogue is hand-crafted as a finite state machine (FSM). This approach has been widely adopted in a limited task domain with a definite task goal, such as filling forms and making transactions. It can also be extended to scenario-based dialogue such as tourist guide. Although it allows for deep interaction, it works only in narrow domains and cannot cope beyond the prepared scenario.
- Question-Answering
  The question-answering (QA) technology is widely used in smartphone assistants and smart speakers. It handles a wide variety of user queries such as weathers and news, and searches the relevant database or Internet to generate an answer. This approach provides wide coverage, but only short interaction. Moreover, it cannot cope beyond the prepared database; in this case, the system simply outputs web search results.
- Statement-Response
  The statement-response function is incorporated in ChatBot and smartphone assistants. There are several approaches to realize this function. One is to prepare a set of statement-response pairs, which can generate relevant responses to a limited number of patterns. Alternatively, the system tries to generate a sentence based on the current focus words, or simply outputs formulaic responses such as "Okay".

The above-mentioned approaches have different advantages and disadvantages, and a combination of them would make a complex system, which a humanoid robot should be equipped with. Figure 3 depicts a hybrid system design based on these modules. For example, a lab guide system can be composed of a hand-crafted flow and some question-answer function. For this architecture, dialogue act recognition and domain recognition are needed to classify user inputs into the appropriate module. The focus word is also detected and tracked to generate relevant responses.

Moreover, we incorporate a backchanneling function as a separate module in order to generate a human-like behavior.

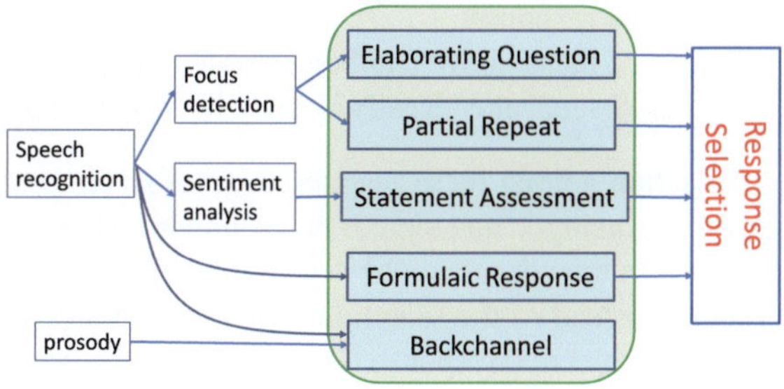

Fig. 4 Architecture of the attentive listening system

4 Attentive Listening System

A major demand for communication robots is to listen to users talking. Talking about troubles occasionally clarifies and solves them. Talking by remembering is important for maintaining the communication ability of senior people. People can talk to pets and dolls instead of a human. That is the reason why many communication robots are designed for attentive listening.

The system should encourage users to talk more smoothly. It needs to respond to any inputs, but does not require a large knowledge base. It is also important to show empathy and entrainment. Therefore, natural and smooth backchanneling is critically important, which is described in the next section.

Based on the techniques of counseling and attentive listening, we have designed an attentive listening system [6], as shown in Fig. 4.

The statement-response function is realized with the four modules.

- Elaborating questions
  Questions are generated based on a focus word. The system tries to combine it with WH phrases (i.e. which, where..) to generate a set of possible questions. The most plausible question is selected based on the N-gram probability [6] computed with a large corpus. For example, given an input "I went to a conference", the system asks "Which conference?"
- Partial repeat
  If none of the elaborating questions is generated, the system can simply respond with the focus word. For example, given an input "I went to Okinawa", the system says "Okinawa?" This simple response shows understanding of the user utterance and encourages the user to talk more. Actually, an effective dialogue strategy is to first output this partial repeat and then generate an elaborating question if the user does not continue to talk.
- Statement assessment
  Sentiment analysis is used to generate an assessment of the user utterances. We prepared specific responses for sentiment categories, for example, "That's nice" for positive and objective facts. These responses should be generated with confidence after listening in a sufficient amount of time.

- Formulaic response
  If none of these above can be generated, the system outputs formulaic responses such as "I see." and "Really?" They have a similar function to backchannels.

Selection among these responses is not so simple. There are many possible responses to an input, and there is no ground truth because responses other than the one observed in the corpus can be acceptable. Therefore, we formulate as the validation problem instead of the selection problem. Namely, we design machine learning to judge if a generated response is acceptable given the linguistic and dialogue context. But this requires extra annotation of all possible responses which are not observed in the corpus. We may turn to crowd-sourcing, but this is the problem to be solved for large-scale training.

Currently, we have made evaluations with the 19 dialogue sessions of attentive listening to senior subjects, and achieved a recall of 70% and a precision of 73% for valid responses on average. This is much better than the case randomly generating according to the distribution in the corpus and the result of training with the corpus occurrence only.

During the corpus collection in the WoZ setting, almost all senior subjects believed that they were talking to ERICA, without awareness of the operator. This means that we can pass a Turing Test if we can generate the same (appropriate) responses for the corpus. But we could achieve 70% at this moment.

We have also conducted trials of the system by asking senior people to talk to the autonomous ERICA. In almost all cases, they can be engaged in dialogue lasting five minutes without serious troubles. Although a quantitative evaluation is yet to be done, ERICA occasionally made occasionally appropriate responses, while she only backchanneled most of the time in some sessions.

5 Generation of Backchannels, Fillers and Laughter

Another factor for human-like conversation is generation of non-lexical utterances. They include backchannels, fillers and laughter.

5.1 Backchannel Generation

Backchannels provide feedback for smooth communication, by indicating the listener is listening, understanding, and agreeing to the speaker. Some types of backchannels are used to express the listener's reactions such as surprise, interest and empathy. Moreover, a series of backchannels produces a sense of rhythm and feelings of synchrony, contingency and rapport [11, 12].

There are three factors in generating backchannels, which are timing (when), lexical form (what), and prosody (how). A majority of the previous works addressed the timing of backchannels [13, 14], but many conventional systems used the same recorded pattern of backchannels, giving a monotonous impression to users.

We investigated the prosody of backchannels, and found their power is correlated with the preceding user utterances [15]. We have also explored for generation of a variety of backchannel forms depending on the dialogue context [16], based on machine learning using linguistic and prosodic features [17–19]. However, the generation of backchannels and choice of their form are arbitrary, and the evaluation with the corpus observation is not meaningful. Thus, we augment the annotation by adding acceptable forms of backchannels for each occurrence point. Similar to the response selection in the attentive listening system, we formulate as the validation problem. The prediction accuracy on average including the not-to-generate case is 64.3%. We also conducted a subjective evaluation by preparing audio files of generated backchannels. The proposed method obtained much higher ratings than the random generation, and to our surprise, it is almost comparable to the original counselor's choice when we use the same voice. This result suggests that we can pass a Turing Test in terms of backchannel form generation. But we need to tune the prosody to be more natural and of variation.

5.2 *Filler Generation*

Conventionally, fillers have been regarded as redundant and thus those to be removed. On the other hand, an utterance of a long sentence without any fillers generated by TTS is not natural [20]. In fact, humans use fillers, not only due to disfluency, but also for showing politeness and for attracting attention.

Fillers are also used for smooth turn-taking, namely either holding the current turn or taking a turn. We are investigating a method to generate fillers at the beginning of the system utterances to indicate an intention of turn-taking or turn-holding just like human conversation [21]. They can be effective for avoiding speech collision, because the collision of fillers with the user utterances is not so harmful and TTS usually cannot cancel the speech output once generated.

5.3 *Laughter Generation*

Laughter is also an important component in human-human dialogue. It is used for ice-breaking and socializing each other. Recently, there are several studies to investigate laughter for humanoid robots [22]. Analysis of our corpus shows that a large majority of laughter is (1) speech laughter rather than stand-alone laughter, and (2) breath laughter rather than obvious laughter. These suggest that laughter is generated not because the subjects feel funny. On the other hand, we observed many negative laughter samples, which follow negative sentences annotated by the sentiment analysis.

Similar to backchannels, there are three factors for laughter generation, but none of them is not an easy problem and has not been seriously investigated. Moreover, inappropriate laughter would make a very bad impression compared with backchannels.

6 Conclusions and Future Directions

This article has introduced our ERICA project, which aims at human-like conversation capability. We have examined realistic social roles, which would bring users into realistic situated dialogue. They are attentive listening, job interview, and speed dating. We are collecting realistic dialogue data, and designing and implementing systems for these three tasks. Then, we try to evaluate the system by comparing its outputs against the corpus. However, the corpus may not be the ground truth in this kind of social interaction tasks, and thus we augment the annotation.

There are other issues which are not addressed in this article. One is flexible turn-taking [23]. We note again that the corpus does not provide the ground truth for this problem. Modeling non-verbal information such as valence and affect is also needed. Character modeling may be also useful for making the system human-like.

We focus on recognition of user engagement [24], which indicates a positive/negative attribute to keep the current dialogue. This is closely related to the performance of the system. The ultimate dialogue with the humanoid robot should be as engaging as human-human dialogue. Therefore, comparable performance to human-like interaction experience should be measured by the engagement level, which will be used for a Total Turing Test.

Acknowledgements This work was supported by JST ERATO Ishiguro Symbiotic Human-Robot Interaction program (Grant Number JPMJER1401), Japan.

References

1. Glas DF, Minato T, Ishi CT, Kawahara T, Ishiguro H (2016) ERICA: the ERATO intelligent conversational android. In: Proceedings of RO-MAN, pp 22–29
2. Inoue K, Milhorat P, Lala D, Zhao T, Kawahara T (2016) Talking with ERICA, an autonomous android. In: Proceedings of SIGdial meeting discourse & dialogue, volume Demo. Paper, pp 212–215
3. Milhorat P, Lala D, Inoue K, Tianyu Z, Ishida M, Takanashi K, Nakamura S, Kawahara T (2017) A conversational dialogue manager for the humanoid robot ERICA. In: Proceedings of international workshop spoken dialogue systems (IWSDS)
4. Fujie S, Matsuyama Y, Taniyama H, Kobayashi T (2009) Conversation robot participating in and activating a group communication. In: Proceedings of InterSpeech, pp 264–267
5. Bohus D, Horvitz E (2009) Models for multiparty engagement in open-world dialog. In: Proceedings of SIGdial

6. Lala D, Milhorat P, Inoue K, Ishida M, Takanashi K, Kawahara T (2017) Attentive listening system with backchanneling, response generation and flexible turn-taking. In: Proceedings of SIGdial meeting discourse & dialogue, pp 127–136
7. DeVault D, Artstein R, Benn G, Dey T, Fast E, Gainer A, Georgila K, Gratch J, Hartholt A, Lhommet M, Lucas G, Marsella S, Morbini F, Nazarian A, Scherer S, Stratou G, Suri A, Traum D, Wood R, Xu Y, Rizzo A, Morency L-P (2014) SimSensei Kiosk: avirtual human interviewer for healthcare decision support. In: Proceedings of AAMAS
8. Kobori T, Nakano M, Nakamura T (2016) Small talk improves user impressions of interview dialogue systems. In: Proceedings of SIGDial, pp 370–380
9. Ranganath R, Jurafsky D, McFarland D (2009) It's not you, it's me: detecting flirting and its misperception in speed-dates. In: Proceedings of EMNLP
10. Ueno S, Inaguma H, Mimura M, Kawahara T (2018) Acoustic-to-word attention-based model complemented with character-level CTC-based model. In: Proceedings of IEEE-ICASSP, pp 5804–5808
11. Levitan R, Hirschberg J (2011) Measuring acoustic-prosodic entrainment with respect to multiple levels and dimensions. In: Proceedings of InterSpeech, pp 3081–3085
12. Xiao B, Georgiou PG, Imel ZE, Atkins D, Narayanan S (2013) Modeling therapist empathy and vocal entrainment in drug addiction counseling. In: Proceedings of InterSpeech, pp 2861–2864
13. Kitaoka N, Takeuchi M, Nishimura R, Nakagawa S (2005) Response timing detection using prosodic and linguistic information for human-friendly spoken dialog systems. J Jpn Soc Artif Intell 20(3):220–228
14. Kitaoka N, Takeuchi M, Nishimura R, Nakagawa S (2005) Response timing detection using prosodic and linguistic information for human-friendly spoken dialog systems. J Jpn Soc Artif Intell 20(3):220–228
15. Kawahara T, Uesato M, Yoshino K, Takanashi K (2015) Toward adaptive generation of backchannels for attentive listening agents. In: Proceedings of international workshop spoken dialogue systems (IWSDS)
16. Kawahara T, Yamaguchi T, Inoue K, Takanashi K, Ward N (2016) Prediction and generation of backchannel form for attentive listening systems. In: Proceedings of INTERSPEECH, pp 2890–2894
17. Ward N (1996) Using prosodic clues to decide when to produce back-channel utterances. In: Proceedings of ICSLP, pp 1728–1731
18. Ward N, Tsukahara W (2000) Prosodic features which cue back-channel responses in English and Japanese. J Pragmat 32(8):1177–1207
19. Koiso H, Horiuchi Y, Tutiya S, Ichikawa A, Den Y (1998) An analysis of turn-taking and backchannels based on prosodic and syntactic features in Japanese map task dialogs. Lang Speech 41(3–4):295–321
20. Andersson S, Georgila K, Traum D, Aylett M, Clark RAJ (2010) Prediction and realisation of conversational characteristics by utilising spontaneous speech for unit selection. In: Proceedings of speech prosody
21. Nakanishi R, Inoue K, Nakamura S, Takanashi K, Kawahara T (2018) Generating fillers based on dialog act pairs for smooth turn-taking by humanoid robot. In: Proceedings of international workshop spoken dialogue systems (IWSDS)
22. Turker BB, Bucinca Z, Erzin E, Yemez Y, Sezgin M (2017) Analysis of engagement and user experience with a laughter responsive social robot. In: Proceedings of InterSpeech, pp 844–848
23. Skantze G, Hjalmarsson A, Oertel C (2014) Turn-taking, feedback and joint attention in situated human-robot interaction. Speech Commun 65:50–66
24. Inoue K, Lala D, Takanashi K, Kawahara T (2018) Latent character model for engagement recognition based on multimodal behaviors. In: Proceedings of international workshop spoken dialogue systems (IWSDS)

Dialog State Tracking for Unseen Values Using an Extended Attention Mechanism

Takami Yoshida, Kenji Iwata, Hiroshi Fujimura and Masami Akamine

Abstract Recently, discriminative models using recurrent neural networks (RNNs) have shown good performance for dialog state tracking (DST). However, the models have difficulty in handling new dialog states unseen in model training. This paper proposes a fully data-driven approach to DST that can deal with unseen dialog states. The approach is based on an RNN with an attention mechanism. The model integrates two variants of RNNs: a decoder that detects an unseen value from a user's utterance using cosine similarity between word vectors of the user's utterance and that of the unseen value; and a sentinel mixture architecture that merges estimated dialog states of the previous turn and the current turn. We evaluated the proposed method using the second and the third dialog state tracking challenge (DSTC 2 and DSTC 3) datasets. Experimental results show that the proposed method achieved DST accuracy of 80.0% for all datasets and 61.2% for only unseen dataset without hand-crafted rules and re-training. For the unseen dataset, the use of the cosine similarity-based decoder leads to a 26.0-point improvement from conventional neural network-based DST. Moreover, the integration of the cosine similarity-based decoder and the sentinel mixture architecture leads to a further 2.1-point improvement.

T. Yoshida (✉) · K. Iwata · H. Fujimura · M. Akamine
Media AI Laboratory, Corporate Research and Development Center,
Toshiba Corporation, Kawasaki, Japan
e-mail: takami.yoshida@toshiba.co.jp

K. Iwata
e-mail: kenji4.iwata@toshiba.co.jp

H. Fujimura
e-mail: hiroshi4.fujimura@toshiba.co.jp

M. Akamine
e-mail: masa.akamine@toshiba.co.jp

L. F. D'Haro et al. (eds.), *9th International Workshop on Spoken Dialogue System Technology*, Lecture Notes in Electrical Engineering 579,
https://doi.org/10.1007/978-981-13-9443-0_7

1 Introduction

Dialog state tracking (DST) is used to update the state of a user's goal, or a dialog state in a task-oriented spoken dialog system. A dialog state is, in practice, a probability distribution of slots and their values that are provided in an ontology. The dialog state is estimated by using a dialog history including the user's utterances and the system's responses up to the present point in time as shown in Table 1. DST is one of the essential functions of a dialog system because the state directly affects the system's decision on its action with respect to the user.

Many approaches to DST have been proposed. They can be classified into three categories: rule-based, generative model-based, and discriminative model-based approaches. The rule-based approach [15] is simple and intuitive. However, making rules is time-consuming and money-consuming, especially when the task is complicated. The generative model-based [13] and the discriminative model-based [3–6, 9, 12] approaches do not need hand-crafted rules because they are statistical and data-driven. Compared to generative models, discriminative ones have shown better performance in the dialog state tracking challenges, DSTC 2 [1] and DSTC 3 [2].

Discriminative models estimate a probability distribution of pre-defined values. Recent discriminative model-based approaches have used neural network models including a feedforward deep neural network (DNN) [4], a recurrent neural network (RNN) [3, 5, 6], and a convolutional neural network (CNN) [12].

However, the discriminative model-based approach has a drawback. Discriminative models need to be trained for slots and values predefined in the ontology of the system. To deal with a new dialog state, we need to prepare training data and train a new model.

To overcome this drawback, two RNN-based DST methods have been proposed. Henderson et al. proposed an RNN-based method with n-gram type features of delexicalized slots and values [3, 5]. The RNN takes input features of n-grams from hypotheses of automatic speech recognition (ASR) and system actions after delexicalization of words relevant to slots and their values. The method showed high accuracy for unseen values in DSTC 3. However, this delexicalization requires a large list of synonyms for slots and their values. Since the list is hand-crafted, the approach is time-consuming.

Jang et al. proposed a DST method based on an RNN with an attention mechanism [7]. The attention mechanism calculates weight parameters called attention weights that represent the importance of each word. The method calculates a context vector, which is a weighted sum of word vectors using attention weights. The RNN estimates probability of values using cosine similarity between the context vector and a word vector of each value. The method can detect an unseen value in training from a user's utterance. However, [7] does not describe how to integrate dialog states of the previous turn and the current turn because the task does not require such integration.

An interesting variation of an attention mechanism has been proposed for language modeling [10]. Language modeling involves estimating the next word in a sentence from the previous words. The paper [10] introduces a mixture model that combines

Table 1 Example of dialog and dialog states

	System response (*tag*)	User's utterance	Dialog state		
			Area	Food	Price
1	Hello. How may I help you? (*welcomemsg()*)	Cheap restaurant in the south part of town	South	*None*[a]	Cheap
2	What kind of food would you like? (*request(food)*)	Any	South	Dontcare	Cheap
3	How about the XX restaurant. (*offer(name = XX)*)	What is the phone number?	South	Dontcare	Cheap
4	Phone number is 1234. (*inform(phone = 1234)*)	Thank you, good bye	South	Dontcare	Cheap

[a]*None* means no value is specified

an RNN-based language model and a pointer model. The former model estimates probability of the next word over vocabulary words as standard language models and the latter probability of the next word that does not appear in training but in the sentence. The combination is done using an attention mechanism called "*sentinel mixture architecture*."

This paper proposes a fully data-driven approach to DST that can track an unseen value without hand-crafted rules and re-training. The approach uses an RNN with an attention mechanism based on cosine similarity [7] to detect an unseen value from a user's utterance. We introduce a sentinel mixture architecture [10] to the RNN model for integrating dialog states of the previous turn and the current turn. We evaluated the proposed method using the DSTC 2 and DSTC 3 datasets.

This paper is organized as follows: Sect. 2 presents the background, Sect. 3 sets out the proposed approach, Sect. 4 details experimental results, and Sect. 5 concludes the paper.

2 Related Work

Jang et al. proposed DST based on an RNN model with an attention mechanism [7] for the fifth dialog state tracking challenge (DSTC 5) [8]. The main task is to estimate a dialog state for each sub-dialog segment consisted of several turns that show the same topic. The main difficulty is the sparsity of the training data: only 56% of values in the ontology is included in the training data [7]. Therefore, DST is required to estimate values unseen in training. To realize such DST, [7] proposed DST based on an RNN with an attention mechanism. The method first calculates an attention weight for each word of a user's utterance using a long short-term memory (LSTM). Then it calculates cosine similarity between a context vector, i.e., a weighted sum of word vectors using the attention weights, and a word vector of each value in the ontology. Since the method estimates a dialog state based on cosine similarities, the

DST can estimate a dialog state unseen in training to the extent that the word vector of the unseen value is available. This approach is effective in handling unseen values.

However, in contrast with DSTC 5, our DST needs to estimate a dialog state for each turn as in DSTC 2 and DSTC 3, and consider the dialog state of the previous turn. The second turn in Table 1 provides an example. Although the user does not mention the area slot, the ground truth of the dialog state keeps "$area = south$," which is mentioned in the previous turn. This example shows that a dialog state is decided by not only the user's utterance of the current turn, but also by a dialog state of the previous turn. The DST in [7] can not handle such situation.

Merity et al. proposed an RNN with a sentinel mixture architecture for language modeling [10]. As mentioned in [10], recent RNN-based language models show high performance but struggle to predict unseen words. To accurately predict both seen and unseen words, [10] proposed an integration of probability for words from a pre-defined vocabulary and probability for words appeared in input sentences. The former is calculated using an RNN like conventional approaches: the RNN outputs the probability of the next word over vocabulary words given previous words in a sentence. The latter is calculated using an attention mechanism: the model regards an attention weight for each word not seen in training but in the input sentence as the probability of the corresponding word. Therefore, the method can predict a word unseen in training if the word is included in the input. The two probabilities are summed up with a weight parameter called a "*sentinel weight*," which is calculated using the attention mechanism. The sentinel weight makes it possible to accurately predict both seen and unseen words, and their approach achieved state-of-the-art performance.

However, unlike language modeling, DST should bring a value and its synonyms together. For example, both "a restaurant in the center of the town" and "a restaurant in the central part of the town" indicate "$area = center$." Therefore, their method should be modified for DST.

To deal with unseen values as well as seen values in DST, we propose to introduce a sentinel mixture architecture to a cosine similarity-based decoder. We integrate a dialog state of the previous turn and the current turn using the sentinel architecture [10]. To the best of our knowledge, there is no previous study on fully data-driven DST that can track an unseen value.

3 Proposed RNN with Attention Mechanism

Figure 1 shows a schematic block diagram of the proposed model. The model consists of an encoding layer and a decoding layer. The model also includes an attention mechanism and sentinel mixture architecture. The model calculates an attention weight for each word and two sentinel weights. The attention weights are used in the attention mechanism and the sentinel weights are used in the sentinel mixture architecture.

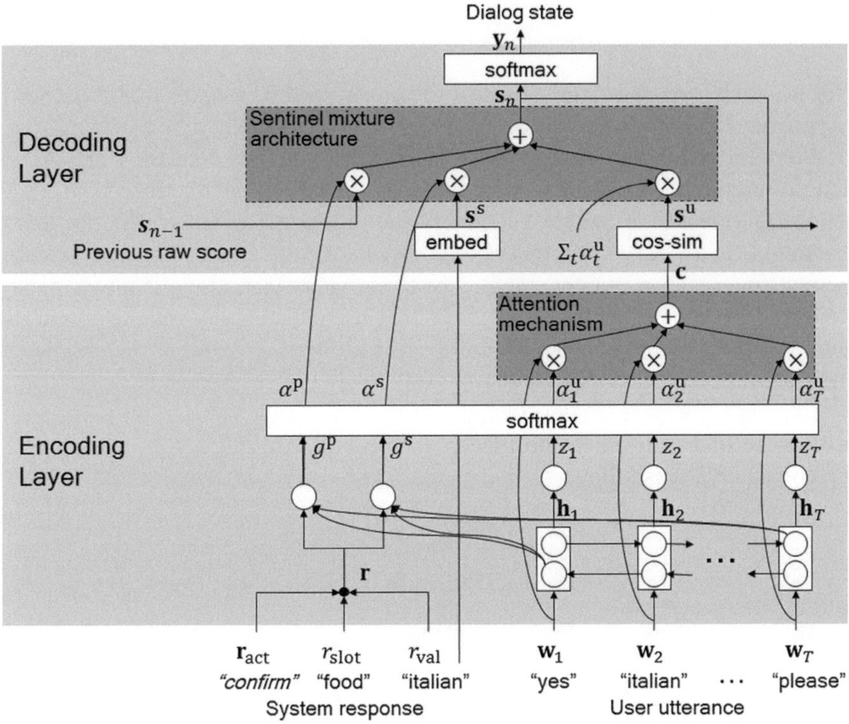

Fig. 1 Model architecture based on the proposed method

The model receives a system response and a user's utterance. The system response consists of a system action tag (e.g., *welcomemsg, confirm, ...*), a target slot (e.g., *area, food, ...*), and a target value (e.g., *north, south, ...*). Examples of the system response include $welcomemsg(), confirm(food = italian), inform(area = north)$. The user's utterance is a word sequence (e.g., "yes italian restaurant please"). The model includes a recurrent connection that receives a raw score. The raw score denotes a reliability that becomes the probability of each value after normalization using a softmax function.

We assume that the probability distribution can be factorized by that of each slot ($P(\text{area, food, price}) = P(\text{area})P(\text{food})P(\text{price})$). We describe how to estimate the probability for a slot in the following sections.

In the following sections, we assume that the DST focuses on the n-th turn, a slot has K values, the utterance consists of T words, and the system can take M actions.

3.1 Encoding Layer

The encoding layer calculates a context vector and sentinel weights from the system response and the user's utterance.

A system action tag, a target slot, and a target value are converted into an action tag vector, a target slot feature, and a target value feature, respectively. The action tag vector is a one-hot vector $\mathbf{r}_{\text{act}}$ whose dimension is M. The target slot feature r_{slot} is a binary value that represents whether the system response includes the slot ($r_{\text{slot}} = 1$) or not ($r_{\text{slot}} = 0$). The target value feature r_{val} is also a binary value that represents whether the system response includes any value ($r_{\text{val}} = 1$) or not ($r_{\text{val}} = 0$). For example, the features $(r_{\text{slot}}, r_{\text{val}})$ for *welcomemsg()*, *request(food)*, and *confirm(food=dontcare)* are (0, 0), (1, 0), and (1, 1), respectively.

The user's utterance is converted into word vectors $(\mathbf{w}_1, \mathbf{w}_2, \ldots, \mathbf{w}_T)$. The word vectors are then converted into hidden state vectors $(\mathbf{h}_1, \mathbf{h}_2, \ldots, \mathbf{h}_T)$ using a bidirectional LSTM as follows:

$$\begin{aligned}
\mathbf{h}_t^{\text{f}} &= \text{LSTM}_{\text{fwd}}\left(\mathbf{h}_{t-1}^{\text{f}}, \mathbf{w}_t\right) , \\
\mathbf{h}_t^{\text{b}} &= \text{LSTM}_{\text{bwd}}\left(\mathbf{h}_{t+1}^{\text{b}}, \mathbf{w}_t\right) , \\
\mathbf{h}_t &= \mathbf{h}_t^{\text{f}} \oplus \mathbf{h}_t^{\text{b}} ,
\end{aligned}$$

where $\text{LSTM}_{\text{fwd}}(\cdot, \cdot)$, $\text{LSTM}_{\text{bwd}}(\cdot, \cdot)$ are forward and backward LSTMs, $\mathbf{h}_t^{\text{f}}$, $\mathbf{h}_t^{\text{b}}$ are hidden states of forward and backward LSTMs, and $\oplus$ is vector concatenation. The hidden state vectors are used to calculate scalar values $(z_1, z_2, \ldots, z_T)$ using a one-layer neural network (NN_{usr}) as follows:

$$z_t = \text{NN}_{\text{usr}}\left(\mathbf{h}_t\right) .$$

The scalar value z_t indicates the importance of the tth word. A larger z_t means the tth word is more important.

To calculate the sentinel weights, two sentinel gate parameters $(g^{\text{p}}, g^{\text{s}})$ are calculated from the system feature vector $(\mathbf{r} = \mathbf{r}_{\text{act}} \oplus r_{\text{slot}} \oplus r_{\text{val}})$ and the hidden state $(\mathbf{h}_{\text{L}} = \mathbf{h}_T^{\text{f}} \oplus \mathbf{h}_1^{\text{b}})$ as follows:

$$\begin{aligned}
g^{\text{p}} &= \text{NN}_{\text{pre}}\left(\mathbf{h}_{\text{L}} \oplus \mathbf{r}\right) , \\
g^{\text{s}} &= \text{NN}_{\text{sys}}\left(\mathbf{h}_{\text{L}} \oplus \mathbf{r}\right) ,
\end{aligned}$$

where $\text{NN}_{\text{pre}}, \text{NN}_{\text{sys}}$ are one-layer neural networks. The sentinel gate parameters indicate the importance of the raw score in the previous turn and that of the system response. A larger gate parameter means the corresponding feature vector is more important. Two sentinel weights $(\alpha^{\text{p}}, \alpha^{\text{s}})$ and attention weighs for the words $(\alpha_1^{\text{u}}, \alpha_2^{\text{u}}, \ldots, \alpha_T^{\text{u}})$ are calculated from g^{p}, g^{s}, and $\mathbf{z} = [z_1, z_2, \ldots, z_T]$ as follows:

$$\boldsymbol{\alpha} = \text{softmax}\left(g^{\text{p}} \oplus g^{\text{s}} \oplus \mathbf{z}\right) , \tag{1}$$

where $\boldsymbol{\alpha} = \left[\alpha^{\mathrm{p}}, \alpha^{\mathrm{s}}, \alpha_1^{\mathrm{u}}, \alpha_2^{\mathrm{u}}, \ldots, \alpha_T^{\mathrm{u}}\right]$.

Using $\boldsymbol{\alpha}$, the model focuses on the corresponding context in estimating dialog states. For example, in estimation of the area slot at the first turn of the dialog shown in Table 1, a dialog state is decided from the fifth word of the user's utterance, "south." The model focuses on the word by setting $\alpha_5^{\mathrm{u}} = 1$. At the second turn, the dialog state is decided from that of the previous turn. Thus, the model focuses on the raw score of the previous turn by setting $\alpha^{\mathrm{p}} = 1$. When the user refers to his or her goal by answering the system's question (e.g., the system asks "Let me confirm. You are looking for a venue in the Newnham area." and the user answers "yes"), the model focuses on the system's action by setting $\alpha^{\mathrm{s}} = 1$.

From the word vector sequence, the context vector $\mathbf{c}$ is calculated using the attention weights $(\alpha_1^{\mathrm{u}}, \alpha_2^{\mathrm{u}}, \ldots, \alpha_T^{\mathrm{u}})$ as follows:

$$\mathbf{c} = \sum_{t=1}^{T} \alpha_t^{\mathrm{u}} \mathbf{w}_t \ . \tag{2}$$

Equation (2) is described as the processing block "*Attention mechanism*" in Fig. 1. When the user's utterance includes a word related to a value, the corresponding attention weight becomes 1 and the context vector is the same as the word vector of the word. Therefore, by comparing the context vector and the word vector of the target value, the system can detect an unseen value in the user's utterance.

3.2 Decoding Layer

The decoding layer estimates a dialog state using the previous raw score, the system response, and the context vector with the sentinel weights. The decoding layer calculates two similarities: a similarity between the context vector and a word vector of each value, and a similarity between the system response and a value. These two similarities and the previous raw score are integrated with the sentinel weights.

From the system response, the decoding layer extracts a binary vector ($\mathbf{s}^{\mathrm{s}} = [s_1^{\mathrm{s}}, s_2^{\mathrm{s}}, \ldots, s_K^{\mathrm{s}}]$) whose k-th component indicates whether the k-th value is included ($s_k^{\mathrm{s}} = 1$) or not ($s_k^{\mathrm{s}} = 0$). This process is described as the processing block "*embed*" in Fig. 1.

Cosine similarity ($\mathbf{s}^{\mathrm{u}}$) between the context vector $\mathbf{c}$ and word vectors of each value $(\mathbf{v}_1, \mathbf{v}_2, \ldots, \mathbf{v}_K)$ is calculated as follows:

$$s_k^{\mathrm{u}} = \frac{\mathbf{c} \cdot \mathbf{v}_k}{\|\mathbf{c}\| \|\mathbf{v}_k\|} \ , \tag{3}$$

$$\mathbf{s}^{\mathrm{u}} = \left[s_1^{\mathrm{u}}, s_2^{\mathrm{u}}, \ldots, s_K^{\mathrm{u}}\right] \ , \tag{4}$$

where $\cdot$ is dot product. If a value consists of more than two words (e.g., *chinese takeaway*), we used the average of the word vectors in Eq. (3). These Eqs. (3) and (4) are represented as the processing block "*cos-sim*" in Fig. 1.

The sentinel weights ($\alpha^{\mathrm{p}}, \alpha^{\mathrm{s}}, \alpha^{\mathrm{u}} = \sum_{t=1}^{T} \alpha_t^{\mathrm{u}}$) control integration of the raw score and the similarities ($\mathbf{s}_{n-1}, \mathbf{s}^{\mathrm{s}}, \mathbf{s}^{\mathrm{u}}$) as follows:

$$\mathbf{s}_n = \alpha^{\mathrm{p}} \mathbf{s}_{n-1} + \alpha^{\mathrm{s}} \mathbf{s}^{\mathrm{s}} + \alpha^{\mathrm{u}} \mathbf{s}^{\mathrm{u}} \ . \tag{5}$$

Equation (5) is described as the processing block "*Sentinel mixture architecture*" in Fig. 1.

A bias b is calculated as $b = \mathrm{NN}_{\mathrm{bias}}\,(\mathbf{h}_{\mathrm{L}} \oplus \mathbf{r})$ where $\mathrm{NN}_{\mathrm{bias}}$ is a one-layer neural network. The bias is used to estimate probability for None that means no value is specified. Note that we omit this bias calculation from Fig. 1 for simplicity. Finally, we calculate output probability $\mathbf{y}_n$ from the concatenation of the current raw score $\mathbf{s}_n$ and the bias b as follows:

$$\mathbf{y}_n = \mathrm{softmax}(\mathbf{s}_n \oplus b) \ . \tag{6}$$

4 Experiments

We evaluated the proposed method through two tasks, "Task 1" and "Task 2." DST performance for seen values was evaluated in Task 1 using the DSTC 2 dataset. In Task 2, DST performance for both seen and unseen values was evaluated using the DSTC 3 dataset. These two datasets were collected in advance, and the system performs DST for the datasets.

4.1 Datasets

The datasets include human–computer dialogs. Users interacted with dialog systems to search for restaurants by specifying constraints. A ground truth of a dialog state was labeled for all turns as the most recently asserted value (which is "*scheme A*" and "*Schedule 1*" in DSTC 3).

Among the slots included in DSTC 2 and DSTC 3, we used the "area," "food," and "price range" slots. A summary of the slots is shown in Table 2. We excluded "hastv" and "childrenallowed" slots because these slots have binary values ("yes" or "no") and no unseen value appeared. We also excluded the "name" slot because word vectors for several values were not obtained.

For Task 1, a model was trained using the dstc2_train dataset and evaluated using the dstc2_test dataset. The datasets include 1,612 and 1,117 sessions. Note that these two datasets use the same ontology.

Table 2 Values included in DSTC 2 and DSTC 3 datasets

	Training	DSTC 2 test set		DSTC 3 test set		
	Values	Seen	Unseen	Seen	Unseen	Examples of unseen
Area	7	7	0	3	14	Girton, arbury, …
Food	92	92	0	20	10	American, cafe food, …
Price range	5	5	0	6	1	Free

For Task 2, a model was trained using the dstc3_train dataset and evaluated using the dstc3_test dataset. The dstc3_train dataset is the sum of the dstc2_train and the dstc2_test datasets. The dstc3_test dataset includes 2,264 sessions.

4.2 System Implementation

We implemented a prototype DST system based on the proposed method using the neural network framework Chainer [14]. One-best ASR results were used as inputs to the encoding layer described in Sect. 3. Contractions were converted to their original forms (e.g., i'm to i am), then each word was converted to a 300-dimensional word vector using a GloVe model [11] distributed on the GloVe website.[1] The word vectors were processed by the RNN as described in Sect. 3. The cell size of the bi-directional LSTM and NNs (NN_{usr}, NN_{sys}, NN_{pre}, NN_{bias}) was 32, and the cells of the LSTM were initialized to 0 before training.

For training the model, we used Adam with a learning rate of 0.001, gradient clipping of 1.0, mini batch size of 32, 50% dropout for the bi-directional LSTM. We used word dropout by randomly replacing a word vector with a zero vector for attention weight calculation. We trained each model by using word dropout ratios from 0% to 50% in 10 increments. Each model was trained with 200 epochs and the best parameter selected based on the accuracy for the development dataset, i.e., the dstc2_dev dataset.

For evaluation, we used a scaling parameter β for the bias b to prevent false negatives of unseen values. We replaced b with $\tilde{b} = b/\beta$ in Eq. (6). We selected the scaling parameter for each model from [1, 2, 4, 8, 16] based on the performance in the DSTC 3 test dataset.

DST based on the proposed method is called as "*Cos with sentinel*" in the following parts.

[1] https://nlp.stanford.edu/projects/glove/.

4.3 Comparative Methods

We implemented three comparative methods: *"Cos without sentinel," "NN,"* and *"RNN."* Cos without sentinel used the cosine similarity-based decoder but did not use the sentinel weights to integrate the similarities. The similarities were updated by the rule $s_{n,k} = 1 - (1 - s_{n-1,k})(1 - s^{u}_{n,k})$ described in [15]. NN used an NN for similarity calculation $s^{u}_{k} = \mathrm{NN_{dec}}\,(\mathbf{c} \oplus \mathbf{v}_k)$. RNN used an LSTM for turn-level recurrent connection in the decoding layer as $s_{n,k} = \mathrm{LSTM_{dec}}\left(\mathbf{c} \oplus \mathbf{r}_k \oplus \mathbf{v}_k, s_{n-1,k}\right)$. Note that the scaling parameter was also used for all methods.

We also show two additional results of the DSTC 2 and DSTC 3 participants: "*focus baseline*" which uses rules, and "*RNN with rule*" which is the DST proposed in [3, 5]. Note that RNN with rule showed the best accuracy in DSTC 3 [2]. Results for the methods were extracted from the dstc2_results and dstc3_results distributed on the website.[2]

4.4 Results

Table 3 shows the accuracy of DST in area, food and price slots in Task 1 and Task 2. We see from this table that the proposed method "Cos with Sentinel" performs better than RNN, NN, and Cos without sentinel models except for the price slot in Task 1 and Task 2. Cos without sentinel performs bad compared with other models. This is because the model failed to merge previous turns and system responses.

Table 4 shows the accuracy only for unseen values. The DST methods using the cosine similarity-based decoder (Cos without sentinel and Cos with sentinel) achieved better accuracy than those using the neural network-based decoder (NN with sentinel and RNN). This shows that the cosine similarity-based attention works for estimating values unseen in training while the neural network-based methods have difficulty in generalizing to unseen values.

The performance of the DSTC 3 participants are higher than that of the proposed method. RNN with rule performs good in particular for unseen values in the slots except the food slot in Task 2 at a cost of hand-crafting rules. The proposed method achieved a good performance without any rules.

4.5 Discussion

To understand the strengths and weaknesses of the proposed method, we analyzed performance on individual slots and values. Among the three slots, accuracy of the price range slot was 52.7%, which is lower than that of the area slot (71.5%) and the food slot (59.5%) as shown in Table 4. This result might be the effect of an incorrect

[2] http://camdial.org/~mh521/dstc.

Table 3 DST accuracy for Task 1 and Task 2

	Task1			Task2			Tasks 1 and 2
	Area	Food	Price	Area	Food	Price	Average
Focus baseline	90.8	83.7	92.9	81.1	90.5	88.4	87.9
RNN with rule	**92.4**	**85.6**	**93.0**	**88.5**	**91.0**	**93.2**	**90.6**
RNN	64.2	69.8	85.1	63.1	67.8	**82.9**	72.2
NN	64.5	79.7	**88.4**	57.3	68.5	79.7	73.0
Cos without sentinel	56.1	52.3	65.5	57.5	38.7	61.1	55.2
Cos with sentinel	**84.7**	**84.4**	83.7	**80.6**	**79.6**	66.9	**80.0**

Table 4 DST accuracy for unseen values

	Task2				Task2 using other unseen values			
	Area	Food	Price	Average	Inexpensive	Affordable	Costly	Pricey
Focus baseline	67.8	**88.1**	87.6	81.2	–	–	–	–
RNN with rule	**85.3**	82.3	**92.3**	**86.6**	–	–	–	–
RNN	21.4	33.1	31.1	28.5	46.2	34.5	26.9	71.7
NN	13.3	31.3	54.6	33.1	0.0	22.1	19.6	0.0
Cos without sentinel	**73.2**	31.1	**73.0**	59.1	81.9	78.5	**89.0**	**89.0**
Cos with sentinel	71.5	**59.5**	52.7	**61.2**	**83.2**	**79.3**	83.8	84.2

word vector of the unseen value "*free*." In fact, the GloVe model gives the 5-top-most similar words for "free" as "Free," "download," "downloads," "downloadable," and "online."

To further validate the proposed method for the price range slot, we evaluated DST performance using other price-related words as unseen values: "*inexpensive*," "*affordable*," "*costly*," and "*pricey*." We replaced "free" with one of the words in the dataset and evaluated the results using the replaced dataset. The right side of Table 4 shows evaluation results for the replaced values. The replacement indicates significant improvement of the performance. We see from this that good word vectors are essential to the proposed DST method.

One of the weaknesses is detection of values that include the same word (e.g., *chinese* and *chinese takeaway*). The cosine similarity-based decoders (both Cos without sentinel and Cos with sentinel) are confused between two values and its accuracy of chinese takeaway was 0. To handle such values, we will take account of the number of words composing values as future work.

5 Conclusion

We proposed a fully data-driven approach to DST using an RNN model with an attention mechanism. The approach integrates a decoder that detects an unseen value from a user's utterance using cosine similarity between a context vector of the user's utterance and that of the unseen value, and a sentinel mixture architecture that merges an estimated dialog state of the previous turn and that of the current turn. We evaluated a DST model based on the proposed approach using the DSTC 2 and DSTC 3 datasets. For unseen values, the results showed that the use of the cosine similarity-based decoder led to a 26.0-point improvement from conventional NN-based DST, and the integration of the cosine similarity-based decoder and sentinel mechanism led to a further 2.1-point improvement. For all values, the accuracy of the conventional cosine similarity-based decoder was 17.8-point lower than that of conventional NN-based DST, but the proposed integration achieved 7.0-point improvement from the NN-based DST.

Future work includes improving accuracy for both seen and unseen values, extending the proposed approach to handle unseen slots, and handling values having no word vectors such as proper nouns.

References

1. Henderson M, Thomson B, Williams J (2014) The second dialog state tracking challenge. In: Proceedings of the 15th annual meeting of the special interest group on discourse and dialogue (SIGDIAL), p 263
2. Henderson M, Thomson B, Williams J (2014) The third dialog state tracking challenge. In: Spoken language technology workshop (SLT). IEEE, pp 324–329
3. Henderson M, Thomson B, Young S (2004) Word-based dialog state tracking with recurrent neural networks. In: Proceedings of the 15th annual meeting of the special interest group on discourse and dialogue (SIGDIAL), pp 292–299
4. Henderson M, Thomson B, Young S (2013) Deep neural network approach for the dialog state tracking challenge. In: Proceedings of the 14th annual meeting of the special interest group on discourse and dialogue (SIGDIAL), pp 467–471
5. Henderson M, Thomson B, Young S (2014) Robust dialog state tracking using delexicalised recurrent neural networks and unsupervised adaptation. In: Spoken language technology workshop (SLT). IEEE, pp 360–365
6. Hori T, Wang H, Hori C, Watanabe S, Harsham B, Roux JL, Hershey JR, Koji Y, Jing Y, Zhu Z et al (2016) Dialog state tracking with attention-based sequence-to-sequence learning. In: Spoken language technology workshop (SLT). IEEE, pp 552–558

7. Jang Y, Ham J, Lee BJ, Chang Y, Kim KE (2016) Neural dialog state tracker for large ontologies by attention mechanism. In: 2016 IEEE spoken language technology workshop (SLT), pp 531–537
8. Kim S, D'Haro LF, Banchs RE, Williams JD, Henderson M, Yoshino K (2016) The fifth dialog state tracking challenge. In: Proceedings of the 2016 IEEE workshop on spoken language technology (SLT)
9. Lee S (2013) Structured discriminative model for dialog state tracking. In: Proceedings of the 14th annual meeting of the special interest group on discourse and dialog (SIGDIAL), pp 442–451
10. Merity S, Xiong C, Bradbury J, Socher R (2016) Pointer sentinel mixture models. CoRR arXiv:abs/1609.07843
11. Pennington J, Socher R, Manning CD (2014) Glove: global vectors for word representation. In: Empirical Methods in Natural Language Processing (EMNLP), pp 1532–1543
12. Shi H, Ushio T, Endo M, Yamagami K, Horii N (2017) Convolutional neural networks for multi-topic dialog state tracking. In: Dialogues with social robots. Springer, Berlin, pp 451–463
13. Thomson B, Young S (2010) Bayesian update of dialogue state: a POMDP framework for spoken dialogue systems. Comput Speech Lang 24(4):562–588
14. Tokui S, Oono K, Hido S, Clayton J (2015) Chainer: a next-generation open source framework for deep learning. In: Proceedings of workshop on machine learning systems in the 29th annual conference on neural information processing systems (NIPS)
15. Wang Z, Lemon O (2013) A simple and generic belief tracking mechanism for the dialog state tracking challenge: on the believability of observed information. In: Proceedings of the 14th annual meeting of the special interest group on discourse and dialog (SIGDIAL), pp 423–432

Generating Fillers Based on Dialog Act Pairs for Smooth Turn-Taking by Humanoid Robot

Ryosuke Nakanishi, Koji Inoue, Shizuka Nakamura, Katsuya Takanashi and Tatsuya Kawahara

Abstract In spoken dialog systems for humanoid robots, smooth turn-taking function is one of the most important factors to realize natural interaction with users. Speech collisions often occur when a user and the dialog system speak simultaneously. This study presents a method to generate fillers at the beginning of the system utterances to indicate an intention of turn-taking or turn-holding just like human conversations. To this end, we analyzed the relationship between a dialog context and fillers observed in a human-robot interaction corpus, where a user talks with a humanoid robot remotely operated by a human. At first, we annotated dialog act tags in the dialog corpus and analyzed the typical type of a sequential pair of dialog acts, called a DA pair. It is found that the typical filler forms and their occurrence patterns are different according to the DA pairs. Then, we build a machine learning model to predict occurrence of fillers and its appropriate form from linguistic and prosodic features extracted from the preceding and the following utterances. The experimental results show that the effective feature set also depends on the type of DA pair.

1 Introduction

A number of spoken dialog systems have been developed and used with smartphones and appliances. The majority of these systems assume that the user utters a query made of single sentence, to which the system responds. In these systems, turn-taking

R. Nakanishi · K. Inoue · S. Nakamura · K. Takanashi · T. Kawahara (✉)
School of Informatics, Kyoto University, Kyoto, Japan
e-mail: kawahara@i.kyoto-u.ac.jp

R. Nakanishi
e-mail: nakanisi@sap.ist.i.kyoto-u.ac.jp

K. Inoue
e-mail: inoue@sap.ist.i.kyoto-u.ac.jp

S. Nakamura
e-mail: shizuka@sap.ist.i.kyoto-u.ac.jp

K. Takanashi
e-mail: takanasi@sap.ist.i.kyoto-u.ac.jp

L. F. D'Haro et al. (eds.), *9th International Workshop on Spoken Dialogue System Technology*, Lecture Notes in Electrical Engineering 579,
https://doi.org/10.1007/978-981-13-9443-0_8

is explicit; the user explicitly signals the start of the utterance with the "push-to-talk" interface or a predefined magic word, and the system indicates when it can accept input with an LED or GUI. However, this is much different from the turn-taking manner in human-human dialog, and is difficult to be applied for humanoid robots designed to be engaged in natural interaction with human. The goal of this study is to realize natural conversational behavior of an autonomous android [7] including backchannels [11] and turn-taking [20].

In fact, without the explicit interfaces mentioned above, speech collisions often occur when a user and the dialog system speak simultaneously, and they usually result in speech recognition errors and dialog breakdown. For natural and smooth turn-taking, this study investigates generation of fillers at the beginning of the system utterances to indicate an intention of turn-taking or turn-holding. For example, by placing a filler, the current speaker can hold the turn while thinking the next utterance, or the other participant can take a turn smoothly before speaking the main utterance. Thus, fillers have an important role and effect in human-human conversations.

In spoken language processing, however, fillers have been regarded as redundant segments, which must be removed in the transcript and are not usually generated by the system. There are several studies to predict fillers [1], but their major aim is to detect and remove them in the speech input. While there are a number of works on prediction and generation of backchannels [11], there are only a limited trials on filler generation [2, 19–21]. Fillers have a variety of forms especially in Japanese, and they are used in different contexts. Watanabe [22] investigated the occurrence ratio of fillers based on the complexity of the following utterance and the syntactic boundary. In this work, we present a method to predict occurrence of fillers and its appropriate form based on dialog act pairs using the linguistic and prosodic features in adjacent utterances. In the remaining of the paper, we describe the corpus and the annotation in Sect. 2, analysis on dialog act pairs and fillers in Sect. 3. Prediction of fillers and its evaluation is conducted in Sect. 4.

2 Corpus and Annotation

2.1 Corpus

We use a human-robot interaction corpus, where a subject talks with an android ERICA [14, 15] remotely operated by a human operator, who talks with a subject and controls non-verbal behaviors. The recording was done in September 2016. There were 39 sessions and each session lasted around 10 min. Engaged in these sessions were 6 operators (female from 20's to 30's) and 39 subjects (male from 20's to 60's and female from 10's to 70's). The android was given a role of a laboratory secretary and the subjects were asked to talk with her as a visitor.

2.2 Annotation

We define fillers as filled pauses which is one of spontaneous speech disfluency, such as "*uh, um, oh*" in English and "*ano:, etto, ma*" in Japanese [13]. For dialog act (DA) annotation, we adopt the following simple four classes based on the classification of general-purpose functions proposed by Bunt et al. [3].

- *Question* (Q): Utterances which function as information-seeking
- *Statement* (S): Utterances which have a role of *Inform / Offer / Promise / Request / Instruct*
- *Response* (R): Utterances which respond to a specific DA such as *Answer*, *Accept Offer* and *Decline Offer*
- *Other* (O): Utterances which do not belong to either Q, R or S such as *Greeting* and *Apology*

Utterances corresponding to *Feedback* in the dimension-specific function among those of O were classified into R because a number of those are lexical responses such as "*I see*". Backchannels are excluded from this annotation. To define the unit of DA, we adopt long utterance units (LUU) [6] which are defined as syntactic and pragmatic disjuncture based on clause units.

In order to validate the reliability of the DA annotation (Q, R, S and O), we calculate Cohen's kappa value [4] and evaluate agreement between annotators. The obtained kappa value was high ($\kappa = 0.799$) from the result of two annotators engaged on three sessions (495 DAs).

2.3 Annotation Results

The total number of DA units was 6441 in all 39 sessions. The occurrence counts of Q, R, S and O by the operators are 758, 1064, 779 and 706, and the occurrence counts of Q, R, S and O by the subjects are 267, 1687, 477 and 703, respectively. In the following, we do not distinguish DAs by the role.

Then, the concept of adjacency pairs of DA units [18] is introduced. We extracted 5080 DA pairs by eliminating DA pairs with an overlap between the preceding DA and the following DA. The total number of fillers observed in the corpus is 4292. In this study, we focus on 1460 fillers (Operator's: 875, Subject's: 585) which occur at the beginning of the following utterance. It is presumed that these fillers are related with the turn management. Table 1 shows the ratio of DA pairs which have a filler between the preceding DA and the following DA. It is observed that fillers are more likely to be used in turn-switching or turn-taking than in turn-keeping cases.

Table 1 Ratio of filler occurrence within DA pairs

DA pairs in turn-switch	33.2% (836/2516)
DA pairs in turn-keep	24.3% (624/2564)
Total DA pairs	28.7% (1460/5080)

Table 2 Bigrams of DA

			Preceding DA			
			Q	R	S	O
Following DA	Turn-holding	Q	0.09 (75)	**0.08 (187)**	0.09 (97)	0.11 (108)
		R	0.01 (6)	**0.33 (731)**	0.00 (5)	0.03 (26)
		S	0.04 (30)	**0.12 (266)**	**0.34 (365)**	0.20 (197)
		O	0.02 (19)	0.07 (146)	0.06 (69)	0.24 (237)
	Turn-taking	Q	0.02 (15)	**0.12 (257)**	0.08 (82)	0.06 (64)
		R	**0.79 (627)**	**0.15 (336)**	**0.26 (274)**	0.02 (22)
		S	0.01 (11)	**0.07 (164)**	0.06 (59)	0.07 (71)
		O	0.02 (14)	0.06 (134)	0.11 (117)	0.27 (269)

3 Analysis Based on Dialog Act (DA) Pairs

3.1 *Typical DA Pairs and Possible Speech Collisions*

Table 2 shows the bigram statistics of DA pairs with classification of turn-holding/taking. Each column represents the preceding DA and each row represents the following DA. Turn-holding means both the preceding DA and the following DA are spoken by the same speaker, and turn-taking means the speaker of the following DA differs from the speaker of the preceding DA. Bigrams are normalized so that the sum over each preceding DA (=column) becomes 1. The numbers in parentheses indicate the occurrence counts of DA pairs. Bigram patterns with frequencies larger than 152 (3% of all DA pairs) are written in bold, and these are focused in the analysis and prediction experiment in this study. The DA pairs related with Other (O) are excluded because most of the utterances tagged with O are the typical expressions like "*sorry*" and the turn-taking behavior in these cases should be different. There are many patterns of typical DA pairs in the corpus, due to a mixed-initiative dialog characteristic between a secretary and a visitor.

First, it is confirmed in the first row that, after questions (Q) by one participant, responses (R) by the other participant are dominant. In this case, there should be a consensus of turn-switching by the both participants. However, as we see in the second column, after responses (R), there is much ambiguity in the following DA. There is ambiguity in turn-management, too. The same participant can continue his utterances (R or S), or the other participant can take back a turn. In the latter

Table 3 Filler class and its definition

Class and ratio of occurrence	Definition
Proper (p) 7%	Forms only used as fillers ("*um*" in English / "*etto*" in Japanese)
Demonstrative (d) 6%	Same forms as demonstrative adjectives ("*so*" in English / "*ano*" in Japanese)
Adverbial (a) 2%	Same forms as adverbs ("*well*" in English / "*ma-*" in Japanese)
Notice (n) 12%	Used to indicate a reaction ("*oh*" in English / "*a*" in Japanese)
No filler (nf) 71%	Not to generate fillers

case, he/she can either respond to the previous utterance (R) or ask a new question (Q). After statements (S), there is not a large variation in the following DA, but big ambiguity in turn-management. The same speaker can continue a statement (S) or the other participants make a response (R) to the statement. It is expected that speech collisions are likely to occur when it is ambiguous which participant should take a turn. The above-mentioned cases are typical. In summary, speech collisions will often occur in the transition from R or S.

3.2 Typical DA Pairs and Filler Patterns

Next, we investigate typical filler patterns for each of the DA pairs which are identified as important in the previous sub-section. We classify fillers into six classes defined by the function and the expression similarity as shown in Table 3.

The typical forms and their occurrence ratio for DA pairs focused in this study are shown in Table 4. We focus on the filler occurrence or not between the preceding DA and the following DA. The upper part and the lower part show the DA pairs of *Keep* (turn-holding) and *Switch* (turn-taking), respectively. The table also gives the most typical or dominant form of fillers used in the DA pair, and then the frequency ratio of that form and other forms as well as no-filler occurrence.

As shown in the upper parts, when the speaker tries to hold the turn, he does not use the notice form, but demonstrative or proper forms. They are used to hold the turn to take time before speaking the next utterance. And the ratio of no-filler occurrence is large in turn-keeping. On the other hand, the lower part suggests that when the speaker is changed, the notice form is most frequently used. The form indicates a response to the preceding utterance and causes natural turn-switching. In QR *Switch* where the turn-switching is most apparent, the ratio of filler usage is smaller and the notice form is not so dominant. It is found out that the tendency of filler usage and its typical form is different depending on the DA pair and in particular turn-switching. Based on this observation, we design prediction and generation of fillers.

Table 4 Typical form of each DA pair

DA pair	Typical form	Ratio of occurrence (tf[a]/o[b]/nf[c])
RQ Keep	Proper	14%/13%/73%
RR Keep	Demonstrative	12%/12%/76%
RS Keep	Demonstrative	11%/21%/68%
SS Keep	Demonstrative	11%/14%/75%
QR Switch	**Notice**	16%/17%/67%
RQ Switch	**Notice**	24%/17%/59%
RR Switch	**Notice**	39%/2%/59%
RS Switch	Proper	10%/20%/70%
SR Switch	**Notice**	30%/7%/63%

[a,b,c] Typical form, other fillers, and no filler, respectively

4 Prediction of Fillers

4.1 Category for Prediction

We assume that the DA of the previous utterance (of either participant) is given and the DA of the next utterance (either holding or taking the turn) is determined. Then, we want to predict the occurrence of a filler and its form. Since there is a different typical form depending on the DA pair as shown in the previous section, the filler form to be predicted is limited to the typical form and other forms collectively. Thus, the target of prediction is reduced to typical form, other forms (o) and no-filler (nf). Moreover, if the ratio of the other forms is small (smaller than 10%) or the number of filler samples is small, they are merged into the typical filler making a single filler category (f).

The baseline single model to predict only occurrence of fillers is also trained by using all data without consideration of DA pairs. This single model outputs the typical filler depending on the DA pair if it predicts filler occurrence.

4.2 Classifier and Features

We use the Random Forest classifier from Scikit-learn [17] and evaluate its performance in 5-fold cross-validation. The number of decision trees which are built by bootstrap is set to ten. Since the number of samples is very different according to the class, we conduct normalization by sub-sampling in training, but the evaluation via cross-validation is conducted for the entire set. The evaluation measures are precision, recall and F-measure (their harmonic mean).

We incorporate linguistic (L) and prosodic (P) features extracted from the preceding utterance (pLUU: preceding LUU) and linguistic (L) features extracted from the following utterance (fLUU: following LUU). They are listed in Table 5.

We adopt features of last words and the boundary type as Japanese has a characteristic in the end-of-sentence expressions [5]. As the length of the preceding utterance is also an important feature, we include the number of words and chunks. Moreover, prosodic features are related with the distinction between turn-taking and turn-holding [12, 16]. Therefore, we extract them from the end of the preceding utterance. F0 and power are extracted by STRAIGHT [9, 10] (XSX [8]). We calculate the regression coefficient, the mean value, the maximum value, and the minimum value of F0 and power. The speech rate is approximately computed by dividing the number of characters by the duration, and the pause is defined as the time (ms) from the end of the pLUU.

The feature extracted from the following utterances (fLUU) would be useful for filler prediction, but we need to conduct prediction before the following utterance. In fact, people generate fillers while thinking of the next utterances. It is unrealistic to get exact information of the following utterance. In this work, however, we assume that the beginning word and the approximate length (two classes of short or long) is determined, and these features are used for prediction. The assumption of availability of these features will stand at least for the system side, and the goal of this study is generation of fillers by the robot after deciding what to speak.

4.3 Prediction Performance

Table 6 shows the prediction result (F-measure) for each DA pair. We tested all possible combinations of three feature sets (pLUU(L), pLUU(P) and fLUU(L)) and the results with the most effective feature sets are presented in the table.

From the upper part of the table, the features including prosody from the preceding utterance (pLUU) are effective in the case of turn-keeping. This is because the speaker suggests turn-holding with the prosody of the previous utterances and that affects the filler generation. On the other hand, in the case of turn-switching, we see that the linguistic features (L) of the following utterance (fLUU) are indispensable. This suggests that the speaker generates a filler after deciding to take a turn, and the form of the filler is determined based on the DA and the approximate length of the following utterance.

The filler prediction model individually trained for each DA pair achieves better performance than the single model for all and assigns the typical form. The overall performance is not necessarily high, but this is due to arbitrary characteristics of fillers; fillers may be placed or not depending on the person and at different times.

Table 5 Feature set

Utterance	Type	Feature	
Preceding utterance (pLUU)	Linguistic (L)	– DA – POS of the last word – Surface of the last word (if POS is an auxiliary verb or a post positional particle) – Clause boundary – # of words – # of chunks	
	Prosodic (P)	F0 power	– Regression coefficient – Maximum value – Minimum value – Mean value
		– Duration – Speech rate – Pause	
Following utterance (fLUU)	Linguistic (L)	– DA – POS of the beginning word – Surface of the beginning word (if POS is a conjunction) – # of words (quantized value) – # of chunks (quantized value)	

4.4 Prediction in Speech Collision Cases

Next, we conduct an experiment whether we can generate fillers to avoid speech collisions between two speakers using the model developed in the last subsection.

Speech collisions are defined as below.

- Both speeches are overlapped for over 500 ms after the beginning of the following utterance.
- The speaker who does not stop speaking after the speech collision takes the turn and the DA of his/her utterance is treated as the following DA.

There are 95 speech collisions in the relevant DA pairs in the corpus and the ratio of occurrence of fillers among these cases is 25.3% (=24/95). As the result of prediction, the proposed model can generate fillers in 57.9% (=55/95) of all cases in the speech collisions. This figure is around 2.3 times larger than the original number in the corpus, and covers 41.7% of them. Note that our model predicts fillers in 34.6% of DA pairs on average, close to the ratio of filler occurrence shown in Table 1.

Table 6 Prediction performance (F-measure)

DA pair	Feature set	Class	Individual	Single
RQ Keep	pLUU + fLUU LP + L	p	0.26	0.26
		o	0.32	0.00
		nf	0.50	**0.67**
		Avg.	**0.36**	0.31
RR Keep	pLUU + fLUU LP + L	d	**0.31**	0.25
		o	0.23	0.00
		nf	0.65	**0.74**
		Avg.	**0.40**	0.33
RS Keep	pLUU + fLUU L + L	d	0.17	0.17
		o	0.36	0.00
		nf	0.45	**0.69**
		Avg.	**0.33**	0.29
SS Keep	pLUU + fLUU LP + L	f	0.35	**0.36**
		nf	0.67	0.67
		Avg.	0.51	**0.52**
QR Switch	fLUU L	n	**0.33**	0.32
		o	0.50	0.00
		nf	0.75	**0.78**
		Avg.	**0.53**	0.37
RQ Switch	fLUU L	n	**0.53**	0.40
		o	0.41	0.00
		nf	0.37	**0.56**
		Avg.	**0.44**	0.32
RR Switch	pLUU + fLUU L + L	f	**0.59**	0.57
		nf	0.70	**0.71**
		Avg.	0.64	0.64
RS Switch	pLUU + fLUU LP + L	p	**0.20**	0.18
		o	0.44	0.00
		nf	**0.60**	0.58
		Avg.	**0.41**	0.25
SR Switch	fLUU L	f	**0.63**	0.50
		nf	**0.75**	0.70
		Avg.	**0.69**	0.60

p: proper form, d: demonstrative form, n: notice form
f: filler, o: other form, nf: no filler

The fillers generated by the proposed model can potentially avoid speech collisions. Even if the filler collides with the utterance by the dialog partner, it does not cause serious harm. This is particularly important for spoken dialog systems which cannot usually cancel the speech output command once generated.

5 Subjective Evaluation

Finally, we conduct a subjective evaluation experiment on the fillers generated by the proposed method. We prepare ten audio samples with inserted fillers (with different frequencies) and no filler, respectively. These dialog segments are extracted from the corpus, and we had 20 people listen to the audio and make an evaluation on the questionnaire regarding to the naturalness and likability.

The result of multiple test shows that a significant difference is confirmed between the no-filler samples and the samples in which the frequency of fillers is low ($t(28) = 4.62$ $p < 0.01$). In the case that the frequency of fillers is high, however, the difference is not significant ($t(28) = 5.16$ $p < 0.1$). The result suggests that the user feels positive towards the system which generates fillers, but it might not be good to generate fillers too much.

6 Conclusions

We have proposed a prediction and generation mechanism of fillers in spoken dialog, which can suggest turn-holding or turn-taking. First, we found out that tendency of filler occurrence and its typical form is different according to the DA pairs. Based on this observation, we prepared a model to predict fillers for each DA pair, using both linguistic and prosodic features of the preceding utterance and the approximate linguistic features of the following utterance. It is shown that the effective feature set for prediction is different according to the DA pairs, in particular turn-holding or turn-taking. It is also shown that the DA features of the following utterance are useful, but this can be considered in the system design so that it determines the DA of the next utterance and then decide to generate a filler. Moreover, it is confirmed that the proposed model successfully generates fillers in more than half cases of speech collision, thus potentially avoid them or mitigate their side effect. We plan to implement this model in the spoken dialog system by an autonomous android ERICA [14, 15], and evaluate the effectiveness in real user experiences.

Acknowledgements This work was supported by JST ERATO Ishiguro Symbiotic Human-Robot Interaction program (Grant Number JPMJER1401), Japan.

References

1. Akita Y, Kawahara T (2010) Statistical transformation of language and pronunciation models for spontaneous speech recognition. IEEE Trans Audio Speech Lang Process 18(6):1539–1549
2. Andersson S, Georgila K, Traum D, Aylett M, Clark R (2010) Prediction and realisation of conversational characteristics by utilising spontaneous speech for unit selection. In: Proceedings of the Speech Prosody

3. Bunt H, Alexandersson J, Carletta J, Chae JW, Fang AC, Hasida K, Lee K, Petukhova O, Popescu-Belis A, Romary L et al (2010) Towards an ISO standard for dialogue act annotation. In: proceedings of the LREC 2010, Malta
4. Cohen J (1960) A coefficient of agreement for nominal scales. Educ Psychol Meas 20(1):37–46
5. Den Y (2015) Some phonological, syntactic, and cognitive factors behind phrase-final lengthening in spontaneous Japanese: a corpus-based study. Lab Phonol 6(3–4):337–379
6. Den Y, Koiso H, Maruyama T, Maekawa K, Takanashi K, Enomoto M, Yoshida N (2010) Two-level annotation of utterance-units in Japanese dialogs: an empirically emerged scheme. In: LREC
7. Inoue K, Milhorat P, Lala D, Zhao T, Kawahara T (2016) Talking with ERICA, an autonomous android. In: Proceedings of the SIGdial meeting discourse and dialogue, pp 212–215
8. Itagaki H, Morise M, Nisimura R, Irino T, Kawahara H (2009) A bottom-up procedure to extract periodicity structure of voiced sounds and its application to represent and restoration of pathological voices. In: MAVEBA, pp 115–118
9. Kawahara H, Masuda-Katsuse I, De Cheveigne A (1999) Restructuring speech representations using a pitch-adaptive time-frequency smoothing and an instantaneous-frequency-based F0 extraction: possible role of a repetitive structure in sounds. Speech Commun 27(3):187–207
10. Kawahara H, Morise M, Takahashi T, Nisimura R, Irino T, Banno H (2008) Tandem-straight: a temporally stable power spectral representation for periodic signals and applications to interference-free spectrum, F0, and aperiodicity estimation. In: IEEE international conference on acoustics, speech and signal processing, 2008. ICASSP 2008. IEEE, pp 3933–3936
11. Kawahara T, Yamaguchi T, Inoue K, Takanashi K, Ward N (2016) Prediction and generation of backchannel form for attentive listening systems. In: Proceedings of the INTERSPEECH, vol 2016
12. Koiso H, Horiuchi Y, Tutiya S, Ichikawa A, Den Y (1998) An analysis of turn-taking and backchannels based on prosodic and syntactic features in Japanese map task dialogs. Lang Speech 41(3–4):295–321
13. Koiso H, Nishikawa K, Mabuchi Y (2006) Construction of the corpus of spontaneous Japanese
14. Lala D, Milhorat P, Inoue K, Ishida M, Takanashi K, Kawahara T (2017) Attentive listening system with backchanneling, response generation and flexible turn-taking. In: Proceedings of the SIGdial meeting discourse and dialogue, pp 127–136
15. Milhorat P, Lala D, Inoue K, Tianyu Z, Ishida M, Takanashi K, Nakamura S, Kawahara T (2017) A conversational dialogue manager for the humanoid robot ERICA. In: Proceedings of the international workshop spoken dialogue systems (IWSDS) (2017)
16. Ohsuga T, Horiuchi Y, Nishida M, Ichikawa A (2006) Prediction of turn-taking from prosody in spontaneous dialogue. Trans Jpn Soc Artif Intell 21:1–8
17. Pedregosa F, Varoquaux G, Gramfort A, Michel V, Thirion B, Grisel O, Blondel M, Prettenhofer P, Weiss R, Dubourg V et al (2011) Scikit-learn: machine learning in python. J Mach Learn Res 12:2825–2830
18. Schegloff EA, Sacks H (1973) Opening up closings. Semiotica 8(4):289–327
19. Shiwa T, Kanda T, Imai M, Ishiguro H, Hagita N (2008) How quickly should communication robots respond? In: 2008 3rd ACM/IEEE international conference on human-robot interaction (HRI). IEEE, pp 153–160
20. Skantze G, Hjalmarsson A, Oertel C (2014) Turn-taking, feedback and joint attention in situated human-robot interaction. Speech Commun 65:50–66
21. Sundaram S, Narayanan S (2002) Spoken language synthesis: experiments in synthesis of spontaneous monologues. In: Proceedings of the IEEE workshop on speech synthesis, pp 203–206
22. Watanabe M (2009) Features and roles of filled pauses in speech communication: a corpus-based study of spontaneous speech. Hitsuji Syobo Publishing

Testing Strategies For Bridging Time-To-Content In Spoken Dialogue Systems

Soledad López Gambino, Sina Zarrieß and David Schlangen

Abstract What should dialogue systems do while looking for information or planning their next utterance? We conducted a study in which participants listened to (constructed) conversations between a user and an information system. In one condition, the system remained silent while preparing a reply, whereas in the other, it "bought time" conversationally, using strategies from previously recorded human interactions. Participants perceived the second system as better at responding within an appropriate amount of time. Additionally, we varied between mid- and high-quality voices, and found that the high-quality voice time-buying system was also seen as more willing to help, better at understanding and more human-like than the silent system. We speculate that participants may have perceived this voice as a better match for the more human-like behavior of the second system.

1 Introduction

A common pattern in spoken human-machine interaction consists of a request for information by the human followed by presentation of this information by the system. Retrieval of this information may take time (e.g., for queries to remote databases). What should a system do while it prepares its reply?

A simple approach would be to remain silent until it can present information. However, this is not what humans do in such a situation. Lopez Gambino et al. [9] show that people have a variety of resources available for "buying time", such as producing fillers (*uhm* or *uh*) [6], repeating parts of the interlocutor's request [5], explaining the reasons for the delay, etc.

S. López Gambino (✉) · S. Zarrieß · D. Schlangen
CITEC Bielefeld University, Universitätsstraße 25, 33615 Bielefeld, Germany
e-mail: m.lopez_gambino@uni-bielefeld.de

S. Zarrieß
e-mail: sina.zarriess@uni-bielefeld.de

D. Schlangen
e-mail: david.schlangen@uni-bielefeld.de

L. F. D'Haro et al. (eds.), *9th International Workshop on Spoken Dialogue System Technology*, Lecture Notes in Electrical Engineering 579,
https://doi.org/10.1007/978-981-13-9443-0_9

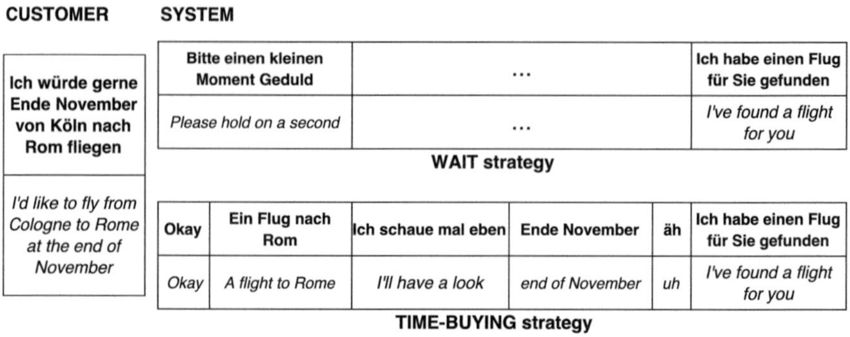

Fig. 1 Example dialogue for each of the two experiment conditions (original utterances in German in bold; English translation provided below in italics)

How would users perceive an automatic system which produces such an array of resources instead of adopting a more traditional "please hold the line" type of approach? Would this system be viewed as a more human-like conversational partner? Or, to the contrary, would this behavior strike listeners as too unusual for an automatic system? To answer these questions, we conducted an overhearer study in which participants compared two (simulated) systems: The WAIT system asked users to wait, and then remained silent until it was able to present information, whereas the TIME-BUYING system produced behaviors similar to those observed in humans (see Fig. 1).

Results showed that participants perceived the TIME-BUYING system as capable of finding a result within a more appropriate time period than the WAIT system, even though the actual time elapsed was the same for both conditions. Furthermore, as long as the system's voice was high quality, the TIME-BUYING system was also perceived as more willing to help, better at understanding and more human-like than the WAIT system. This, however, was not the case when the system used a mid-quality voice (see Sect. 3).

2 Method

DESIGN The main factor was WAIT versus. TIME-BUYING (see above). We conducted two runs of the study, with two different speech synthesizers, the first more easily identifiable as a machine and the second sounding more natural (see *MATERIALS* below). Participants listened to four recordings, two for each condition, in random order.

PARTICIPANTS Recruitment was carried out on the crowdsourcing platforms Amazon Mechanical Turk and Crowdflower and limited to workers in Germany. Forty-two subjects participated in the first run (16 female and 26 male, aged 20 to

69) and 39 in the second run (15 female and 24 male, aged 21 to 63). The study was published in the form of a questionnaire on the online platform SoSciSurvey.[1]

MATERIALS For the first run, the system's utterances were synthesized using MaryTTS, whereas Cereproc was used for the second run.[2] For MaryTTS we chose an HSMM voice, which resulted in (subjectively) less natural sound than the second one, a commercial professional voice. We used a male voice and the same utterances in both runs. The utterances were also the same for all participants. In order to produce them, we implemented a simple "time-buying generator" which produced a sequence of five time-buying utterances and then announced having found a flight. The system used the time-buying categories described in [9]. Some examples are **filler** (*uh, uhm*), **echoing** (A: *I need a flight to Bristol.* B: *Okay, a flight to Bristol…*) and **justification** (*The system is very slow today.*). At each step, the system chose one of these categories and produced one out of a set of canned utterances belonging to that category. The choice of category depended on: (a) the previous system utterance and (b) the time elapsed since the beginning of the time-buying stretch. Given these two parameters, the system selected a category by sampling from a probability distribution over all possible categories. The probabilities were trained on the DSG-Travel Corpus, a corpus of human interactions simulating a travel agency scenario [9]. The full recordings, as presented to the participants, consisted of a customer's request for a flight, followed by the system's time-buying utterances and final announcement of having found a result (as illustrated in Fig. 1).

PROCEDURE The participants first provided some demographic data, did a brief German language check, and read the task instructions. Participants then listened to recordings of enacted phone conversations between a human customer and an automatic system at a travel agency.[3] The human customer asked for a flight meeting certain criteria and the system pretended to look for an option which satisfied the customer's needs (see Fig. 1). After a while, the system announced having found an appropriate flight. The time between the end of the customer's request and the system's announcement was approximately 12 s.[4] The behavior of the system during this period varied according to the experimental condition:

- WAIT: The system asks the customer to wait by producing an utterance such as *Bitte einen kleinen Moment Geduld* (Please be patient for a moment), and then remains silent until it announces having found the flight.
- TIME-BUYING The system produces a variety of utterances separated by short pauses, thus "buying time" until it has found a flight.

After each recording, participants rated the corresponding system on a 1–5 scale (5 meaning "strongly agree") with respect to five statements (here in translation):

[1] URLs: https://www.mturk.com/, https://www.crowdflower.com, https://www.soscisurvey.de/.

[2] http://mary.dfki.de/, https://www.cereproc.com/.

[3] The customers' utterances were taken from the DSG-Travel corpus [9].

[4] We considered 12 seconds to be a realistic waiting period a relatively lengthy lookup might take, yet not so long that the WAIT strategy would obviously be disadvantaged.

1. The system understood the caller well.
2. The system took an appropriate amount of time to find a flight.
3. The system sounds as if willing to help.
4. The system acts the way I would expect a person to act.
5. If I had to buy a flight on the phone, I would use this system.

3 Results

We compared the ratings between the WAIT and the TIME-BUYING strategy. We test significance of differences through a paired-samples t-test and Wilcoxon signed-rank test, using Bonferroni adjusted alpha levels (0.05/5 = **0.01**, 0.01/5 = **0.002**, 0.001/5 = **0.0002**). In the first run (Mary-TTS voice is used), mean ratings for TIME-BUYING are higher than for WAIT, for all five statements. However, the difference only proved significant in the case of statement 2, "The system took an appropriate amount of time to find a flight" ($t(83) = 3.22$, $p < 0.002$; $W = 244.5$, $p < 0.002$) (Table 1).

In the second run (Cereproc Text-to-Speech is used), the TIME-BUYING strategy was rated better for each of the five statements, and differences were highly significant in all cases (see Table 2).

Table 1 Mean ratings, standard deviations and medians for both conditions in statement 2, in the first run of the study

Condition	Mean	Std. Dev.	Median
WAIT	3.7	0.99	4
TIME-BUYING	4.07	0.94	4

Table 2 Statistics for the statements (see Sect. 2); high-quality voice run

Statement	*M* WAIT	*M* TB	*Mdn* WAIT	*Mdn*TB	t-test	Wilcoxon
1	3.91 (*SD* = 0.85)	4.47 (*SD* = 0.71)	4	5	$t(77) = 6.11$, $p < 0.0002$	$W = 111$, $p < 0.0002$
2	3.21 (*SD* = 1.17)	4.38 (*SD* = 0.77)	3	5	$t(77) = 9.38$, $p < 0.0002$	$W = 52$, $p < 0.0002$
3	3.33 (*SD* = 1.02)	3.98 (*SD* = 0.91)	4	4	$t(77) = 5.67$, $p < 0.0002$	$W = 163.5$, $p < 0.0002$
4	3.03 (*SD* = 1.03)	3.7 (*SD* = 1.09)	3	4	$t(77) = 5.03$, $p < 0.0002$	$W = 248$, $p < 0.0002$
5	2.85 (*SD* = 1.04)	3.42 (*SD* = 1.17)	3	4	$t(77) = 5.5$, $p < 0.0002$	$W = 132$, $p < 0.0002$

4 Discussion

The results presented above show that an information-providing dialogue system which can use speech to avoid long gaps after a user's request—similarly to what humans usually do—can make a better impression on overhearers than a system which asks the user to wait and then remains silent until it can provide an answer. In the first run of our study, participants found waiting times to be more appropriate in the TIME-BUYING system than in the WAIT one, even though the actual times remained constant across conditions. Additionally, the second run revealed that overhearers also perceived the TIME-BUYING system as more willing to help, better understanding of the user's request, and more human-like than the WAIT system. Finally, participants preferred the former over the latter for their own use. These results suggest that dialogue systems could benefit from the incorporation of time-buying capabilities.

Additionally, the differences between the results of both study runs open up questions regarding the interplay of voice quality and time-buying strategy. One possible interpretation is that participants may have found the more human-like voice in the second run a better match for the more human-like behavior of the TIME-BUYING system. This could be connected to the idea of the *metaphors* involved in humans' perception of dialogue systems. Edlund et al. [7] draw a distinction between the *interface metaphor*, in which the system is perceived as a machine, and the *human metaphor*, in which the system is viewed as an interlocutor with whom speech is the natural interaction channel, and highlight the need for internal coherence between the metaphor selected and the behavior of the system.

From this perspective, one could argue that a system seeking to buy time like humans should use a voice as similar as possible to that of a human. However, deciding what kind of voice is best for a dialogue system is not always so straightforward, and other considerations also need to be taken into account. One of them is flexibility. Many commercial TTS systems sound relatively human-like but do not offer many options for acoustic modification (other than general emotion tags, etc). Systems like MaryTTS, on the other hand, offer both unit selection and HSMM voices, and the latter grant the possibility, for example, to adjust the frequency and duration of each phone to specific values [12]. It is therefore necessary to take this trade-off between human-likeness and flexibility into account, and prioritize depending on the aims and specificities of the dialogue system under construction.

5 Related Work

Our results for both study runs are compatible with the idea that "filled time" is perceived as shorter than "unfilled time". This is, however, a somewhat contested assumption: Although there is research suggesting its validity [8, 14], it has also been postulated that what creates a perception of shorter waiting time is not the fact that the time is filled, but rather the nature of the information which is used to fill it. An example could be information about the waitee's place in the queue, which may convey a feeling of progressing towards the goal [10] or information about the estimated total duration of the wait [1].[5] This seems to be connected to a need for transparency regarding the state of the interaction. Such considerations are highly relevant when it comes to incorporating more conversational time-buying utterances in a system, since these utterances may also enable the system to provide justification for the wait and convey a sense of progress towards the desired goal. Finally, we highlight the importance of time-buying mechanisms within the area of incremental speech processing, since a number of studies have shown the benefits of systems with the ability to start producing some speech even before they have a full plan of the information to present [2–4, 11, 13].

6 Conclusion and Future Work

We have presented an overhearer study in which participants rated two information systems: one which asked the interlocutor to wait and remained silent while looking for the information to present, and another one which produced utterances during the wait. We found that participants perceive the time elapsed between the interlocutor's request and the system's response as longer in the first condition. Additionally, if the synthesized voice is relatively human-like, the system producing utterances is also perceived as more willing to help, better understanding of the user's request, and more human-like. In the future, we plan to incorporate time-buying capabilities into an actual dialogue system and explore the effects of different time-buying strategies in an interactive scenario, with regard to users' preferences as well as to more objective measures of task performance [3, 15, 16].

Acknowledgements This work was supported by the Cluster of Excellence Cognitive Interaction Technology 'CITEC' (EXC 277) at Bielefeld University, which is funded by the German Research Foundation (DFG).

References

1. Antonides G, Verhoef P, van Aalst M (2002) Consumer perception and evaluation of waiting time: a field experiment. J Consum Psychol 12(3):193–202
2. Baumann T, Schlangen D (2013) Open-ended, extensible system utterances are preferred, even if they require filled pauses. In: Proceedings of short papers at SIGdial 2013
3. Betz S, Carlmeyer B, Wagner P, Wrede B (2017) Interactive hesitation synthesis and its evaluation. https://www.preprints.org/manuscript/201712.0058/v1

[5] In this study, information about duration of the wait did not make perceived waiting time shorter than actual waiting time, but it did reduce overestimation of its length in comparison to other experimental conditions.

4. Buschmeier H, Baumann T, Dosch B, Kopp S, Schlangen D (2012) Combining incremental language generation and incremental speech synthesis for adaptive information presentation. In: Proceedings of the 13th annual meeting of the special interest group on discourse and dialogue, pp 295–303
5. Byron D, Heeman P (1997) Discourse marker use in task-oriented spoken dialog. In: Proceedings of Euro speech 97
6. Clark H, Fox Tree J (2002) Using uh and um in spontaneous speaking. Cognition 84(1):73–111
7. Edlund J, Gustafson J, Heldner M, Hjalmarsson A (2008) Towards human-like spoken dialogue systems. Speech Commun 50:630–645
8. Hirsch I, Bilger R, Heatherage B (1950) The effect of auditory and visual background on apparent duration. Am J Psychol, 69
9. Lopez Gambino S, Zarrieß S, Schlangen D (2017) Beyond on-hold messages: conversational time-buying in task-oriented dialogue. In: Proceedings of SIGdial 2017
10. Munichor N, Rafaeli A (2007) Numbers or apologies? customer reactions to telephone waiting time fillers. J Appl Psychol 92(2):511–518
11. Schlangen D, Skantze G (2011) A general, abstract model of incremental dialogue processing. Dialogue Discourse 2(1):83–111
12. Schröder M, Trouvain J (2003) The German text-to-speech synthesis system MARY: a tool for research, development and teaching. Int J Speech Technol 6:365–377
13. Skantze G, Hjalmarsson A (2010) Towards incremental speech generation in dialogue systems. In: Proceedings of the 11th annual meeting of the special interest group on discourse and dialogue, SIGDIAL '10. Association for Computational Linguistics, Stroudsburg, PA, USA , pp 1–8
14. Tom G, Burns M, Zeng Y (1997) Your life on hold: the effect of telephone waiting time on customer perception. J Direct Mark 11(3):25–31
15. Walker M, Kamm C, Litman D (2000) Towards developing general models of usability with PARADISE. Nat Lang Eng 6:3–4
16. Whittaker S, Walker M (2005) Evaluating dialogue strategies in multimodal dialogue systems. In: Minker W, Bühler D (eds) Spoken multimodal human-computer dialogue in mobile environments. text, speech and language technology, vol 28

Faster Responses Are Better Responses: Introducing Incrementality into Sociable Virtual Personal Assistants

Vivian Tsai, Timo Baumann, Florian Pecune and Justine Cassell

Abstract Speech-based interactive systems, such as virtual personal assistants, inevitably use complex architectures, with a multitude of modules working in series (or less often in parallel) to perform a task (e.g., giving personalized movie recommendations via dialog). Add modules for evoking and sustaining sociability with the user and the accumulation of processing latencies through the modules results in considerable turn-taking delays. We introduce incremental speech processing into the generation pipeline of the system to overcome this challenge with only minimal changes to the system architecture, through partial underspecification that is resolved as necessary. A user study with a sociable movie recommendation agent objectively diminishes turn-taking delays; furthermore, users not only rate the incremental system as more responsive, but also rate its recommendation performance as higher.

1 Introduction

We present a way to improve turn-taking responsiveness in a social, multimodal dialog system [9] that builds a relationship with users while recommending movies, and we show that increased responsiveness is also perceived as improved perfor-

V. Tsai (✉)
Johns Hopkins University, Maryland, USA
e-mail: viv@jhu.edu

T. Baumann
Universität Hamburg, Hamburg, Germany
e-mail: mail@timobaumann.de

F. Pecune
University of Glasgow, Glasgow, Scotland
e-mail: florian.pecune@glasgow.ac.uk

J. Cassell
Carnegie Mellon University, Pittsburgh, USA
e-mail: justine@cs.cmu.edu

L. F. D'Haro et al. (eds.), *9th International Workshop on Spoken Dialogue System Technology*, Lecture Notes in Electrical Engineering 579,
https://doi.org/10.1007/978-981-13-9443-0_10

mance regarding those recommendations. The base movie agent first prompts users to specify preferred genres, directors, and actors; a knowledge graph-based recommendation system [7] consequently produces suitable movie titles and explanations of their relevance. The agent delivers not only this task-based information, but also conversational utterances intended to build rapport with the user (Table 3).

When comparing human-dialog system interactions with human-human interactions, Ward et al. [12] found that a crucial issue with the former is "responsiveness," exceeded in importance only by "recognition and understanding" and "time-outs." Indeed, the response delay in our legacy application frequently disrupts interactions: users become frustrated when the agent fails to respond promptly and consequently end the interaction prematurely or even abandon the interaction mid-exchange. We therefore seek to improve the responsiveness of the movie agent.

An analysis of our legacy system showed that the most significant source of delay is the recommendation system, which takes 1.7 s on average to return the requested movie content. A user study demonstrates that our incorporation of incrementality not only objectively improves response time by eliminating this delay, but also positively impacts users' evaluations of their interactions with the movie agent in interesting ways.

2 Related Work

Incremental speech analysis and speech generation have previously been used as effective means of ensuring fast responses for spoken dialog systems [5, 6]. Specifically, the required processing times in a dialog system can take place *while* speech is delivered by the user or the system. Skantze and Hjalmarsson [11] implemented this strategy via a system that played a filler ('uhm') to give the appearance of reduced turn-taking delays (by 0.6 s on average); the resulting system was rated by users as significantly more efficient than a non-incremental version. However, the integration of incremental processing into larger existing dialog system architectures has been limited (e.g., requiring switching between different modes for certain dialog states [3]).

The incremental speech synthesis system in [2] uses only partially specified descriptions of utterances when initiating the delivery of those utterances: the beginning of each utterance must always be known for the system to start speaking it, but later parts may be underspecified as long as they are fully specified by the time the system requires them for synthesis. This concept of underspecification has not yet been applied in the context of full incremental dialog systems, to the best of our knowledge. Bąk et al. [1] present the concept of partial specification for object-oriented modeling, which we use similarly in our implementation.

3 Implementation

Our system is split into a frontend client, for speech recognition and synthesis, and a server backend (Fig. 1), which involves a multiuser framework for managing multiple users and a pipeline architecture for dialog processing. Pipeline modules include a natural language understanding (NLU) module, a dialogue manager (DM), a social reasoner (SR), and a natural language generator (NLG).

For each user-agent exchange, the NLU/DM interprets user input, then returns a response intent for the agent (e.g., *ask for favorite genre*, or *give movie rec.*) and, if applicable, a movie recommendation object (comprised of suitable movie titles and reasonings behind these selections) from the recommendation system. The SR [10] selects a conversational strategy based on the DM response intent, and the NLG uses both the response intent and the SR strategy to formulate an appropriate response. Finally, the NLG replaces any variables in this response (i.e., [*movie title*]) with the corresponding content, thus producing a complete response to be output to the frontend.

Our incremental solution (Fig. 2) integrates the concept of partial specification while retaining the pipeline of the legacy system. Since the recommendation system creates the most significant delay, we adapt the architecture so that during interactions involving recommendations, processes for which that recommendation content is unnecessary can continue while the delay occurs. Rather than waiting for the recommendation system's output, the DM immediately outputs the response intent and an *underspecified* recommendation object. The former allows the SR to produce a strategy, which in turns allows the NLG to generate its response.

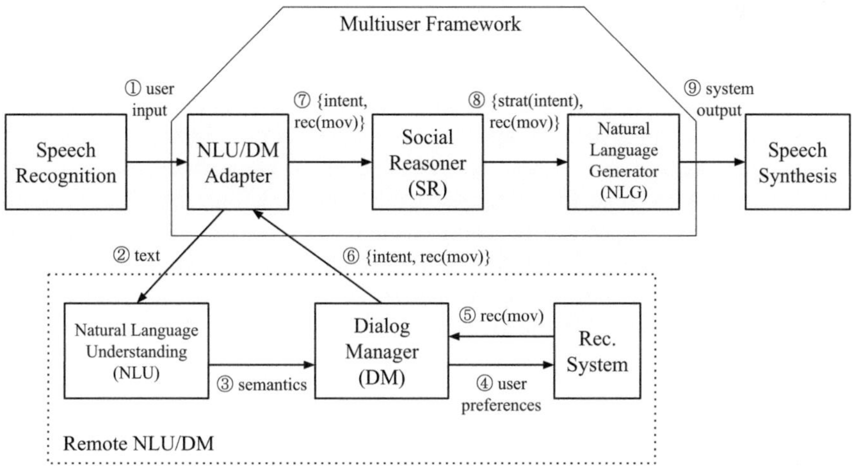

Fig. 1 Task flow for legacy, non-incremental system: modules are arranged in a pipeline; NLU, DM, and recommendation are outsourced to a separate service. The order of the information flow is shown by the circled numbers

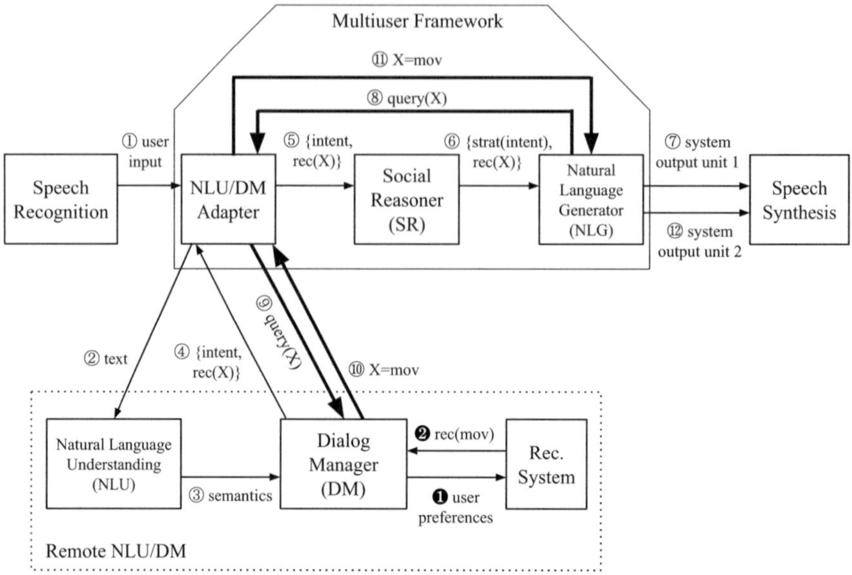

Fig. 2 Task flow for incremental system, where the response shown is split into two separate units. Movie recommendation querying is performed concurrently (inversely circled numbers). The first unit of system output is spoken while details of the second are being determined (see strong arrows)

While the legacy system considers each NLG response as a whole, the incremental system splits it at unit markets, then considers each unit in sequence and incrementally sends results to the frontend. (On the frontend side, these units are placed in a queue and synthesized only when all preceding units have already been verbalized.) If a unit contains a variable and the value for that variable is currently underspecified, the NLG waits until that value can be provided; in the case of an underspecified recommendation object, the NLG queries the DM for the fully specified recommendation object (and is blocked until it can be provided).

By incrementally generating responses, our solution allows the agent to begin speaking as soon as the first unit is complete and thus before the movie recommendation is available. In other words, if a response contains enough units prior to a unit with the movie title variable, the recommendation system latency can be folded into the time required to verbalize the preceding units.

4 User Study

To evaluate the efficacy of our incremental solution, we compared users' interactions with the agent under both non-incremental (N) and incremental (I) conditions. Three dependent variables were assessed: *responsiveness of agent*, *attentiveness of agent*, and *enjoyability of conversation*. We also assessed the *recommendation quality* (i.e., quality of the agent's movie recommendations) as a presumably unaffected control variable. This gave us the following hypotheses:

- **H-Inc**. The incremental system (I) will positively impact the dependent variables: under the incremental system (I), users will rate the agent as more responsive, more attentive, and more enjoyable than they will under the non-incremental system (N).
- **H-Rec**. The movies recommended are not affected by the I/N condition and their perceived quality should thus remain the same.

Our system's incrementality is presently limited to whole sentences as units. We thus structured NLG response patterns to ensure that all responses are comprised of two units: a "social" sentence (e.g., "Wow, here's one I'd love to go to!") and a "recommendation" sentence, the latter of which contained the movie title variable. Each social sentence was constructed such that its verbalization by the agent took more than 1.7 s, thus covering the average latency of the recommendation system.

4.1 Procedure

The experiment was conducted with a total of 24 subjects (12 female, 12 male; mean age 20–25) recruited from the university community and randomly split into four equally sized groups. Each subject held two conversations, one with the non-incremental (N) system and one with the incremental (I), with two different scenarios (each giving a specific genre, director, and actor to feed to the agent). The system-scenario, as well as N/I ordering combinations, was counter-balanced for the four groups to remove ordering and/or scenario preference effects. Subjects were informed only that the conversations would differ in scenario; no information about the difference in the systems was given.

For each conversation, subjects began by using scenario content to answer the movie agent's initial questions; continued the conversation by commenting on the agent's responses ("I've already seen that movie") and/or specifying new preferences ("Actually, I'd prefer comedies"); and ended the conversation once they were satisfied with the quantity/quality of the recommendations and the overall interaction. They then evaluated their experience by completing a 7-point Likert questionnaire (eight randomly ordered statements, two (one original, one reverse-coded) for each of the four variables) on the conversation. To conclude, subjects completed an additional final questionnaire on whether or not they noticed a difference between the two conversations; afterwards, they were debriefed on the differing N/I system conditions.

Data from all conversations were collected in the form of log files, which included system internals, conversation transcripts, and timestamps for both frontend and backend processes, as well as audio files.

Table 1 Mean durations (in seconds) for user-agent exchanges

	Non-incremental system	Incremental system
Non-recommendation exchanges	0.7	0.7
Recommendation exchanges	2.2	0.5

Table 2 P-values of sign tests for each variable from non-incremental versus incremental conditions

Variable	All subjects (N = 24)	≥7 years of English (N = 18)
Responsiveness	0.038*	0.033*
Attentiveness	0.058	0.038*
Enjoyability	0.084	0.038*
Recommendation quality	0.017*	0.015*

*Statistical significance ($p < 0.05$)

4.2 Results

The average duration of subjects' conversations with the agent was 200 s, with a mean of 6 movie recommendations given per conversation. We used timestamps from the log file data to calculate the latency between each pair of the user's request and the agent's subsequent response.

While the differing systems had no impact on exchanges that did not involve movie recommendations (Table 1), the incremental system saved an overall average of 1.7 s per user-agent exchange in which a movie was recommended (cutting down the response time to nearly 20%) and was thus objectively more responsive than the non-incremental system. To measure participants' subjective evaluations of the non-incremental and incremental system, we performed sign tests on each of the four dependent variables (see Tables 2 and 3).

The significant advantage of the incremental system supports our hypothesis **H-Inc**, particularly when focusing on users proficient in English. Most subjects maintained during their debriefings that they did not notice any time differences between the two systems, but many noted in their final questionnaire that their conversation with the incremental system went "more smoothly" and seemed to "flow better."

The incremental system was also rated higher for recommendation quality (the quality of movie recommendations), even though we used the same recommendation system for both conditions. We therefore reject hypothesis **H-Rec**. This result indicates that turn-taking delays have a significant influence on perceived quality, pointing to the need to attend to turn-taking latencies in system implementation.

In sum, the study results support our hypothesis that an incremental system is preferred and more positively rated by users. We also find that incrementality improves the perceived recommendation quality.

Table 3 Conversation example

	Speaker	Utterance
Part 1	*Movie Agent*	*Hi, I'm Sara. I can help you find a good movie to watch. Would you mind telling me your favorite genres?*
	User	my favorite genres are comedy and action
	Movie Agent	*I like the way you think! Who are your favorite directors?*
	User	Christopher Nolan and Christopher McQuarrie
	Movie Agent	*Okay! Who are your favorite actors?*
	User	Christian Bale and Amy Adams
Part 2	*Movie Agent*	*I think this movie fits your tastes. How about Interstellar (2014)?*
	User	oh that's a good movie but I've already watched it do you want to recommend another one for me
	Movie Agent	*Wow, here is one I'd love to go to. It's called The Dark Knight Rises (2012).*
	User	yeah that's a good movie too I've seen that too so do you have another recommendation
	Movie Agent	*I think this movie fits your tastes. How about American Hustle (2013)?*
	User	oh that's a good recommendation I haven't watched that before
	Movie Agent	*If I can be of further assistance, I'll be here*

5 Conclusion and Future Work

Our findings indicate that the impact of incremental processing extends beyond latencies and can significantly influence users' perception of a system's capabilities. However, our current solution is only effective when responses are structured such that social utterances are long enough to cover the recommendation system delay; furthermore, when no variables exist, these social utterances unnecessarily prolong user-agent interactions.

In the future, we envision an incremental generation system that delivers information to the user as quickly as is reasonable and otherwise buys time [8] using social talk until information becomes available. Using anytime processing [14], we could additionally consider expected recommendation performance to determine whether or not to buy time (i.e., the system could delay its response if it believes that a better recommendation may still be produced through a more extensive search). We also envision using incremental speech synthesis [4, 13] in order to break down the granularity of speech output to smaller units than sentences for higher flexibility.

Appendix

The following table shows an audio transcription of a conversation between a user and our movie agent. During the first part of this conversation, the movie agent asks for the user's preferences; during the second part, the movie agent suggests relevant movie titles (adding any new specifications to the list of preferences) until the system determines that the user is satisfied.

Note that each of the movie agent's recommendation sentences is preceded by a "social" sentence, which is generated in accordance with a conversational strategy in order to build rapport with the user.

References

1. Bąk K, Diskin Z, Antkiewicz M, Czarnecki K, Wąsowski A (2013) Partial instances via subclassing, pp 344–364
2. Baumann T (2014) Partial representations improve the prosody of incremental speech synthesis. In: Proceedings of interspeech
3. Baumann T, Paetzel M, Schlesinger P, Menzel W (2013) Using affordances to shape the interaction in a hybrid spoken dialogue system. In: Proceedings of ESSV, pp 12–19
4. Baumann T, Schlangen D (2012) INPRO_iSS: a component for just-in-time incremental speech synthesis. In: Proceedings of ACL system demonstrations
5. Buschmeier H, Baumann T, Dorsch B, Kopp S, Schlangen D (2012) Combining incremental language generation and incremental speech synthesis for adaptive information presentation. In: Proceedings of SigDial, pp 295–303
6. Buß O, Baumann T, Schlangen D (2010) Collaborating on utterances with a spoken dialogue system using an ISU-based approach to incremental dialogue management. In: Proceedings of SigDial, pp 233–236
7. Kanjirathinkal RC, Cohen W (2016) Personalized recommendations using knowledge graphs: a probabilistic logic programming approach. In: Proceedings of ACM RecSys, pp 325–332
8. López Gambino S, Zarrieß S, Schlangen D (2017) Beyond on-hold messages: conversational time-buying in task-oriented dialogue. In: Proceedings of SigDial, pp 241–246
9. Matsuyama Y, Bhardwaj A, Zhao R, Romeo O, Akoju S, Cassell J (2016) Socially-aware animated intelligent personal assistant agent. In: Proceedings of the 17th annual meeting of the special interest group on discourse and dialogue, pp 224–227
10. Papangelis A, Zhao R, Cassell J (2014) Towards a computational architecture of dyadic rapport management for virtual agents. In: Proceedings of the international conference on intelligent virtual agents, pp 320–324
11. Skantze G, Hjalmarsson A (2013) Towards incremental speech generation in conversational systems. Comput Speech Lang 27(1):243–262
12. Ward NG, Rivera AG, Ward K, Novick DG (2005) Root causes of lost time and user stress in a simple dialog system. In: Proceedings of interspeech, pp 1565–1568. ISCA
13. Wester M, Braude DA, Potard B, Aylett MP, Shaw F (2017) Real-time reactive speech synthesis: incorporating interruptions. Proceedings of Interspeech 2017:3996–4000
14. Zilberstein S (1996) Using anytime algorithms in intelligent systems. AI Mag 17(3):73

Latent Character Model for Engagement Recognition Based on Multimodal Behaviors

Koji Inoue, Divesh Lala, Katsuya Takanashi and Tatsuya Kawahara

Abstract Engagement represents how much a user is interested in and willing to continue the current dialogue and is the important cue for spoken dialogue systems to adapt the user state. We address engagement recognition based on listener's multimodal behaviors such as backchannels, laughing, head nodding, and eye gaze. When the ground-truth labels are given by multiple annotators, they differ according to each annotator due to the different perspectives on the multimodal behaviors. We assume that each annotator has a latent character that affects its perception of engagement. We propose a hierarchical Bayesian model that estimates both the engagement level and the character of each annotator as latent variables. Furthermore, we incorporate other latent variables to map the input feature into a sub-space. The experimental result shows that the proposed model achieves higher accuracy than other models that do not take into account the character.

1 Introduction

A number of spoken dialogue systems have been developed and practically used in various kinds of contexts such as user assistants and conversational robots. The systems interact with the user in certain tasks such as question answering [10] and medical diagnoses [6]. In most cases, however, the interaction is human-machine

K. Inoue (✉) · D. Lala · K. Takanashi · T. Kawahara
Graduate School of Informatics, Kyoto University, Kyoto, Japan
e-mail: inoue@sap.ist.i.kyoto-u.ac.jp

D. Lala
e-mail: lala@sap.ist.i.kyoto-u.ac.jp

K. Takanashi
e-mail: takanasi@sap.ist.i.kyoto-u.ac.jp

T. Kawahara
e-mail: kawahara@sap.ist.i.kyoto-u.ac.jp

L. F. D'Haro et al. (eds.), *9th International Workshop on Spoken Dialogue System Technology*, Lecture Notes in Electrical Engineering 579,
https://doi.org/10.1007/978-981-13-9443-0_11

specific and much different from the case of human-human dialogue. Our ultimate goal is to realize conversational robots which behave like human beings and pervade many aspects of our daily lives in a symbiotic manner. To this end, it is needed for the systems to recognize and understand the conversational scene such as the user state. In this paper, we focus on user engagement in human-robot interaction. Engagement represents the process by which dialogue participants establish, maintain, and end their interaction [25]. Practically, it has been defined as the user state which represents how much a user is interested in and willing to continue the current dialogue [20, 29]. By recognizing user engagement in dialogue, the system can generate adaptive behaviors, which contributes to smooth and natural interaction.

In this study, we address engagement recognition based on listener's multimodal behaviors such as verbal backchannels, laughing, head nodding, and eye gaze. Since these behaviors are used by listeners to express responses toward speakers, it is presumed that these are related to engagement. To obtain the ground-truth labels of engagement, we ask third-party people (annotators) to judge the user engagement of dialogue data. Since the perception of engagement is subjective, the annotation result often depends on each annotator. Previous studies integrated engagement labels among annotators like majority voting to train recognition models [15, 16, 28].

The difference among annotators suggests that each annotator has different perspective on multimodal behaviors and engagement. We assume that each annotator has its latent character, and the character affects his/her perspective for engagement. The latent character represents a kind of template for the perspective on engagement. We propose a latent character model which estimates not only the engagement level but also the character of each annotator as latent variables. The model can simulate each annotator's perception more precisely. This study contributes to a variety of recognition tasks containing subjectivity such as emotion recognition in that the proposed model takes into account the differences and commonalities of multiple annotators.

2 Related Work

Engagement has been variously defined in different kinds of studies [8]. The definitions are mainly classified into two types. The first one focuses on the start and end of the interaction. For example, it is defined as "*the process by which two (or more) participants establish, maintain, and end their perceived connection*" [25]. This type is related to other concepts such as *attention* and *involvement* [19, 30]. The second type focuses on the quality of interaction. For example, engagement was defined as "*how much a participant is interested in and attentive to a conversation*" [29] and "*the value that a participant in an interaction attributes to the goal of being together with the other participant(s) and of continuing the interaction*" [20]. This type is related to *interest* and *rapport*. In this study, we consider engagement in the context of the latter type.

Engagement recognition has been widely studied in previous studies. It was formulated as a binary classification problem: engaged or not (disengaged), or a category classification problem [1]. The used features were based on non-linguistic multimodal behaviors. Non-linguistic information is commonly used as features because linguistic information is specific to the dialogue domain and content, and speech recognition is error-prone. Previous studies investigated the relationship between engagement and multimodal behaviors such as spatial information (e.g. location, trajectory, distance) [2, 14, 28], eye gaze (e.g. looking at a robot, mutual gaze) [1, 3, 15, 16, 22, 28, 31], facial information (e.g. facial movement, expression, head pose) [3, 4, 31], verbal backchannels [22, 28], head nodding [16], laughing [27], and posture [7, 23]. Additionally, low-level signals such as acoustic and visual features were considered [4, 11, 29]. The initial recognition models were based on heuristic rules [14, 19, 24]. The recent approach is based on machine learning techniques such as support vector machines (SVM) [7, 16, 28, 31], hidden Markov model (HMM) [29], and convolutional neural networks (CNN) [11]. Recently, some researchers have undertaken a study on system behaviors after recognizing user engagement. They found that the engagement level is related to turn-taking behaviors [12, 28]. Other researchers investigated how to handle user disengagement by changing the dialogue policy or changing the system responses [26, 31]. Our purpose of engagement recognition is similar to those of these studies.

In this study, we address a problem of subjectivity on the annotation of engagement. The perception of engagement is subjective and thus often results in disagreement among annotators. Earlier studies took an approach to train a few annotators to avoid disagreement [1, 11, 28, 31]. When the annotators have to consider multimodality, the annotation becomes more complicated and diverse. Besides, it is natural that there are various perspectives for understanding multimodal behaviors. To collect the various perspectives, we can use another approach based on "wisdom of crowds" where many annotators are recruited and asked to annotate user engagement. Previous study integrated the various labels given by the multiple annotators using majority voting [15, 16, 28].

We take into account the various perspectives of the annotators. We assume that each annotator has a latent character that affects his/her perception of engagement. Our proposed model estimates not only user engagement but also the character of each annotator. The model can simulate each annotator's perception of engagement. It is expected that we can understand the differences and common points among the annotators from the annotation data. The similar model considering the difference of annotators is a two-step conditional random fields (CRF), which was proposed for a backchannel prediction task [17, 18]. The prediction model was trained for each annotator, and the final result is determined by voting from the individual models. On the other hand, we train the model based on the character, not for each annotator. Therefore, more robust estimation is expected even if the amount of data for each annotator is small.

3 Annotation of Listener's Engagement

We have collected a human-robot interaction corpus where an autonomous android robot, named ERICA [13], interacted with a human subject. ERICA was operated by another human subject, called an operator, who was in a remote room. Figure 1 shows a snapshot of the dialogue. The dialogue scenario was as follows. ERICA works in a laboratory as a secretary, and the subject visited the professor. Since the professor was absent for a while, the subject talked with ERICA until the professor would come back. Each dialogue lasted about 10 min. The voice uttered by the operator was directly played with a speaker placed on ERICA in real time. We recorded the dialogue with directed microphones, a 16-channel microphone array, RGB cameras, and Kinect v2. We manually annotated utterances, turn units, and dialogue acts. From this corpus, we used 20 sessions for the annotation of subject engagement. The subjects were 12 females and 8 males, with ages ranging from teenagers to over 70 years old. The operators were 6 actresses in their 20 and 30 s. One of the 6 actresses was assigned to each session. All the participants were native Japanese speakers.

We annotated subject engagement by recruiting other 12 females who had not participated in the above dialogue experiment. Note that we considered other methods asking the subjects or the operators to annotate subject engagement by themselves right after the dialogue. However, it was hard to make them annotate it due to a time constraint. Besides, we sometimes observe a bias where the subjects tend to give positive evaluations of themselves [21]. Each dialogue session was randomly assigned to 5 annotators. The instructions given to the annotators was as follows. Engagement was defined as "How much the subject is interested in and willing to continue the current dialogue with ERICA". We also explained a list of listener's behaviors which could be related to engagement, with example descriptions. This list included facial expression, laughing, eye gaze, backchannels, head nodding,

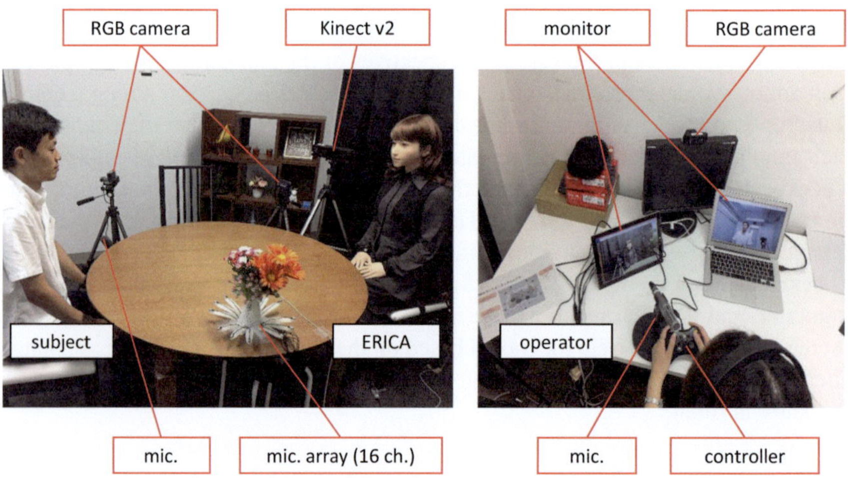

Fig. 1 Setup for conversation

body pose, moving of shoulders, and moving of arms or hands. The annotators were asked to watch the dialogue video from ERICA's viewpoint, and to judge subject engagement based on the subject's behaviors. Specifically, the annotators had to press a button when the following three conditions were being met: (1) the subject was being a listener, (2) the subject was expressing any listener's behaviors, and (3) the behavior shows the high level of engagement.

In this study, we use ERICA's conversational turns as a unit for engagement recognition. When an annotator pressed the button during an ERICA's turn more than once, we regarded the turn was annotated as *engaged* by the annotator. Therefore, each turn has binary labels: engaged or not. There were 433 turns in the 20 sessions, and the number of the engaged labels was 894, and that of the not-engaged ones was 1,271. We investigated the agreement score among the annotators. The average value of Cohen's kappa on every pair of two annotators was 0.291 with a standard deviation of 0.229. Figure 2 shows the matrix of the Cohen's kappa values on each pair among the annotators. Some pairs showed scores which were higher than the moderate agreement (larger than 0.4). This result suggests that the annotators could be clustered into some groups based on their perspectives on multimodal behaviors and engagement.

We also investigated which listener's behaviors were related to engagement. After the annotation work, we asked each annotator to select all meaningful behaviors to annotate subject engagement. As a result, the annotators mostly selected facial expression, laughing, eye gaze, backchannels, head nodding, and body pose. Among them, we use four behaviors, backchannels, laughing, head nodding, and eye gaze in the following experiment. We manually annotated the occurrence of these behaviors. The definition of backchannels was responsive interjections (such as "*huh*" in English and "*un*" in Japanese) and expressive interjections (such as "*oh*" in English and "*he-*" in Japanese) [5]. The laughing was defined as vocal laughing, not just smiling without any vocal utterance. The occurrence of head nodding was judged by the vertical movement of the head. The eye gaze of the subject was annotated as a binary state: the subject was gazing at ERICA's face. We defined the occurrence of eye-gaze

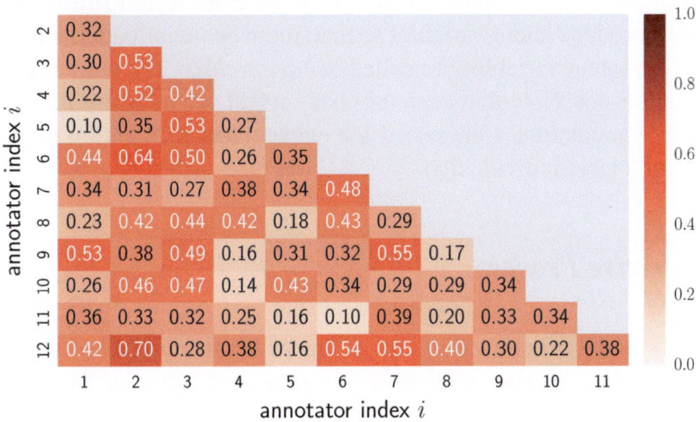

Fig. 2 Inter-annotator agreement scores on each pair of the annotators (Cohen's kappa)

behaviors as the event when the subject was gazing at ERICA's face continuously more than 10 s. We confirmed the histogram of the continuous gazing times, and then made sure of the certain number of occurrences with this criteria. It was difficult to annotate other behaviors such as facial expression and body pose due to its ambiguity. Note that these behaviors will be considered in the future work.

4 Latent Character Model Using Different Annotator Perspectives

It is essential for engagement recognition to consider various annotator perspectives. The annotation result suggests that each annotator has a different perspective on the multimodal behaviors for the perception of engagement. We assume that the different perspectives can be interpreted by a latent variable called *character*. This character represents a template for the perception of engagement. For example, annotators with one character tend to regard the laughing behavior as the engagement indicator. On the other hand, other annotators with another character tend to regard backchannels as the indicator. We introduce a hierarchical Bayesian model to estimate not only the engagement level but also the latent character from the annotation data. This model is called latent character model and enables us to simulate the various perspectives by considering the different characters.

4.1 Problem Formulation

Engagement recognition is done on each system's dialogue turn. The input is based on the occurrences of the four behaviors: laughing, backchannels, head nodding, and eye gaze. Specifically, the input feature is a four-dimensional binary vector corresponding to the combination of the occurrences of the four behaviors, and this is called *behavior pattern*. Therefore, the number of possible states is exponential to the input behaviors ($16 = 2^4$ states in this case). Since this leads to the data sparseness problem, we introduce latent variables to map these behavior patterns into a smaller dimension. The latent variables are called *behavior class*. The output is also binary state: engaged or not. Note that each turn has several ground-truth labels annotated by the multiple annotators. Concretely, the engagement recognition model predicts each annotator's label individually.

4.2 Generative Process

The graphical model is depicted in Fig. 3. The generative process is as follows. For each annotator, the character distribution is generated from the Dirichlet distribution as

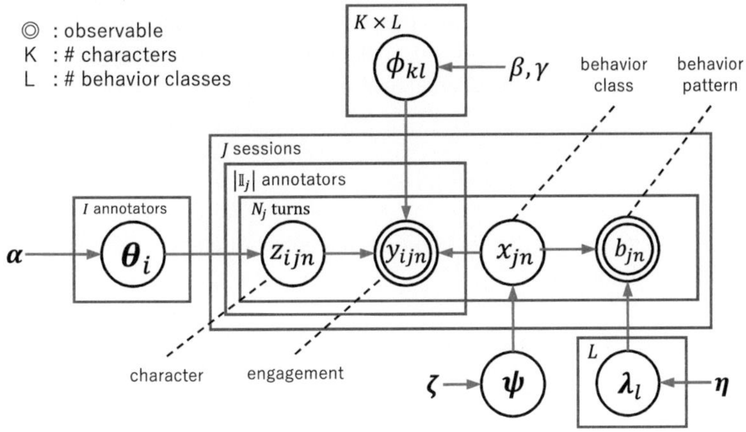

Fig. 3 Graphical model of the proposed model

$$\boldsymbol{\theta}_i = (\theta_{i1}, \cdots, \theta_{ik}, \cdots, \theta_{iK}) \sim \text{Dirichlet}(\boldsymbol{\alpha}), 1 \le i \le I, \tag{1}$$

where i, I, K denote the annotator index, the number of annotators, and the number of characters, respectively, and $\boldsymbol{\alpha} = (\alpha_1, \cdots, \alpha_k, \cdots, \alpha_K)$ is a hyperparameter. The model parameter θ_{ik} represents the probability that the i-th annotator has the k-th character. The behavior-class distribution is generated from the Dirichlet distribution as

$$\boldsymbol{\psi} = (\psi_1, \cdots, \psi_l, \cdots, \psi_L) \sim \text{Dirichlet}(\boldsymbol{\zeta}), \tag{2}$$

where l, L denote the behavior-class index, the number of behavior classes, respectively, and $\boldsymbol{\zeta} = (\zeta_1, \cdots, \zeta_l, \cdots, \zeta_L)$ is a hyperparameter. The model parameter ψ_l represents the probability that the l-th behavior class is generated. For each combination of the character and the behavior class, the engagement distribution is generated from the beta distribution as

$$\phi_{kl} \sim \text{Beta}(\beta, \gamma), 1 \le k \le K, 1 \le l \le L, \tag{3}$$

where β and γ are hyperparameters. For example, the parameter ϕ_{kl} represents the probability that annotators with k-th character give the engaged label when they observe the l-th behavior class. For each behavior class, the behavior-pattern distribution is generated from the Dirichlet distribution as

$$\boldsymbol{\lambda}_l = (\lambda_{l1}, \cdots, \lambda_{lm}, \cdots, \lambda_{lM}) \sim \text{Dirichlet}(\boldsymbol{\eta}), 1 \le l \le L, \tag{4}$$

where m, M denote the behavior-pattern index, the number of behavior patterns, respectively, and $\boldsymbol{\eta} = (\eta_1, \cdots, \eta_m, \cdots, \eta_M)$ is a hyperparameter. For example, the parameter λ_{lm} represents the l-th behavior class generates the m-th behavior pattern. In the case of the current setting, the number of behavior patterns (M) is 16.

There are J dialogue sessions, and the set of annotator indices who annotated the j-th session is represented as $\mathbb{I}_j$. Besides, there are N_j system's dialogue turns in the j-th session. For each turn, the character of the i-th annotator is generated from the categorical distribution as

$$z_{ijn} \sim \text{Categorical}(\boldsymbol{\theta}_i), i \in \mathbb{I}_j, 1 \le j \le J, 1 \le n \le N_j, \tag{5}$$

where n denotes the turn index. In addition, the behavior class is generated from the categorical distribution as

$$x_{jn} \sim \text{Categorical}(\boldsymbol{\psi}), 1 \le j \le J, 1 \le n \le N_j. \tag{6}$$

Based on the generated behavior class, the behavior pattern is observed from the categorical distribution as

$$b_{jn} \sim \text{Categorical}(\boldsymbol{\lambda}_{x_{jn}}), 1 \le j \le J, 1 \le n \le N_j. \tag{7}$$

The behavior patterns correspond to the input features. When the i-th annotator with the character z_{ijn} perceives the behavior class x_{jn}, the engagement label is observed based on the Bernoulli distribution as

$$y_{ijn} \sim \text{Bernoulli}(\phi_{z_{ijn}x_{jn}}), i \in \mathbb{I}_j, 1 \le j \le J, 1 \le n \le N_j. \tag{8}$$

The engagement labels correspond to the outputs of the model. Among the above variables, the characters and the behavior classes are latent variables, and the engagement labels and the behavior patterns are observable.

Given the data set of the above variables and parameters, the joint distribution is represented as

$$\begin{aligned} &p(\boldsymbol{X}, \boldsymbol{Y}, \boldsymbol{Z}, \boldsymbol{B}, \boldsymbol{\Theta}, \boldsymbol{\Phi}, \boldsymbol{\Psi}, \boldsymbol{\Lambda}) = \\ &\quad p(\boldsymbol{X}|\boldsymbol{\Psi})p(\boldsymbol{Y}|\boldsymbol{X}, \boldsymbol{Z}, \boldsymbol{\Phi})p(\boldsymbol{Z}|\boldsymbol{\Theta})p(\boldsymbol{B}|\boldsymbol{X}, \boldsymbol{\Lambda})p(\boldsymbol{\Theta})p(\boldsymbol{\Phi})p(\boldsymbol{\Psi})p(\boldsymbol{\Lambda}), \end{aligned} \tag{9}$$

where the bold capital letters represent the data sets of the variables written by those small letters. Note that $\boldsymbol{\Theta}$, $\boldsymbol{\Phi}$, $\boldsymbol{\Psi}$, and $\boldsymbol{\Lambda}$ are the model parameters.

4.3 Training

In the training phase, the model parameters $\boldsymbol{\Theta}$, $\boldsymbol{\Phi}$, $\boldsymbol{\Psi}$, and $\boldsymbol{\Lambda}$ are estimated. We use the collapsed Gibbs sampling which marginalizes the model parameters and iteratively and alternatively samples the latent variables. Here, we sample the character z_{ijn} and the behavior class x_{jn} from those conditional probability distributions as

$$z_{ijn} \sim p(z_{ijn}|\boldsymbol{X}, \boldsymbol{Y}, \boldsymbol{Z}_{\backslash ijn}, \boldsymbol{\alpha}, \beta, \gamma), \tag{10}$$

$$x_{jn} \sim p(x_{jn} | \boldsymbol{X}_{\backslash jn}, \boldsymbol{Y}, \boldsymbol{Z}, \boldsymbol{B}, \beta, \gamma, \boldsymbol{\zeta}, \boldsymbol{\eta}), \tag{11}$$

where the model parameters $\boldsymbol{\Theta}$, $\boldsymbol{\Phi}$, $\boldsymbol{\Psi}$, and $\boldsymbol{\Lambda}$ are marginalized. Note that $\boldsymbol{Z}_{\backslash ijn}$ and $\boldsymbol{X}_{\backslash jn}$ are the set of characters without z_{ijn} and the set of behavior classes without x_{jn}, respectively. The detail of sampling formulas is omitted here, but it can be obtained in the same manner as the other work [9]. After sampling, we select one of the sampling results as $\boldsymbol{X}^*$ and $\boldsymbol{Z}^*$ where the joint probability $p(\boldsymbol{X}, \boldsymbol{Y}, \boldsymbol{Z}, \boldsymbol{B} | \boldsymbol{\alpha}, \beta, \gamma, \boldsymbol{\zeta}, \boldsymbol{\eta})$ is maximized. The model parameters $\boldsymbol{\Theta}$, $\boldsymbol{\Phi}$, $\boldsymbol{\Psi}$, and $\boldsymbol{\Lambda}$ are estimated based on the sampling result $\boldsymbol{X}^*$ and $\boldsymbol{Z}^*$ as

$$\theta_{ik} = \frac{D_{ik} + \alpha_k}{\sum_{k'=1}^{K} (D_{ik'} + \alpha_{k'})}, \tag{12}$$

$$\phi_{kl} = \frac{N_{kl1} + \beta}{N_{kl1} + N_{kl0} + \beta + \gamma}, \tag{13}$$

$$\psi_l = \frac{T_l + \zeta_l}{\sum_{l'=1}^{L} (T_{l'} + \zeta_{l'})}, \tag{14}$$

$$\lambda_{lm} = \frac{S_{lm} + \eta_m}{\sum_{m'=1}^{M} (S_{lm'} + \eta_{m'})}. \tag{15}$$

Note that D_{ik} is the number of turns where the i-th annotator has the k-th character. N_{kl1} is the number of times when annotators with the k-th character gave the engaged labels for the l-th behavior class. Similarly, N_{kl0} is the number of the not-engaged labels. T_l is the number of times when the l-th behavior class was generated. Finally, S_{lm} is the number of times when the m-th behavior pattern was observed from the l-th behavior class. These numbers are counted up among the sampling results $\boldsymbol{X}^*$ and $\boldsymbol{Z}^*$ and also the observable datasets $\boldsymbol{Y}$ and $\boldsymbol{B}$

4.4 Testing

In the testing phase, the unseen engagement label given by a target annotator is predicted by using the estimated model parameters. Specifically, the model is given the estimated model parameters $\boldsymbol{\Theta}$, $\boldsymbol{\Phi}$, $\boldsymbol{\Psi}$, and $\boldsymbol{\Lambda}$, the input behavior pattern b_t, and the target annotator index i. Note that t represents the turn index in the test data. Given the input behavior pattern, the probability of each behavior class is calculated as

$$p(l | b_t, \boldsymbol{\Psi}, \boldsymbol{\Lambda}) = \frac{1}{\Xi} \psi_l \lambda_{lb_t}, \tag{16}$$

where Ξ is the partition function. The probability that the target annotator gives the engaged label is calculated by marginalizing both the characters and the behavior classes as

$$p(y_{it} = 1|b_t, i, \boldsymbol{\Theta}, \boldsymbol{\Phi}, \boldsymbol{\Psi}, \boldsymbol{\Lambda}) = \sum_{k=1}^{K} \theta_{ik} \sum_{l=1}^{L} \phi_{kl}\, p(l|b_t, \boldsymbol{\Psi}, \boldsymbol{\Lambda}). \quad (17)$$

The t-th turn is recognized as *engaged* by the target annotator when this probability is higher than a threshold.

5 Experimental Evaluation

We compared the proposed model with other methods which do not consider the different annotator perspectives. We conducted the cross-validation with the 20 dialogue sessions: 19 for training and the rest for testing. In this experiment, we used the input behavior patterns which were manually annotated. The output ground-truth labels were the annotation results described in Sect. 3. In the proposed model, the number of sampling was 3,000, and all prior distributions were the uniform distribution. The number of characters (K) was changed from 1 to 5 on a trial basis. The unique character ($K = 1$) means the case where we do not consider the different perspectives. Besides, the number of behavior classes (L) was chosen from $\{2, 4, 8, 12, 16\}$.

The evaluation metric is as follows. Each session has different ground-truth labels given by 5 annotators. We evaluated each annotator's labels individually. Given the target annotator index i, the engaged probability (Eq. 17) was calculated for each turn. Setting the threshold at 0.5, we calculated the accuracy which is a ratio of the number of the correct turns to the total number. We averaged the accuracy scores for all five annotators and also among the cross-validation. The chance level was 0.579 (=1,271/2,165).

Compared methods are based on the logistic regression. We considered two types of training: *majority* and *individual*. In the *majority* type, we integrated the training labels of the five annotators by the majority voting and trained an unique model which is independent of the annotators. In the *individual* type, we trained an individual model for each annotator with his/her data only and used each model according to the target annotator index i in the test phase. Although the *individual* type can learn the different perspective of each annotator, the amount of training data is much smaller.

Table 1 summarizes the recognition accuracy with the proposed model. Note that the accuracies of the compared methods are 0.670 and 0.681 for *majority* and

Table 1 Recognition result (average accuracy)

K (#character)	L (#behavior class)				
	2	4	8	12	16
1	0.669	0.667	0.667	0.662	0.667
2	0.695	0.698	0.702	0.705	0.702
3	0.698	0.714	0.708	0.702	0.712
4	0.697	0.709	0.705	0.712	0.707
5	0.689	0.707	0.708	0.711	0.703

individual types, respectively. The proposed method achieves the accuracy by 0.714 which is higher than those of the compared methods. The best accuracy was achieved when the number of characters (K) is 3, and the number of behavior classes (L) is 4. This result suggests that the character can be substantially represented in 3 dimensions and the behavior patterns are potentially classified by 4 variables.

6 Conclusion

We have addressed engagement recognition from listener's multimodal behaviors in spoken dialogue. The different perspectives of multiple annotators are represented by the latent characters in the proposed model. Besides, the input behavior pattern is classified into smaller meaningful classes. The proposed latent character model achieved the higher accuracy than the compared methods which do not consider the character. In future work, we will implement a spoken dialogue system utilizing the engagement recognition model. To this end, the engagement recognition model will be integrated with automatic behavior detection methods. Furthermore, we will design the system behaviors after the system recognizes the user engagement.

Acknowledgements This work was supported by JSPS KAKENHI (Grant Number 15J07337) and JST ERATO Ishiguro Symbiotic Human-Robot Interaction program (Grant Number JPMJER1401), Japan.

References

1. Bednarik R, Eivazi S, Hradis M (2012) Gaze and conversational engagement in multiparty video conversation: an annotation scheme and classification of high and low levels of engagement. In: Proceedings of the ICMI workshop on eye gaze in intelligent human machine interaction
2. Bohus D, Horvitz E (2009) Learning to predict engagement with a spoken dialog system in open-world settings. In: Proceedings of the SIGDIAL, pp 244–252
3. Castellano G, Pereira A, Leite I, Paiva A, McOwan PW (2009) Detecting user engagement with a robot companion using task and social interaction-based features. In: Proceedings of the ICMI, pp 119–126
4. Chiba Y, Ito A (2016) Estimation of users willingness to talk about the topic: analysis of interviews between humans. In: Proceedings of the IWSDS
5. Den Y, Yoshida N, Takanashi K, Koiso H (2011) Annotation of Japanese response tokens and preliminary analysis on their distribution in three-party conversations. In: Proceedings of the oriental COCOSDA, pp 168–173
6. DeVault D, Artstein R, Benn G, Dey T, Fast E, Gainer A, Georgila K, Gratch J, Hartholt A, Lhommet M, Lucas G, Marsella S, Morbini F, Nazarian A, Scherer S, Stratou G, Suri A, Traum D, Wood R, Xu Y, Rizzo A, Morency LP (2014) SimSensei kiosk: a virtual human interviewer for healthcare decision support. In: Proceedings of the autonomous agents and multi-agent systems, pp 1061–1068
7. Frank M, Tofighi G, Gu H, Fruchter R (2016) Engagement detection in meetings. arXiv preprint arXiv:1608.08711
8. Glas N, Pelachaud C (2015) Definitions of engagement in human-agent interaction. In: Proceedings of the international workshop on engagement in human computer interaction, pp 944–949

9. Griffiths TL, Steyvers M (2004) Finding scientific topics. Proc Natl Acad Sci 101(suppl 1):5228–5235
10. Higashinaka R, Imamura K, Meguro T, Miyazaki C, Kobayashi N, Sugiyama H, Hirano T, Makino T, Matsuo Y (2014) Towards an open-domain conversational system fully based on natural language processing. In: Proceedings of the COLING, pp 928–939
11. Huang Y, Gilmartin E, Campbell N (2016) Conversational engagement recognition using auditory and visual cues. In: Proceedings of the INTERSPEECH
12. Inoue K, Lala D, Nakamura S, Takanashi K, Kawahara T (2016) Annotation and analysis of listener's engagement based on multi-modal behaviors. In: Proceedings of the ICMI workshop on multimodal analyses enabling artificial agents in human-machine interaction
13. Inoue K, Milhorat P, Lala D, Zhao T, Kawahara T (2016) Talking with ERICA, an autonomous android. In: Proceedings of the SIGDIAL, pp 212–215
14. Michalowski MP, Sabanovic S, Simmons R (2006) A spatial model of engagement for a social robot. In: Proceedings of the international workshop on advanced motion control, pp 762–767
15. Nakano YI, Ishii R (2010) Estimating user's engagement from eye-gaze behaviors in human-agent conversations. In: Proceedings of the IUI, pp 139–148
16. Oertel C, Mora KAF, Gustafson J, Odobez JM (2015) Deciphering the silent participant: on the use of audio-visual cues for the classification of listener categories in group discussions. In: Proceedings of the ICMI
17. Ozkan D, Morency LP (2011) Modeling wisdom of crowds using latent mixture of discriminative experts. In: Proceedings of the ACL, pp 335–340
18. Ozkan D, Sagae K, Morency LP (2010) Latent mixture of discriminative experts for multimodal prediction modeling. In: Proceedings of the COLING, pp 860–868
19. Peters C (2005) Direction of attention perception for conversation initiation in virtual environments. In: Proceedings of the international workshop on intelligent virtual agents, pp 215–228
20. Poggi I (2007) Mind, hands, face and body: a goal and belief view of multimodal communication. Weidler
21. Ramanarayanan V, Leong CW, Suendermann-Oeft D (2017) Rushing to judgement: how do laypeople rate caller engagement in thin-slice videos of human-machine dialog?. In: INTERSPEECH, pp 2526–2530
22. Rich C, Ponsler B, Holroyd A, Sidner CL (2010) Recognizing engagement in human-robot interaction. In: Proceedings of the HRI, pp 375–382
23. Sanghvi J, Castellano G, Leite I, Pereira A, McOwan PW, Paiva A (2011) Automatic analysis of affective postures and body motion to detect engagement with a game companion. In: Proceedings of the HRI, pp 305–311
24. Sidner CL, Lee C (2003) Engagement rules for human-robot collaborative interactions. In: Proceedings of the ICSMC, pp 3957–3962
25. Sidner CL, Lee C, Kidd CD, Lesh N, Rich C (2005) Explorations in engagement for humans and robots. Artif Intell 166(1–2):140–164
26. Sun M, Zhao Z, Ma X (2017) Sensing and handling engagement dynamics in human-robot interaction involving peripheral computing devices. In: CHI, pp 556–567
27. Türker, B.B., Buçinca Z, Erzin E, Yemez Y, Sezgin M (2017) Analysis of engagement and user experience with a laughter responsive social robot. In: INTERSPEECH, pp 844–848
28. Xu Q, Li L, Wang G (2013) Designing engagement-aware agents for multiparty conversations. In: Proceedings of the CHI, pp 2233–2242
29. Yu C, Aoki PM, Woodruff A (2004) Detecting user engagement in everyday conversations. In: Proceedings of the ICSLP, pp 1329–1332
30. Yu Z, Nicolich-Henkin L, Black AW, Rudnicky AI (2016) A Wizard-of-Oz study on a non-task-oriented dialog systems that reacts to user engagement. In: Proceedings of the SIGDIAL, pp 55–63
31. Yu Z, Ramanarayanan V, Lange P, Suendermann-Oeft D (2017) An open-source dialog system with real-time engagement tracking for job interview training applications. In: Proceedings of the IWSDS

Utilizing Argument Mining Techniques for Argumentative Dialogue Systems

Niklas Rach, Saskia Langhammer, Wolfgang Minker and Stefan Ultes

Abstract This work presents a pilot study for the application of argument mining techniques in the context of argumentative Dialogue Systems. We extract the argument structure of an online debate and show how it can be utilized to generate artificial persuasive dialogues in an agent-agent scenario. The interaction between the agents is formalized as argument game and the resulting artificial dialogues are evaluated in a user study by comparing them to human generated ones. The outcomes indicate that the artificial dialogues are logically consistent and thus show that the use of the employed argument annotation scheme was successful.

1 Introduction

Argumentation is an essential part of human conversation and is often employed in order to resolve conflicts or persuade an opponent. Enabling virtual agents to engage with humans (and each other) in a similar way (i.e. by exchanging arguments) is crucial for tasks such as reasoning, understanding new concepts and deducing new knowledge [9]. However, implementations of such argumentative systems have to overcome different barriers [21] and rely on knowledge about existing arguments and their formal relations to each other [3, 13, 15, 20]. With the large amount of arguments present online in different forms and from numerous sources, approaches to harness the same for the before mentioned systems are of particular interest. The field of argument mining [7–9] is concerned with the automatic extraction and analysis of argumentative structures from natural language sources and thus provides promising approaches for this task. In this work, we employ the annotation scheme

N. Rach (✉) · S. Langhammer · W. Minker
Institute of Communication Engineering, Ulm University, Ulm, Germany
e-mail: niklas.rach@uni-ulm.de

W. Minker
e-mail: wolfgang.minker@uni-ulm.de

S. Ultes
Department of Engineering, University of Cambridge, Cambridge, UK

L. F. D'Haro et al. (eds.), *9th International Workshop on Spoken Dialogue System Technology*, Lecture Notes in Electrical Engineering 579,
https://doi.org/10.1007/978-981-13-9443-0_12

introduced in [17] to extract the argument structure of an online debate and show how it can be utilized in an argumentative Dialogue System.

We choose an agent-agent scenario as testbed and generate artificial dialogues from the resulting argument structure in order to analyze and evaluate our approach. Following the classification of [14], we focus on persuasive dialogues meaning that each agent has the goal to establish a convincing line of argumentation and to weaken the one of the opponent. To this end, the dialogue is formalized as dialogue game for argumentation (in short argument game, see [11, 19] for an overview) based on the formal system presented in [10]. The evaluation is done by comparing the resulting artificial dialogues to human generated ones in a user study. The present work reports first steps in the implementation of the argumentative Dialogue System discussed in [12] and builds on the work presented in [5].

The remainder of the paper is as follows: Sect. 2 examines related work on argumentative Dialogue Systems. Section 3 introduces the employed argument mining scheme and discusses the textual source as well as the annotation. In Sect. 4, we introduce the architecture of the system in combination with the theoretical background of the respective components. Subsequently, in Sect. 5, the outcome and the evaluation of the corresponding user study are discussed. We close with a conclusion and a brief discussion of future work in Sect. 6.

2 Related Work

In this section we summarize related work on argumentative (Dialogue) Systems. Implementations of this kind are comparatively scarce due to several issues [21] that have to be solved or bypassed.

Two examples that are also based on argument games are presented in [3, 20]. The first one implements a specification of the Toulmin dialogue game, whereas the latter one is based on the dialogue model DE. Both systems assume a certain structure for the required database of arguments but in contrast to the present work, the focus lies not on its generation. Instead, exemplary argument structures are employed to illustrate the underlying principles. The Arvina system [6] allows users to exchange arguments present in the Argument Web [4] following the rules of a previously selected argument game. In addition, virtual agents are available to represent authors in the Argument Web by reflecting their arguments on a certain topic. Thus, these agents do not establish an individual line of argumentation, as in the herein discussed case.

Rosenfeld and Kraus [15] introduced a persuasive agent capable of learning an optimal strategy by means of Reinforcement Learning (RL) in a bipolar weighted argument framework. In contrast to the above mentioned systems, the interaction in this case is restricted to the exchange of available arguments, meaning that strategical moves (for instance questioning the validity of a previous argument) are not possible. The corpus of arguments is derived from human-human dialogues about the same topic by mapping utterances to arguments in the framework. Thus, an inclusion of arguments from external sources as proposed in this work is not intended. Rakshit et al. [13] recently introduced an argumentative chatbot that relies on a corpus of annotated arguments generated from web resources. The system responses are derived by means of similarity measures between the user utterance and responses available in the corpus. Hence, each system response has to be included explicitly in the data. In contrast, the herein presented approach addresses the system response in the argument game framework, allowing the agents to respond to earlier utterances (and not just the latest) and to employ additional moves that have no corresponding instance in the data (for example challenging the validity of an argument).

3 Data and Annotation

In this section we discuss the annotation scheme and the textual source it is applied to. Our source of choice is a sample debate from the *Debatabase* of the idebate.org[1] website. The reasons for this choice are as follows: Firstly, idebate.org is operated by the International Debate Education Association (IDEA), a global network of organizations devoted to debating education. Hence, the debates offered here can be expected to meet certain quality standards regarding both form and content. Secondly, all debates presented here explore both sides of their respective topic. Lastly, all *Debatabase* debates adhere to a specific structure which both facilitates the quick screening for suitable candidates and potentially aids the argument annotation process later on. The sample debate employed in the scope of this work is concerned with the topic *Marriage is an outdated institution*. This choice is mostly due to the high amount of arguments provided by the *Debatabase* for this topic.

The employed annotation scheme was introduced by Stab et al. [17] for the analysis of written essays. It includes three argument components (*major claim, claim, premise*) and two directed relations (*support* and *attack*) between these components and is not tied to a specific domain. Thus, it provides every aspect required for the herein considered textual source. In addition, the resulting argument structure is compatible with the employed argument game (see Sect. 4) which makes the annotation scheme a reasonable choice for our task.

If a component ϕ_1 *supports* or *attacks* another component ϕ_2, we say that ϕ_2 is the target of ϕ_1 (or that ϕ_1 targets ϕ_2, respectively). A debate usually has one *major claim*, which formulates the overall topic around which the debate is built (here *Marriage is an outdated institution*) and is the only component that has no target. *Claims* are statements or assertions that express a certain opinion towards the *major claim* but require additional argumentative justification. Thus, *claims* can only target the *major claim*, not other *claims* or *premises*. An example *claim* from the herein examined debate is:

> Marraige does not provide any more of a stable environment for child rearing than a regular monogamous relationship.

[1]https://idebate.org/debatabase (last accessed 16 March 2019).

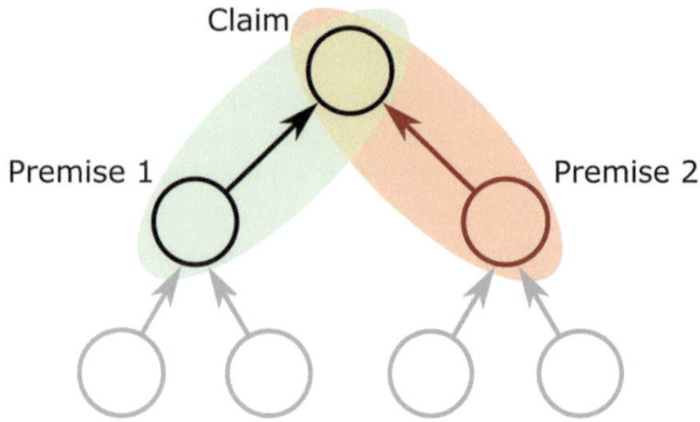

Fig. 1 Construction of arguments from the graph representing the annotated argument components. The green circle indicates an argument of the form *(premise* 1, *so claim)* the red circle a counterargument with the form *(premise* 2, so $\neg claim$)

A *premise* on the other hand provides reason for or against a *claim* or extends an already existing line of argumentation, meaning a *premise* can target the *major claim*, a *claim* or another *premise*. An example *premise* supporting the above mentioned *claim* is:

So many marriages end in divorce with the resulting splits affecting the children.

All argument components can target no more than one other component, but can be targeted by more than one and no argument component can target itself. This hierarchical structure of components allows for a representation of the annotated argument structure as a tree, where the argument components constitute the nodes and the argument relations constitute the edges. The arguments employed in the artificial dialogue can be constructed from this graph as a pair of nodes and their relation to each other. We denote the set of all arguments of this kind with *args*. The elements of *args* have the form $a = (\phi_1, \text{so}\, \phi_2)$ if ϕ_1 supports ϕ_2 and $b = (\phi_1, \text{so}\, \neg\phi_2)$ if ϕ_1 attacks ϕ_2. The construction of arguments from the graph is depicted in Fig. 1. It is important to note that it is generally possible to build arguments with more than two components. Throughout this work we limit ourselves to the above described arguments consisting of two components, in order to ensure the best compatibility with the employed argument game framework.

The annotation was done by an expert based on the guidelines of [17] by first identifying argument components in the debate and secondly annotating the relations between them. The identification of overall topic and stance of the author included in the original work as a separate step are not required here, as both aspects are brought out by the structure of the debates. The annotation resulted in a total of 72 argument components (1 *major claim*, 10 *claims* and 61 *premises*) and their corresponding relations and was encoded in an OWL ontology [2] for further use. In order to facilitate the Natural Language Generation of arguments in the artificial dialogues, the original annotated sentences were modified slightly to form a complete and reasonable utterance. Thus, implications are made explicit, references and citations were reformulated and expressions that are exclusively used in the debate format were adapted to the dialogue context (for example "*[...] as the opposition claims*" was changed to "*[...] as you claim*").

4 The System Architecture

In the following, we describe the architecture of our system as well as the theoretical foundations it is based on. The core of the system consists of two agents (Alice and Bob) that argue about a certain topic. The interaction follows rules defined in the argument game [10] which determines the player to move and the available moves in each state of the dialogue. The set of available arguments for both agents is provided by the argument tree discussed in Sect. 3. In the current state of our work, each dialogue has a fixed length, i.e. a maximal number of turns after which the interaction terminates and thus, no termination criterion is employed, yet. After the game is finished, each move is subject to a template based Natural Language Generation (NLG) transforming the game moves into a natural language utterance. In the subsequent sections we discuss three aspects of the system in more detail.

4.1 The Argument Game

To structure the interaction between the agents, we utilize the formal system for computational persuasion of Prakken [10] that formalizes utterances in the dialogue as moves in a game. The possible moves for each agent have the form *claim*(ϕ), *argue*(a), *why*(ϕ), *retract*(ϕ), *concede*(ϕ) with ϕ an argument component in the graph discussed above and $a \in args$. It is worth noting that only two of these types introduce new content, i.e. new argument components to the dialogue (*claim* and *argue*), whereas the remaining three deal with components introduced earlier. In addition, the formalism introduces a protocol to determine the outcome of the game, the player to move and the list of available moves at each state. The outcome is tied to the termination criterion which is not employed here due to the finite length of the dialogues. Thus, we do not determine a winner (in the sense of the dialogue game) in the herein discussed scenario.

The list of available moves is determined by means of a relevance criterion for previous moves defining which of the latter ones can be addressed in the current state of the game. Thus, a response to earlier moves (and not just the latest one) is possible. We refer to responses of this kind as *topic switch*, since the focus is switched from the latest move to an earlier one. This allows to respond to a move more than once (if the required conditions are met) and gives the agents the possibility to explore different branches of the argument tree in the same dialogue. Apart from the opening move (*claim*), each move either attacks a previous move (*argue*, *why*) or surrenders to a previous move (*concede*, *retract*).

For a detailed discussion of the framework we refer the interested reader to [10, 11]. It should be noted that the *claim* move is not to be confused with the *claim* component of the annotation scheme. In fact, the *claim* move is only employed to open the game and is thus always introduces the *major claim* component to the dialogue.

4.2 Agent Strategy

The second issue is the agent strategy. As the framework restricts the agents only as much as necessary, a strategy to select the next move from the list of allowed ones is required as different moves lead to different outcomes. Throughout this work, we employ rules that lean to the argumentative agent profile described in [1]. The key assumption is, that the agent attacks whenever possible. Consequently, he only surrenders to an opponent move if there is no other option left. This choice is reasonable for the herein discussed case, as each participant has the goal to convince the opponent and thus to strengthen the own and weaken the opponents line of argumentation whenever possible. To keep the dialogue focused on the current topic, we add a preference of moves that respond to the latest opponent move over a topic switch. In addition, we add a preference of *argue* over *why* moves in order to prevent an extensive use of the latter one. The rules thus read as:

- Attack if possible. If you do so,
  - If possible, attack the previous utterance of the opponent.
  - Prefer *argue* moves over *why* moves.
- If no attack is possible, *surrender*. If possible, *surrender* to the latest opponents move.

By using these rules, each agent identifies the next move from the set of possible moves. If there is more than one move fulfilling the same conditions, the next move is picked from this subset randomly.

4.3 Natural Language Generation

The NLG of the system relies on the original textual representation of the argument components. As discussed in Sect. 3, the annotated sentences were slightly modified to form a stand-alone utterance which serves as a template for the respective *argue* (and *claim*) move. In addition, a list of natural language representations for each additional type of move was defined. The explicit formulation is chosen from this list randomly during the generation of the utterances. Moreover, topic switches are indicated by referencing formulations, followed by the textual representation of the referenced argument component as for example: *I'd like to go back to something you mentioned earlier. I think you said: [...]*. Again, the explicit formulation is picked randomly for each utterance.

Finally, we added additional connecting phrases like *That's not a valid point.* to some argue moves in order to increase the naturalness of the dialogue.

5 Outcome and Evaluation

In this section, we discuss the outcome of the overall system, i.e. the artificial dialogues between the two agents Alice and Bob, and evaluate the annotation in this context. To this end, we first examine one of the artificial dialogues as well as an excerpt of a human generated discussion and highlight structural differences. Subsequently, we discuss the findings of a user study comparing both types of dialogues. The focus herein lies on the logic consistency of the lines of argumentation established by the agents as this consistency depends directly on the annotated argument structure. From our perspective, this is a fair evaluation of both the annotation and application, as errors in one of the two or even both aspects would be reflected in the outcome.

5.1 Artificial Dialogue

In the following, we first show an exemplary artificial dialogue generated by the system by means of the annotated argument structure.[2] The utterances are generated by the NLG, whereas the content of the brackets indicates the respective move (*claim: c, argue: a, why: w*) including its index, followed by its target. For instance, (a2,a1) denotes *argue* move 2, addressing (the previous) *argue* move 1.

Alice: "*Marriage is an outdated institution.*" (c1)
Bob: "*Why do you think that?*" (w1,c1)
Alice: "*The frequency and accessibility of divorce undermines the entire purpose of marriage.*" (a1,w1)
Bob: "*Could you please elaborate?*" (w2,a1)
Alice: "*Marriage no longer leads to a stable or permanent relationship.*" (a2,w2)
Bob: "*That's not a valid point. The purpose of marriage is not an eternal, unrelenting union, whether it is wanted or not.*" (a3,a2)
Alice: "*Could you be more specific?*" (w3,a3)
Bob: "*The purpose of marriage is to foster a more stable relationship than would be possible without marital vows.*" (a4,w3)
Alice: "*Could you please elaborate?*" (w4,a4)
Bob: "*Well, maybe we can come back to this point later. You said earlier: Marriage no longer leads to a stable or permanent relationship. I'm not sure I understand what you're getting at.*" (w5,a2)
Alice: "*Our society no longer respects marriage as a permanent institution.*" (a5,w5)
Bob: "*Could you be more specific?*" (w6,a5)

[2]Material reproduced from www.iedebate.org with the permission of the International Debating Education Association.

Alice: "*Serial monogamy is becoming ever more common.*" (a6,w6)
Bob: "*Could you please elaborate?*" (w7,a6)
Alice: "*50% of all divorces in the UK go on to remarry according to the Office for National Statistics.*" (a7,w7)

It should be noted that this dialogue does not include any surrendering move, as both agents prefer attack moves over surrender moves and the dialogue has fixed length, so no agent has run out of attack options yet. We recognize one topic switch (w5), that makes a new branch of the argument tree available for both agents and allows Alice to support one of her points (a2) with additional arguments. Moreover, a frequent use of *why* moves can be observed, due to the fact that the system up to now does not allow for chained arguments, i.e. an extended utterance covering multiple argument components. This is particularly evident in comparison with the excerpt of a human generated dialogue shown below,[3] where the opposite can be observed:

Alice: "*Marriage is an outdated institution.*"
Bob: "*Why do you think that?*"
Alice: "*According to the Office for National Statistics the average age, in the UK, to get married is approximately 30 years old. Life expectancy in the UK is approximately 80 years. This means the average marriage expects people to commit to maintain a certain way of life for a period that is longer than they have actually been alive. This goes hand in hand with the rise of social acceptability of people having more than one life partner in their life to show that either marriage is an unreasonable expectation of someone or a meaningless charade that is not actually expected to be maintained.*"
Bob: "*This argument only works under the assumption that we live in a society where divorce does not exist. If a person enters into a marriage without full awareness of what they have committed to and later need to get out of that marriage, they are free to. Being able to leave a marriage, though, does not make marriage a meaningless charade, as you claim. It is still more difficult to leave a marriage than it is to leave a non-marital committed relationship and so it makes a big difference.*"

We see that in this case, a single utterance encapsulates multiple argument components and that the dialogue as a whole consists of less turns. Moreover, only one *why* move is present in this case. A last difference in view of the line of argumentation is the fact, that in the human generated discussion some premises occur before the corresponding conclusion is present. Up to now this is not possible for virtual agents in our system as it could lead to inconsistent lines of argumentation and is only reasonable in combination with utterances containing multiple arguments. In order to enable the system to consistently argue this way, a modification of the employed argument game is required.

---

[3]Material reproduced from www.iedebate.org with the permission of the International Debating Education Association.

5.2 *Survey*

To validate the annotation and the herein discussed use of the resulting argument structure, we compared the two kinds of dialogues discussed above in a user study realized by clickworker.[4] Each participant was given one random instance out of the two possible categories (agent-agent or human dialogue). To include all aspects of the original debate, five human generated dialogues and 20 agent-agent dialogues were utilized as the argument density was higher in the human case. The 122 participants were from the UK and assigned randomly to one instance, resulting in a splitting of 54 participants rating the agent-agent case and 68 rating the human generated case. The rating was done on a five point scale from *completely disagree* (1) to *completely agree* (5) and 10 questions about the persuasiveness of the involved parties, logical consistency of the argumentation and an overall impression of the dialogue. The questions distinguish between the two agents Alice and Bob and ask about both of them separately:

- I was not convinced by Bob/Alice and how he/she presented his/her case. (Strat. Bob/Alice)
- It was always clear which previous utterance Bob/Alice addressed in his/her turn. (Prev. Bob/Alice)
- The arguments presented by Alice/Bob are logically consistent responses to the utterances they refer to. (Arg. Bob/Alice)
- Alice's/Bob's line of argumentation is not logically consistent. (Arg. line Bob/Alice)
- It was difficult to follow the line of argumentation throughout the debate. (Arg. line diff.)
- The whole debate is natural and intuitive. (Nat. and int.)

It should be noted that due to the different formulations the best ranking is not always the highest. Table 1 shows the corresponding statistical results for all questions. Each line includes the median for the artificial dialogues (Agent), the human dialogues (Human) and the corresponding p value achieved with a Mann-Whitney-U test and all 122 ratings. As mentioned in the beginning, our focus in the context of this work lies on the questions assessing the logical consistency of the argumentation. These are in particular the questions asking for appropriateness of the arguments (Arg. Bob/Alice) and the questions that asses the complete line of argumentation (Arg. line Bob/Alice). The questions related to the agent strategy (Strat. Bob/Alice) were posed in order to decouple the rating of the dialogical behavior form the rating of the lines of argumentation. Thus, the corresponding results are not discussed further, here. We see that in the case of the overall consistency (Arg. line Bob/Alice) the ratings for the different scenarios are close to each other as the median is equal for both agents. Moreover, the case of Bob yields no significant difference whereas Alice is on the threshold of $p = 0.05$. For the single step rating (Arg. Bob/Alice), we see in both cases a significant difference between the human generated and the agent-agent

[4]https://marketplace.clickworker.com (last accessed 16 March 2019).

Table 1 Median and p value for both agent-agent (Agent) and human-generated (Human) dialogues. Bold lines indicate questions related to the logic consistency of the argumentation

	Agent	Human	p
Strat. Bob	3.0	2.5	0.11
Strat. Alice	3.0	3.0	0.26
Prev. Bob	4.0	4.0	0.06
Prev. Alice	4.0	4.0	0.29
Arg. Bob	**4.0**	**4.0**	**0.02**
Arg. Alice	**3.5**	**4.0**	**$\leq$ 0.01**
Arg. line Bob	**2.0**	**2.0**	**0.72**
Arg. line Alice	**2.0**	**2.0**	**0.05**
Arg. line diff.	3.0	2.0	$\leq$ 0.01
Nat. and int.	2.0	4.0	$\leq$ 0.01

dialogue. Nevertheless, the median is the same for the case of Bob and still above the neutral value of 3.0 for Alice. As mentioned earlier, we experienced a frequent use of *why* moves as well as some unintuitive changes of topic that may have lead to distraction and irritation of the participant. This is mostly reflected in the different ratings for the two last questions which asses the overall impression of the dialogue. As the argumentation was nevertheless in each case rated as consistent, we value the annotation scheme as an adequate approach to collect the data for systems of the herein discussed kind.

6 Conclusion and Outlook

We have presented a Dialogue System application for argument mining techniques. We discussed the annotation of a written debate and how the resulting argument structure can be employed by virtual agents to play an argument game. Moreover, we have evaluated the resulting artificial dialogues in a user study. The results indicate that the consistency is prevailed in the artificial dialogues, although there is room for improvement in view of the naturalness of the same. We conclude that the use of the employed annotation scheme was successful but additional effort is required in order to enable a more natural and intuitive interaction.

Thus, future work will focus on multiple aspects. First of all, we aim for an extended database by including both additional topics and additional arguments for the present topic. The long term goal in this context is to employ automatic identification of argument components and relations as investigated in [16, 18] and thus utilize the full potential of argument mining in the herein discussed context. A second direction of interest is the dialogue management, i.e. the selection of arguments which will be optimized by means of Reinforcement Learning to provide a more intuitive and natural line of argumentation. The naturalness of the dialogue can in our opinion also be increased by a more advanced NLG. Finally an interaction of one of the agents with a human user is of interest, as this is eventually the goal of our system.

Acknowledgements This work has been funded by the Deutsche Forschungsgemeinschaft (DFG) within the project "How to Win Arguments—Empowering Virtual Agents to Improve their Persuasiveness", Grant Number 376696351, as part of the Priority Program "Robust Argumentation Machines (RATIO)" (SPP-1999).

References

1. Amgoud L, Parsons S (2001) Agent dialogues with conflicting preferences. In: ATAL, vol 1. Springer, pp 190–205
2. Bechhofer S (2009) Owl: web ontology language. In: Encyclopedia of database systems. Springer, pp 2008–2009
3. Bench-Capon TJ (1998) Specification and implementation of Toulmin dialogue game. In: Proceedings of JURIX, vol 98, pp 5–20
4. Bex F, Lawrence J, Snaith M, Reed C (2013) Implementing the argument web. Commun ACM 56(10):66–73
5. Langhammer S (2017) A debating ontology for argumentative dialogue systems. Bachelor's thesis, In: Institute of communication engineering. Ulm University
6. Lawrence J, Bex F, Reed C (2012) Dialogues on the argument web: mixed initiative argumentation with arvina. In: COMMA, pp 513–514
7. Lippi M, Torroni P (2016) Argumentation mining: state of the art and emerging trends. ACM Trans Internet Technol (TOIT) 16(2):10
8. Moens MF (2013) Argumentation mining: where are we now, where do we want to be and how do we get there?. In: Post-proceedings of the 4th and 5th workshops of the forum for information retrieval evaluation. ACM, p 2
9. Palau RM, Moens MF (2009) Argumentation mining: the detection, classification and structure of arguments in text. In: Proceedings of the 12th international conference on artificial intelligence and law. ACM, pp 98–107
10. Prakken H (2000) On dialogue systems with speech acts, arguments, and counterarguments. In: JELIA. Springer, pp 224–238
11. Prakken H (2006) Formal systems for persuasion dialogue. knowl Eng Rev 21(2):163–188
12. Rach N, Minker W, Ultes S (2017) Towards an argumentative dialogue system. In: Bex F, Grasso F, Green N (eds) Proceedings of the 17th workshop on computational models of natural argument co-located with ICAIL 2017; 2017 Jul 16; London, UK, London: CEUR Workshop Proceedings, 27–29 p
13. Rakshit G, Bowden KK, Reed L, Misra A, Walker M (2017) Debbie, the debate bot of the future. arXiv preprint arXiv:1709.03167
14. Reed C, Norman T (2003) Argumentation machines: new frontiers in argument and computation, vol. 9. Springer Science & Business Media
15. Rosenfeld A, Kraus S (2016) Strategical argumentative agent for human persuasion. In: ECAI, pp 320–328
16. Stab C, Daxenberger J, Stahlhut C, Miller T, Schiller B, Tauchmann C, Eger S, Gurevych I (2018) Argumentext: searching for arguments in heterogeneous sources. In: Proceedings of the 2018 conference of the North American chapter of the association for computational linguistics: demonstrations, pp 21–25

17. Stab C, Gurevych I (2014) Annotating argument components and relations in persuasive essays. In: COLING, pp 1501–1510
18. Stab C, Gurevych I (2014) Identifying argumentative discourse structures in persuasive essays. In: EMNLP, pp 46–56
19. Wells S, Reed CA (2012) A domain specific language for describing diverse systems of dialogue. J Appl Logic 10(4):309–329
20. Yuan T, Moore D, Grierson A (2008) A human-computer dialogue system for educational debate: a computational dialectics approach. Int J Artif Intell Educ 18(1):3–26
21. Yuan T, Moore D, Reed C, Ravenscroft A, Maudet N (2011) Informal logic dialogue games in human-computer dialogue. Knowl Eng Rev 26(2):159–174

Part III
Dialogue Evaluation and Analysis

Multimodal Dialogue System Evaluation: A Case Study Applying Usability Standards

Andrei Malchanau, Volha Petukhova and Harry Bunt

Abstract This paper presents an approach to the evaluation of multimodal dialogue systems, applying usability metrics defined in ISO standards. Users' perceptions of effectiveness, efficiency and satisfaction were correlated with various performance metrics derived from system logfiles and reference annotations. Usability experts rated questions from a preliminary 110-items questionnaire, and an assessment of their agreement on usability concepts has led to a selection of eight main factors: task completion and quality, robustness, learnability, flexibility, likeability, ease of use and usefulness (value) of an application. Based on these factors, an internally consistent and reliable questionnaire with 32 items (Cronbach's alpha of 0.87) was produced. This questionnaire was used to evaluate the Virtual Negotiation Coaching system for metacognitive skills training in a multi-issue bargaining setting. The observed correlations between usability perception and derived performance metrics suggest that the overall system usability is determined by the quality of agreements reached, by the robustness and flexibility of the interaction, and by the quality of system responses.

1 Introduction

Modern digital services and devices get more and more interconnected and integrated in everyday human activities. They often come each with their own interfaces, which users need to learn. Multimodal natural-language based dialogue is becoming a

A. Malchanau · V. Petukhova (✉)
Spoken Language Systems Group, Saarland University, Saarbrücken, Germany
e-mail: v.petukhova@lsv.uni-saarland.de

A. Malchanau
e-mail: andrei.malchanau@lsv.uni-saarland.de

H. Bunt
Department of Cognitive Science and Artificial Intelligence, Tilburg University, Tilburg, The Netherlands
e-mail: harry.bunt@uvt.nl

L. F. D'Haro et al. (eds.), *9th International Workshop on Spoken Dialogue System Technology*, Lecture Notes in Electrical Engineering 579,
https://doi.org/10.1007/978-981-13-9443-0_13

feasible and attractive human-machine interface which can be used to provide a universal, accountable and personalized form of access to a variety of products and contents. Such interfaces offer a mode of interaction that has certain similarities with natural human communication, in using a range of input and output modalities that people normally employ in communication, achieving a certain level of 'digital immersion' which boosts user acceptance and enriches user experience. As a part of the interactive application design, evaluations are performed in order to assess the success of the developed solutions. Evaluation results serve to inform designers about the functional and non-functional aspects of the system performance.

Commonly, dialogue systems are evaluation by asking users to fill in a questionnaire after interacting with the system. It is still largely an open question which parameters should be taken into account when designing a satisfaction questionnaire, and which of these may correlate well with user satisfaction. Qualitative and quantitative measures are often automatically computed from test interactions with real or simulated users. Most existing evaluation metrics are designed for task-oriented information-seeking spoken dialogue systems and do not apply well to complex multimodal interactions. In this paper we propose to assess multimodal dialogue system performance by relating various performance metrics and subjective perception of *usability* factors as defined by the ISO 9241-11 and ISO/IEC 9126-4 standards. This enables usability quantification in a meaningful and systematic way.

This paper is structured as follows. Section 2 discusses existing approaches to the evaluation of interactive conversational systems. Section 3 presents the ISO 9241-11 usability definition and metrics for effectiveness, efficiency and satisfaction. We adapt these metrics to the multimodal dialogue system evaluation task by specifying factors that impact usability perception by its users. In Sect. 4 we describe experiments and report results evaluating the Virtual Negotiation Coach application. Section 5 summarises our findings and outlines future research.

2 Related Work

Several dialogue system evaluation approaches have been proposed in the past. PARADISE, one of the most widely-used evaluation models [1], aims at predicting user global satisfaction given a set of parameters related to task success and dialogue costs. Satisfaction is calculated as the arithmetic mean of nine user judgments on different quality aspects rated on 5-point Likert scales. Subsequently, the relation between task success and dialogue cost parameters and the mean human judgment is estimated by means of a multivariate linear regression analysis.

Another approach is to evaluate a dialogue system on the basis of test interactions substituting human users by computer agents that emulate user behaviour, see e.g. [2]. The various types of users and system factors can be systematically manipulated, e.g. using interactive, dialogue task and error recovery strategies.

As for system performance metrics, several sets have been recommended for spoken dialogue system evaluation ranging from 7 parameters defined in [16] to 52 in [17] related to the entire dialogue (duration, response delay, number of turns), to meta-communication strategies (number of help requests, correction turns), to the system's cooperativity (contextual appropriateness of system utterances), to the task which can be carried out with the help of the system (task success, solution quality), as well as to the speech input performance of the system (word error rate, understanding error rate).

When evaluating an interactive application, users' judgments provide valuable insights into how well the application meets their expectations and needs. One of the methods to measure users' attitudes is to observe their behaviour and establish links between their emotions and actions [20]. Current technical possibilities in the tracking and analysis of large amounts of logged user-generated multimodal data enable such observations [19]. For instance, gaze re-direction, body movements, facial muscle contraction, skin conductivity and heart rate variance may serve as a source of information for analysing a user's affective state and deliver complementary valuable evaluation data.

The most common practice is to solicit users' judgments on different system quality aspects with the help of a questionnaire. The absence of standard questionnaires for dialogue system evaluation makes it difficult to compare the results from different studies, and the various existing questionnaires exhibit great differences:

- The PARADISE questionnaire has nine user satisfaction related questions [12].
- The Subjective Assessment of Speech System Interfaces (SASSI) questionnaire contains 44 statements rated by respondents on 7-point Likert scales [13].
- The Godspeed questionnaire comprises 24 bipolar adjective pairs (e.g. fake-natural, inert-interactive, etc.) related to (1) anthropomorphism, (2) animacy, (3) likeability, (4) perceived intelligence and (5) perceived safety to evaluate human-robot interactions on 5-point Likert scales [18].
- The REVU (Report on the Enjoyment, Value, and Usability) questionnaire was developed to evaluate interactive tutoring applications and comprises 53 statements rated on 5-point Likert scales divided into three parts: OVERALL, NL (Natural Language), and IT (Intelligent Tutor) [3].
- The Questionnaire for User Interface Satisfaction (QUIS 7.0, [8]) measures satisfaction related (1) overall user reaction, (2) screen, (3) terminology and system information, (4) learnability, (5) system capabilities, (6) technical manuals and on-line help, (7) on-line tutorials, (8) multimedia, (9) teleconferencing, and (10) software installation. A short 6-dimensional form contains 41 statements rated on 9-point Likert scales, a long one has 122 ratings used for diagnostic situations.

The QUIS questionnaire is widely used and is considered as de-facto standard for user satisfaction assessment when performing usability studies. The QUIS forms can be customized by selecting evaluation aspects relevant for a specific application and use case, as we will show in the next sections when evaluating a multimodal dialogue system.

3 Usability Definition

It is common practice to evaluate an interactive system and its interface using a number of observable and quantifiable metrics for effectiveness, efficiency and satisfaction—see the ISO 9241-11 and ISO/IEC 9126-4 standards.

Task completion and the accuracy with which users achieve their goals are associated with the system's *effectiveness*. Task completion is calculated as the proportion of successfully completed tasks given the total number of tasks. To measure success of information retrieval tasks in information seeking dialogues, Attribute Value Matrix (AVM) metrics are used as proposed in PARADISE. In tutoring interactive applications, the task completion rate will depend on the system's ability to provide meaningful feedback [3]. In the next section we will define effectiveness metrics for our negotiation training use case.

Efficiency is associated with the effort that users spend to perform specified tasks and is often correlated with temporal and duration properties of the interaction, e.g. number of turns, pace, reaction times, etc. Measures of efficiency associated with user's cognitive costs relate to [21]:

- *robustness*, referring to the level of support provided to the user in determining achievement and assessment of goals; is related to observability, recoverability, responsiveness and task conformance;
- *learnability*, referring to the ease with which new users can begin effective interaction and then to attain a maximal level of performance; is related to predictability, familiarity and consistency; and
- *flexibility*, referring to the multiplicity of ways in which the user and the system can communicate; is related to initiative, task substitutivity and customisability.

Satisfaction is concerned with user attitudes associated with the product use, and is often assessed with the help of questionnaires. Satisfaction is measured at the task and test levels. Popular post-task questionnaires are After-Scenario Questionnaire (ASQ, [4]), NASA Task Load Index (TLX, [15]) and Single Ease Question (SEQ, [14]). Satisfaction at the test level serves to measure users' impression of the overall ease of use of the system being tested.

In order to develop a reliable questionnaire for assessing user perception of a multimodal dialogue system usability we conducted an online study. QUIS 7.0 served as the basis for respondents to make their selection of aspects they think are important for them when evaluating a multimodal dialogue system. QUIS provides a useful decomposition of the usability concept into several dimensions (factors), enabling a clear mapping of system performance to distinctive usability perception aspects, with the advantage of being able to assess the impact of different items on usability perception instead of simply summing up or averaging to compute an overall satisfaction score (as e.g. in PARADISE or with the System Usability Scale, SUS [5]). Adapting the QUIS questionnaire for the purposes of multimodal dialogue system evaluation, we considered factors assessed by the SAASI and Godspeed questionnaires. Previous studies showed that evaluative adjectives, bipolar adjective pairs and specific evaluative statements appeared to be more accurate than global satisfaction questions and were the most preferred forms for respondents [8, 22]. In our study, 36 evaluative adjectives, 40 bipolar adjective pairs, and 34 evaluative statements were ranked on 5-point Likert scales by 73 respondents, from which 69.6% considered themselves as dialogue researchers or related, and all respondents used dialogue systems at least once in their life. The study showed that important aspects related to user satisfaction are concerned with *task completion, task quality, robustness, learnability, flexibility, likeability, ease of use* and *usefulness/value* of the application. We adopted the QUIS 7.0 structure and populated it with 32 selected items rated the highest (>4.0 points with standard deviation <1.0) in the online study. The resulting questionnaire[1] has six dimensions measuring (1) overall reaction, (2) perceived effectiveness, (3) system capabilities, (4) learnability, (5) visuals/displays and animacy, (6) real-time feedback. The questionnaire allows to evaluate a system's functionality related to multimodality (items in dimension 3 and 5) and tutoring capabilities (dimension 6). The questionnaire is used to perform user-based evaluation as reported in Sect. 4.3 and is evaluated on internal consistency reliability.

4 Experimental Set-Up

The use case considered in this study is concerned with the evaluation of a multimodal Intelligent Tutoring System designed to train metacognitive skills in a multi-issue bargaining setting—the Virtual Negotiation Coach (VNC). The system's goal is to make negotiators aware of and reason about their negotiation behaviour and negotiation strategies and those of their opponent.

4.1 Context and Scenario

The specific setting considered in this study involved a multi-issue bargaining scenario about anti-smoking legislation passed in the City of Athens. The negotiated regulations were concerned with four main *issues*: (1) smoke-free public areas ('smoking ban scope'); (2) tobacco tax increase ('taxation'); (3) effective anti-smoking campaign programs ('campaign'); and (4) enforcement policy and police involvement ('enforcement'), see Fig. 1. Each of these issues involves four to five most important negotiation *values* with preferences representing negotiation positions, i.e. preference profiles. The strength of preferences was communicated to the negotiators through colours. Darker orange colours indicated increasingly negative options; darker blue colours increasingly positive options.

[1]The usability questionnaire for a multimodal dialogue system evaluation is provided in the Appendix.

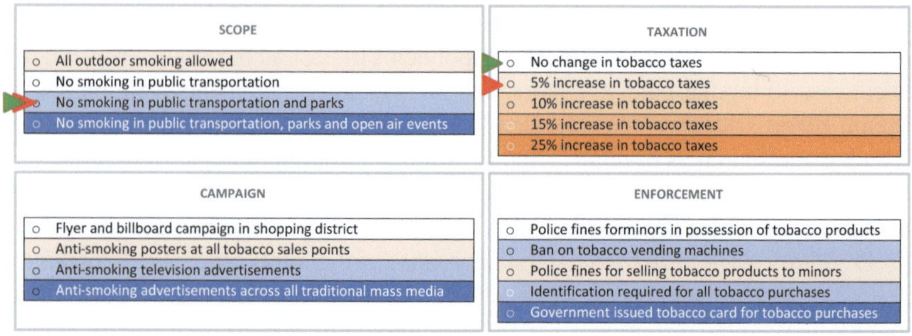

Fig. 1 Preference card: example of values in four negotiated issues presented in colours. Partners' offers visualized with red arrow (system) and green one (user)

In our evaluation experiment, 28 participants aged 25–45, professional politicians or governmental workers, were interacting with the VNC (see Sect. 4.2) for an hour. Nine negotiation scenarios were used, based on different preference profiles. Users ('trainees') were assigned a City Councilor role and a random scenario. All sessions were recorded and the system performance was logged.

The trainees' task was to negotiate with a Small Business Representative (system) an agreement which assigns exactly one value to each issue, exchanging and eliciting offers concerning $\langle ISSUE; VALUE\rangle$ options. The task is considered as completed when for all four issues an agreement is reached. Negotiators were allowed to withdraw and/or re-negotiate previously made agreements within one session. The negotiation task quality is measured by the quality of the agreements reached. In integrative bargaining this can be determined by the number of *Pareto optimal outcomes*.[2]

The negotiation success is influenced by the negotiators' strategies. For integrative negotiations, where the negotiators strive for a balance between cooperation and competition, two main negotiation strategies are observed: cooperative and non-cooperative. Cooperative negotiators share information about their preferences and attempt to find mutually beneficial agreements. They are not engaged in positional bargaining[3] tactics, instead, they try to find issues where a trade-off is possible. Non-cooperative negotiators prefer to withhold their (true) preferences and focus on positional bargaining, rarely asking for or ignoring an opponent's preferences. They threaten to end the negotiation or make very small concessions. In our experiments, we calculated the *cooperativeness level* as the number of cooperative actions given the total number of task-related negotiation actions performed. A third task success metric is related to the the *number of negative deals*, i.e. dispreferred agreements on bright 'orange' options as shown in Fig. 1.

[2]Pareto optimality reflects a state of affairs when there is no alternative state that would make any partner better off without making anyone worse off.

[3]Positional bargaining involves holding on to a fixed preferences set regardless of the interests of others.

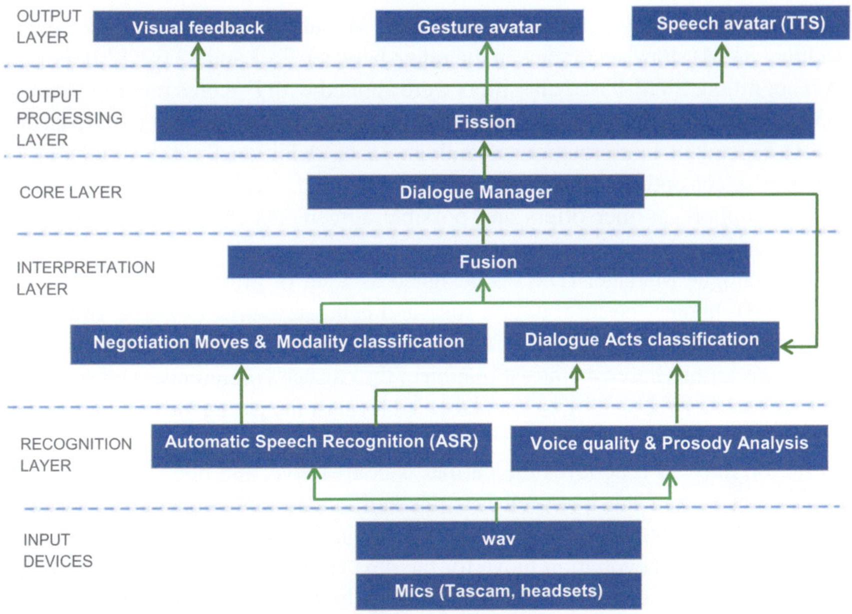

Fig. 2 Architecture of the virtual negotiation coach system

4.2 *Virtual Negotiation Coach*

We designed the Virtual Negotiation Coach (VNC), a multimodal interactive system with the functionality described in the scenario section. The VNC gets a speech signal, recognizes and interprets it, and generates multimodal actions as response, i.e. speech and gestures of a virtual negotiator and positive and negative visual feedback of a virtual tutor. Figure 2 shows the VNC architecture and processing workflows.

Speech signals are recorded by multiple devices: wearable microphones or headsets, and an all-around microphone placed between participants. The speech signals serve as input for Automatic Speech Recognition (ASR). The Kaldi-based ASR component incorporates acoustic and language models developed using 759 hours of data from various available data sources.[4] The collected 'smoking ban' in-domain data [9] is used for language model adaptation. The ASR performance is measured at 34.4% Word Error Rate (WER) [6].

The ASR output is used for lexical, syntactic, and semantic analysis to perform negotiation moves, modality and dialogue act (DA) classification. Negotiation moves specify events and their arguments of $NegotiationMove(ISSUE; VALUE)$ type [23]. Sequence learning Conditional Random Field (CRF) models were trained to predict three types of classes: negotiation move, issue, preference value. A 10-fold cross-

[4]Examples of resources are: the Wall Street Journal WSJ0 corpus, HUB4 News Broadcast data and the VoxForge corpus.

validation yielded an F-score of 0.7 on average. Modality classifiers (Support Vector Machine, SVM) show accuracies in the range between 73.3 and 82.6% [26]. For the DA recognition, SVM-based classifiers were applied with F-scores ranging between 0.83 and 0.86 [7]. Information related to the dialogue history has been used to ensure context-dependent interpretation of dialogue acts. Additionally, the trainee has a choice to select options using a graphical interface as depicted in Fig. 2. As task progress support, partner offers and possible agreements are visualized with red (system) and green (user) arrows.

The Dialogue Manager (DM) is designed as a set of processes ('threads') that receive data, update the information state and generate output [24, 25]. Integrated into the DM is the Negotiation Task Agent (NTA), which interprets and produces negotiation actions based on the estimation of the partner's preferences and goals and adjusts its strategy according to the perceived level of the partner's cooperativeness. It begins neutrally, requesting the partner's preferences, reacts with a cooperative negotiation move if it believes the partner is cooperative, and non-cooperatively if the partner's actions are interpreted as unconditionally non-cooperative, see [23]. The system reasons about the overall state of the negotiation task, and attempts to identify the best negotiation move for the next action. The DM computes (1) the system's counter-move, and (2) feedback sharing the system's beliefs about the user's preferences and the user's negotiation strategy. Additionally, the DM takes care of actions concerning contact and social obligations management, as well as elaborate recovery and error handling actions.

The system includes Fusion and Fission components. The Fusion module currently fuses interpretations from two modules obtaining full semantic representations of user speech contributions. Given the dialogue acts provided by the Dialogue Manager, Fission generates system responses splitting content into different modalities, such as Avatar and Voice (TTS) for negotiation actions, and visual feedback for tutoring actions. The latter includes a representation of the negotiators' current cooperativeness, visualized by happy and sad face emoticons. At the end of each negotiation session, summative feedback is generated specifying the number of points gained or lost for each partner, the number of negative agreements, and the Pareto optimality of the reached agreements. This type of feedback accumulates across multiple consecutive negotiation rounds.

4.3 User-Based Evaluation: Perception Versus Performance

The VNC system was evaluated measuring usability in terms of effectiveness, efficiency and satisfaction. Previous research suggests that there are differences in perceived and actual performance [10]. Performance and perception scores are correlated, but they are different usability metrics and both need to be considered when conducting quantitative usability studies. In our design, subjective perception of effectiveness, efficiency and satisfaction were correlated with various performance metrics to assess their impact on the qualitative usability properties. We computed bi-variate correlations to determine possible factors impacting user perception of system usability and the derived performance metrics from logged and annotated evaluation sessions.

The perceptive assessments come from the user satisfaction judgments on different aspects after interacting with the system. The questionnaire designed for this purpose is, first, evaluated on internally consistency and reliability measuring Cronbach's alpha. The internal consistency of the factors (dimensions) were: (1) overall reaction, alpha = 0.71; (2) perceived effectiveness, alpha = 0.74; (3) system capabilities, alpha = 0.73; (4) learnability, alpha = 0.72; (5) visuals and animacy, alpha = 0.75; and (6) real-time feedback, alpha = 0.82. All alpha values were >0.7, so we can conclude that all factors have sufficient internal consistency reliability.

As part of the user-based evaluation, users were asked to provide an overall rating of the system that they interacted with using six bipolar negative-positive adjective pairs such as frustrating-satisfying, difficult-easy, inefficient-efficient, unnatural-natural, rigid-flexible and useless-useful rated on 5-points Likert scales. Correlations between the mean overall satisfaction (3.64) and each of the other factors was measured as follows: effectiveness, $r = 0.79$; system capabilities, $r = 0.59$; learnability, $r = 0.87$; visuals and animacy, $r = 0.76$; and feedback, $r = 0.48$. Thus, users appreciate when the system effectively meets their goals and expectations and supports them in completing their tasks, is easy to learn how to interact with and offers flexible input and output processing and generation in multiple modalities.

As performance metrics, system and user performance related to task completion rate[5] and its quality[6] were computed. We also compared system negotiation performance with human performance on the number of agreements reached, the ability to find Pareto optimal outcomes, the degree of cooperativeness, and the number of negative outcomes.[7] It was found that participants reached a lower number of agreements when negotiating with the system than when negotiating with each other, 66% versus 78%. Participants made a similar number of Pareto optimal agreements (about 60%). Human participants show a higher level of cooperativity when interacting with the system, i.e. 51% of the actions are perceived as cooperative. This may mean that humans were more competitive when interacting with each other. A lower number of negative deals was observed for human-agent pairs, 21% versus 16%. Users perceived their interaction with the system as effective when they managed to complete their tasks successfully reaching Pareto optimal agreements by performing cooperative actions but avoiding excessive concessions. No significant differences in this respect were observed between human-human and human-system interactions.

---

[5]We consider the overall negotiation task as completed if parties agreed on all four issues or parties came to the conclusion that it is impossible to reach any agreement.

[6]Overall task quality was computed in terms of number of *reward points* the trainee gets at the end of each negotiation round and summing up over multiple repeated rounds; and *Pareto optimality* (see footnote 5).

[7]We considered negative deals as flawed negotiation action, i.e. the sum of all reached agreements resulted in an overall negative value, meaning that the trainee made too many concessions and selected mostly dispreferred bright 'orange' options (see Fig. 1).

As for efficiency, we assessed temporal and duration dialogue parameters, e.g. time elapsed and number of system and/or user turns to complete the task (or subtask) and the interaction as a whole. We also measured the system response time, the silence duration after the user completed his utterance and before the system responded. Weak negative correlation effects have been found between user perceived efficiency and system response delay, meaning that users found the system reaction and the interaction pace too slow. Dialogue quality is often assessed measuring word and sentence error rates [1, 2] and turn correction ratio [11]. Many designers noticed that what contributes to the perceived system quality is not so much how many errors the system makes, but rather the system's ability to recognize errors and recover from them. Users value if they can easily identify and recover from their own mistakes. All system's processing results were visualized to the user in a separate window, which contributes to the system observability. System's and user's applied repair and recovery strategies are evaluated by two expert annotators and agreement was measured in terms of kappa. Repairs were estimated as the number of corrected segments, recoveries as the number of regained utterances which were partially failed at recognition and understanding, see also [11]. While most annotators agreed that repair strategies were applied adequately, longer dialogue sessions due to frequent clarifications seem to be undesirable.

The VNC is evaluated to be relatively easy to interact with (4.2 Likert points). However, users found an instruction round with a human tutor prior to the interaction useful. Most users were confident enough to interact with the system on their own, some of them however found the system too complex and experienced difficulties in understanding certain concepts/actions. A performance metric which was found to negatively correlate with system learnability is user response delay, the silence duration after the system completed its utterance and the user proposed relevant dialogue continuation. Nevertheless, the vast majority of users learned how to interact with the system and complete their tasks successfully in the consecutive rounds. We observed a steady decline in user response delays from round to round.[8]

Users appreciated the system's flexibility. The system offered the option to select continuation task actions using a graphical interface on a tablet in case the system processing failed entirely. The use of concurrent multiple modalities was positively evaluated by the users. It was always possible for users to take initiative in starting, continuing and wrapping up the interaction, or leave these decisions to the system. At each point of interaction, both the user and the system were able to re-negotiate any previously made agreement.[9]

As overall satisfaction, the interaction was judged to be satisfying, rather reliable and useful, however, less natural (2.76 Likert points). The latter is largely attributed to rather tedious multimodal generation and poor avatar performance. System actions

[8]For now, this is only the general observation and the metric will be taken into consideration in future test-retest experiments.

[9]Performance metrics related to initiative and task substituitivity aspects and their impact on the perceived usability will be an issue for future research.

Table 1 Summary of evaluation metrics and obtained results in terms of correlations between subjective perceived system properties and actions, and objective performance metrics. (*R* stands for Pearson coefficient; * = statistically significant ($p < 0.05$)

Usability metric	Perception	Performance		*R*
	Assessment	Metric/parameter	Value	
Effectiveness (task completeness)	Mean rating score effectiveness 4.08	Task completion rate[5]; in %	66.0	0.86*
Effectiveness (task quality)		Reward points[6]; mean, max0.10	5.2	0.19
		User's action error rate (UAER, in %)[7]	16.0	0.27*
		Pareto optimality[6]; mean, between 0 and 1	0.86	0.28*
		Cooperativeness rate; mean, in %	51.0	0.39*
Efficiency (overall)	Mean rating score efficiency 4.28	System response delay (SRD); mean, in ms	243	−0.16
		Interaction pace; utterance/min	9.98	−0.18
		Dialogue duration; in min	9:37	−0.21
		Dialogue duration average, in number of turns	56.2	−0.35*
Efficiency (learnability)	3.3 (mean)	User response delay (URD); mean, in ms	267	−0.34*
Efficiency (robustness)	3.2 (mean)	System recovery strategies (SRS) correctly activated (Cohen's κ)	0.89	0.48*
		User recovery strategies (URS) correctly recognized (Cohen's κ)	0.87	0.45*
Efficiency (flexibility)	3.8 (mean)	Proportion spoken/on-screen actions mean, in % per dialogue	4.3	0.67*
Satisfaction (overall)	Aggregated per user ranging between 40 and 78	ASR word error rate; WER, in %	22.5	−0.29*
		Negotiation moves recognition accuracy, in %	65.3	0.39*
		Dialogue act recognition; accuracy, in %	87.8	0.44*
		Correct responses (CR)[11] relative frequency, in %	57.6	0.43*
		Appropriate responses (AR)[10] relative frequency, in %	42.4	0.29*

were judged by expert annotators as appropriate,[10] correct[11] and easy to interpret. Other module-specific parameters reflecting widely used metrics computed by comparing system performance with reference annotations were various types of error rates, accuracy, and κ scores measuring agreement between the system performance and human annotations of the evaluation sessions. Recognition and interpretation mistakes turned out to have moderate negative effects on the user satisfaction. Table 1 summarizes the results.

[10]System action is appropriate given the context if it introduces or continues a repair strategy.

[11]System action is considered as correct if it addresses the user's actions as intended and expected. These actions exclude recovery actions and error handling.

Satisfaction questionnaires were constructed in such a way that, along with overall user satisfaction, we could also evaluate the system's tutoring performance. Participants indicated that system feedback was valuable and supportive. However, they expected more visual real-time feedback and more explicit summative feedback on their learning progress. Most respondents think that the system presents an interesting application for skills training and would use it as a part of their training routine.

5 Conclusions and Future Research

We have presented an approach to multimodal dialogue system evaluation according to the available ISO standards on usability and qualitative metrics for effectiveness, efficiency and satisfaction. A prototype questionnaire was designed, based on established measures and best practices for the usability evaluation of interactive systems and interfaces. Potential questionnaire items were rated by respondents. Eight factors were selected as having a major impact on the perceived usability of a multimodal dialogue system and related to task success, task quality, robustness, learnability, flexibility, likeability, ease of use and usefulness (value). Performance metrics were either automatically derived from logfiles or computed using reference annotations. Perception and performance were correlated to be able to quantify usability. It was observed that the overall system usability is determined most by the user satisfaction with the task quality, by the robustness and flexibility of the interaction, and by the quality of system responses.

Further efforts will be directed at refining performance metrics and computing advanced performance parameters. We are investigating the ways to better understand and use data coming from modern tracking and sensing devices to determine the affective state of the user. In particular, continuous and non-intrusive monitoring and assessment of the level of user's motivation and engagement, which seems feasible nowadays, will allow optimising the system functional efficiency boosting user acceptance and satisfaction.

Appendix

Usability Perception Questionnaire: Multimodal Dialogue System

1. Overall reaction to the system						
	1	2	3	4	5	
frustrating	○	○	○	○	○	satisfying
difficult	○	○	○	○	○	easy
inefficient	○	○	○	○	○	efficient
unnatural	○	○	○	○	○	natural
rigid	○	○	○	○	○	flexible
unreliable	○	○	○	○	○	reliable
useless	○	○	○	○	○	useful
2. Task effectiveness						
I was successful in completion of all my tasks						
never	○	○	○	○	○	always
To complete tasks was						
difficult	○	○	○	○	○	easy
I knew what to do at each point of interaction						
never	○	○	○	○	○	always
System support in task completion was						
unhelpful	○	○	○	○	○	helpful
3. System capabilities						
Interaction with the system was						
too slow	○	○	○	○	○	fast enough
System performance was mostly						
unreliable	○	○	○	○	○	reliable
System recovered from mistakes						
never	○	○	○	○	○	always
Correcting mistakes was mostly						
difficult	○	○	○	○	○	easy
System informed me about what it is doing						
never	○	○	○	○	○	always
System's behaviour was mostly						
rigid	○	○	○	○	○	flexible
4. Learnability						
Learning how to interact with the system						
difficult	○	○	○	○	○	easy
Navigation through tasks						
confusing	○	○	○	○	○	clear
Exploring new features was						
difficult	○	○	○	○	○	easy
Remembering rules and commands was						
difficult	○	○	○	○	○	easy
System help was						
inadequate	○	○	○	○	○	adequate
5. Visuals, displays and animacy						
Embodied characters were						
artificial	○	○	○	○	○	human-like
Use of multiple input modalities was mostly						
useless	○	○	○	○	○	useful
Position of messages on screen was						
confusing	○	○	○	○	○	clear
Amount of information on screen was						
distracting	○	○	○	○	○	supportive
Sequences of screens were						
inconsistent	○	○	○	○	○	coherent
6. Online help and real-time feedback						
useless	○	○	○	○	○	useful
distracting	○	○	○	○	○	supportive
excessive	○	○	○	○	○	sufficient
delayed	○	○	○	○	○	timely
abrupt	○	○	○	○	○	as expected

References

1. Walker MA, Litman DJ, Kamm CA, Abella A (1997) PARADISE: a framework for evaluating spoken dialogue agents. In: Proceedings of the 8th conference on European chapter of the association for computational linguistics, pp 271–280
2. López-Cózar R, Callejas Z, McTear M (2006) Testing the performance of spoken dialogue systems by means of an artificially simulated user. Artif Intell Rev 26(4):291-323. Springer (2006)
3. Dzikovska M, Moore J, Steinhauser N, Campbell G (2011) Exploring user satisfaction in a tutorial dialogue system. In: Proceedings of the 12th annual meeting of the special interest group on discourse and dialogue (SIGdial 2011), pp 162–172
4. Lewis JR (1991) Psychometric evaluation of an after-scenario questionnaire for computer usability studies: the ASQ. ACM Sigchi Bull 23(1):78–81
5. Brooke J (1996) SUS-A quick and dirty usability scale. Usability Eval Ind 189(194):4–7
6. Singh M, Oualil Y, Klakow D (2017) Approximated and domain-adapted LSTM language models for first-pass decoding in speech recognition. In: Proceedings of the 18th annual conference of the international speech communication association (INTERSPEECH), Stockholm, Sweden
7. Amanova D, Petukhova V, Klakow D (2016) Creating annotated dialogue resources: cross-domain dialogue act classification. In: Proceedings of the 9th international conference on language resources and evaluation (LREC 2016), ELRA, Paris
8. Chin JP, Diehl VA, Norman KL (1988) Development of an instrument measuring user satisfaction of the human-computer interface. In: Proceedings of the SIGCHI conference on human factors in computing systems, pp 213–218, ACM
9. Petukhova V, Stevens CA, de Weerd H, Taatgen N, Cnossen F, Malchanau A (2016) Modelling multi-issue bargaining dialogues: data collection, annotation design and corpus. In: Proceedings of the 9th international conference on language resources and evaluation (LREC 2016), ELRA, Paris
10. Nielsen J (2012) User satisfaction vs. performance metrics. Nielsen Norman Group
11. Danieli M, Gerbino E (1995) Metrics for evaluating dialogue strategies in a spoken language system. In: Proceedings of the 1995 AAAI spring symposium on empirical methods in discourse interpretation and generation, vol 16, pp 34–39
12. Walker M, Kamm C, Litman D (2000) Towards developing general models of usability with PARADISE. Nat Lang Eng 6(3–4):363–377
13. Hone KS, Graham R (2001) Subjective assessment of speech-system interface usability. In: Proceedings of the 7th European conference on speech communication and technology
14. Sauro J, Dumas JS (2009) Comparison of three one-question, post-task usability questionnaires. In: Proceedings of the SIGCHI conference on human factors in computing system, ACM, pp 1599–1608
15. Hart SG, Staveland LE (1988) Development of NASA-TLX (task load index): results of empirical and theoretical research. Adv Psychol 52, 139–183 (Elsevier)
16. Fraser N (1997) Assessment of interactive systems. In: Gibbon D, Moore R, Winski R (eds) Handbook on standards and resources for spoken language systems. Mouton de Gruyter, Berlin, pp 564–615
17. Möller S (2004) Quality of telephone-based spoken dialogue systems. Springer Science & Business Media
18. Bartneck C, Kulić D, Croft E, Zoghbi S (2009) Measurement instruments for the anthropomorphism, animacy, likeability, perceived intelligence, and perceived safety of robots. Int J Soc Robot 1(1):71–81
19. Linek SB, Marte B, Albert D (2008) The differential use and effective combination of questionnaires and logfiles. In: Computer-based knowledge and skill assessment and feedback in learning settings (CAF), Proceedings of the ICL
20. Kooijmans T, Kanda T, Bartneck C, Ishiguro H, Hagita N (2007) Accelerating robot development through integral analysis of human robot interaction. IEEE Trans Robot 23(5):1001–1012

21. Dix A (2009) Human-computer interaction. In: Encyclopedia of database systems, pp 1327–1331, Springer US
22. Root RW, Draper S (1983) Questionnaires as a software evaluation tool. In: Proceedings of the SIGCHI conference on human factors in computing systems, pp 83–87, ACM
23. Petukhova V, Bunt H, Malchanau A (2017) Computing negotiation update semantics in multi-issue bargaining dialogues. In: Proceedings of the SemDial 2017 (SaarDial) workshop on the semantics and pragmatics of dialogue, Germany
24. Malchanau A, Petukhova V, Bunt H, Klakow D (2015) Multidimensional dialogue management for tutoring systems. In: Proceedings of the 7th language and technology conference (LTC 2015), Poznan, Poland
25. Malchanau A, Petukhova V, Bunt H (2018) Towards integration of cognitive models in dialogue management: designing the virtual negotiation coach application. Dialogue Discourse 9(2):35–79
26. Lapina V, Petukhova V (2017) Classification of modal meaning in negotiation dialogues. In: Proceedings of the 13th joint ACL-ISO workshop on interoperable semantic annotation (ISA-13), pp 59–70, Montpellier, France

Toward Low-Cost Automated Evaluation Metrics for Internet of Things Dialogues

Kallirroi Georgila, Carla Gordon, Hyungtak Choi, Jill Boberg, Heesik Jeon and David Traum

Abstract We analyze a corpus of system-user dialogues in the Internet of Things domain. Our corpus is automatically, semi-automatically, and manually annotated with a variety of features both on the utterance level and the full dialogue level. The corpus also includes human ratings of dialogue quality collected via crowdsourcing. We calculate correlations between features and human ratings to identify which features are highly associated with human perceptions about dialogue quality in this domain. We also perform linear regression and derive a variety of dialogue quality evaluation functions. These evaluation functions are then applied to a held-out portion of our corpus, and are shown to be highly predictive of human ratings and outperform standard reward-based evaluation functions.

1 Introduction

The Internet of Things (IoT) is the network of physical devices (e.g., home appliances, health monitoring devices, etc.) connected to the Internet. IoT devices can be controlled each one separately by individual apps, or all together via an integrated

K. Georgila (✉) · C. Gordon · J. Boberg · D. Traum
USC Institute for Creative Technologies, Playa Vista, USA
e-mail: kgeorgila@ict.usc.edu

C. Gordon
e-mail: cgordon@ict.usc.edu

J. Boberg
e-mail: boberg@ict.usc.edu

D. Traum
e-mail: traum@ict.usc.edu

H. Choi · H. Jeon
Samsung Electronics Co., Ltd., Seocho-gu, Seoul, Korea
e-mail: ht777.choi@samsung.com

H. Jeon
e-mail: heesik.jeon@samsung.com

L. F. D'Haro et al. (eds.), *9th International Workshop on Spoken Dialogue System Technology*, Lecture Notes in Electrical Engineering 579,
https://doi.org/10.1007/978-981-13-9443-0_14

(1) User: Play music in 10 min.
(2) System: In which room shall I play music?
(3) User: Bedroom please.
(4) System: As you wish. I will play music in the bedroom in 10 min.
(5) User: Thanks. Please turn on the air-conditioner there too.
The user wants the air-conditioner to be turned on in 10 min but the user request is ambiguous and misenterpreted by the system.
(6) System: Okay, the air-conditioner in the bedroom is now on.
(7) User: Wait, I want the air-conditioner on in 10 min, not now.
(8) System: Oops my mistake. The air-conditioner in the bedroom is now off. I will turn it on in 10 min. Anything else?
(9) User: Yes, turn on the light.
The user has switched to a different task in a new room but the system thinks that the new command is related to the previous context (bedroom).
(10) System: I've turned on the light in the bedroom.
(11) User: Not in the bedroom, in the kitchen. I'm going to the kitchen now.
(12) System: I've turned on the light in the kitchen.
(13) User: Good. Move the TV up one channel.
(14) System: To be sure, you'd like me to move up one channel the TV in the kitchen, correct?
(15) User: Yes, this is exactly what I'd like you to do.
(16) System: Moving the kitchen TV up one channel.
(17) User: No, I mean down one channel.
(18) System: Got it. Moving the TV down one channel.
(19) User: Hmm. There is nothing interesting on.
(20) System: I totally sympathize. The quality of TV has gone downhill.
(21) User: Oh well, just turn it off.
There is a speech recognition error: "on" is recognized instead of "off".
(22) System: Sorry I can't do that. The TV is already on.
(23) User: I said turn the TV off.
(24) System: Done.
(25) User: Thanks.
(26) System: You're very welcome. I'm always here to help.

Fig. 1 Example system-user dialogue in the IoT domain

app. Alternatively, IoT devices can be controlled by a smart assistant via human-system dialogue interaction [11], which is the approach that we follow in this paper.

Building dialogue systems for the IoT domain is an area of research that has only recently emerged, and which poses many challenges. Figure 1 shows an example dialogue between a human user and a future smart home assistant that controls the IoT devices. As we can see, the human-system interaction can be quite complex. An intelligent dialogue system should be able to understand the user's requests and if needed follow up with smart clarification questions. For example, in utterance (2) the system behaves intelligently by asking the user a clarification question about which room it should play music in. Dialogue context is another factor that can affect dialogue quality, both regarding interpreting user requests and responding appropriately. For example, in utterances (5)–(6) the system misinterprets the user's ambiguous request, and instead of scheduling to turn on the air-conditioner in 10 min the air-conditioner is turned on immediately. A better system response would have been a clarification question "Shall I turn on the air-conditioner now or in 10 min?". Furthermore, in utterance (14) the system's confirmation request about the TV in the kitchen makes sense in this dialogue context, but it would sound weird in a different context. Regardless of the dialogue context, speech recognition errors can be another source of noise in the interaction as it is the case for all spoken dialogue systems, e.g., in utterances (21)–(22).

The quality of IoT dialogues may also be influenced by potential side-effects of some actions. Consider utterances (9)–(12). The result of this system-user exchange is that both lights are now on (in the bedroom and the kitchen), even though the user wanted only the light in the kitchen to be on. Unless the bedroom light is turned off, this dialogue has a side-effect as well as the main effect on the light in the kitchen. This side effect might be undesirable, or desirable, or neutral. Furthermore, timing can be an important issue. For example, the system's action in (18) depends on what the system did between (16)–(17). If the system already moved the TV up one channel as described in (16), then in (18) the system must move the TV down two channels. However, if (17) came before the system had a chance to change the channel, it would only need to move down one channel. Thus, not only the linguistic context is important, but also the context of device state and actions performed.

Given that users are bound to interact with this smart home assistant on a regular basis, it may not be enough that the system performs its tasks and behaves in a rational manner. People may have preferences about the system's personality and expect it to be polite, nice, and adapt to their mood, needs, and desires. For example, some users may expect the system to sympathize with their opinions as in utterance (20) or be chatty as in utterances (8) and (26). Also, some users may prefer the system to be explicit about what it has done or is planning to do (e.g., in utterance (4) "I will play music in the bedroom in 10 min.") whereas other users may appreciate brevity (e.g., in utterance (24) "Done.").

As we can see from the above examples, there are similarities between dialogues in the IoT domain and multi-domain task-oriented dialogues. What makes the IoT domain particularly interesting is that the devices involved can work together in a synergistic way rather than being totally separate domains or tasks. For example, users may have a routine when they leave the house in the morning, e.g., locking the windows and doors, turning off the coffee machine and the toaster, etc. Or when they come back in the evening they may want the smart home assistant to create a relaxing atmosphere in the living room, e.g., by playing classical music at an appropriate speaker volume level, dimming the lights, etc.

The example dialogue of Fig. 1 illustrates that it is not clear at all what constitutes a successful system-user dialogue in the IoT domain. Our goal in this paper is to take the first steps toward developing low-cost evaluation metrics that are predictive of user perceptions about dialogue quality in this domain. By low-cost we mean that these metrics should be based on automatically extracted features or, if this is not possible, rely on simple annotations that can be performed by non-experts in linguistics or dialogue. Our contribution is two-fold. First, we provide insights about unique challenges in developing dialogue quality evaluation metrics in the IoT domain. We also present a novel annotation scheme for annotating a variety of both social and task-related aspects in this domain. Second, using a methodology similar to the PARADISE evaluation framework [20], we develop novel evaluation functions for the IoT domain. When applied to a held-out portion of our dialogue corpus, these evaluation functions are shown to be highly predictive of human ratings of dialogue quality, and outperform standard reward-based evaluation functions.

2 Related Work

Over the years, a variety of evaluation frameworks and metrics have been proposed for measuring the quality of human-system dialogue interaction, mainly for task-oriented dialogue systems [8]. Some metrics are subjective (e.g., user satisfaction, perceived task completion, etc.), and others are objective (e.g., word error rate, dialogue length, etc.). Objective measures can be calculated from the interaction logs while subjective assessments can be collected via surveys and questionnaires [10, 15].

PARADISE [20] is perhaps the most well-known framework for evaluating dialogue systems, and an attempt to automate the evaluation process. PARADISE seeks to optimize a desired quality such as user satisfaction by formulating it as a linear combination of a variety of metrics, such as task success and dialogue cost (e.g., dialogue length). The advantage of this method is that once a desired quality has been formulated as a realistic evaluation function, it can be optimized by controlling the factors that affect it. In the example above, user satisfaction can be optimized by increasing task success and minimizing dialogue length.

Reinforcement learning (RL) has become the standard technique for learning dialogue policies from data or simulated users (SUs). In RL, a typical reward function is for the system to earn a number of points for a fully successful dialogue (or for partial success, e.g., when some of the requested information is provided or confirmed), and subtract a penalty per system turn to ensure that the learned dialogue policies will not favor lengthy and tedious dialogues [9]. Note however that longer dialogue lengths are not necessarily indicative of poor dialogue quality but depending on the task they may actually indicate user engagement and satisfaction [2].

A variety of metrics have been employed for measuring the quality of SUs used for training and evaluating dialogue policies. The idea is that the action generated by the SU is compared against the user action in a human-human or human-system reference corpus (in the same dialogue context), and measures such as precision, recall, accuracy, and perplexity are used [5, 6, 18]. However, these metrics can be problematic because if a SU action is not the same as the user action in the reference corpus, this does not necessarily mean that it is a poor action. Also, once a user or system response deviates from the corresponding action in the reference corpus, the remaining dialogue will unfold in an entirely different way than the fixed dialogue in the reference corpus, which will make further comparisons meaningless.

In non-task-oriented dialogue systems (e.g., chatbots) developing robust evaluation metrics can be even harder than for task-oriented dialogue. Here it is not clear what success means and thus task-specific objective metrics are not appropriate. Instead subjective evaluations for appropriateness of responses can be much more meaningful, which has led to the development of coding schemes for response appropriateness and scoring in such cases [17, 19]. Another approach is to evaluate dialogue systems in a semi-formal manner, using human judges to rate the coherence of a conversational agent and correlating these judgements with measures extracted from within the system [1]. Dialogue coherence can also be formulated as an information ordering task [4, 16]. In [16], a binary classifier was built for distinguishing between coherent and incoherent dialogues using local transition patterns that span over adjacent dialogue turns encoding lexical and semantic information. In [4], random permutations of dialogue segments were generated and rated by human judges on a Likert scale in terms of coherence. It was found that Kendall's τ correlated well with human judgements.

Word-overlap similarity metrics such as BLEU, METEOR, and ROUGE (originally employed in machine translation and summarization) are currently widely used for measuring chatbot dialogue quality. However, BLEU, METEOR, and ROUGE suffer from the same problems as the aforementioned SU evaluation metrics. In fact it has been shown that BLEU, METEOR, and ROUGE do not correlate well with human judgements of dialogue quality [13]. Note that BLEU has also been used for evaluating SUs [12]. Discriminative BLEU, a variation of BLEU where reference strings are scored for quality by human raters, was found to correlate better with human judgements than standard BLEU [3]. To address the issues with BLEU, METEOR, and ROUGE, next utterance classification was introduced as a method for evaluating chatbots [14], but the proposed metric recall@k is very similar to the recall metric previously used for evaluating SUs, and consequently has the same limitations. Recently, topic-based metrics for chatbot evaluation (topic breadth and topic depth) were found to correlate well with human judgements [7].

3 Our Dialogue Corpus

Our corpus currently consists of approximately 6200 dialogues in the IoT domain between a smart home assistant and a user. These dialogues were written by several linguists, and obviously they are not as realistic as the dialogues we would get by having humans interact with a real system or in a Wizard of Oz setting, which is something we plan to do for future work. However, despite this limitation, our dialogues are designed to capture a variety of phenomena that we would encounter in real human-system dialogues, including speech recognition errors, misunderstandings, clarification requests, timing, context and scheduling issues, and generally all the phenomena that we discussed in the dialogue example of Fig. 1. For this reason we consider our corpus to be a valuable resource for bootstrapping our research in the IoT dialogue domain.

Our corpus includes system-user conversations regarding one home appliance at a time (washer, speaker, bulb, TV, air-conditioner) or multiple devices at the same time (e.g., the air-conditioner and the TV). Also, in our IoT ontology we have multiple

Table 1 Statistics of our corpus (232 dialogues)

Dialogue feature	Mean	Standard deviation
Number of tasks per dialogue	1.41	0.68
Number of system turns per dialogue	2.80	0.98
Number of user turns per dialogue	2.80	0.98
Number of all turns per dialogue	5.60	1.96
Number of system words per dialogue	14.08	8.01
Number of user words per dialogue	15.32	6.00
Number of all words per dialogue	29.40	12.15
Average number of system words per utterance	5.14	2.61
Average number of user words per utterance	5.69	1.88
Average number of all words per utterance	5.41	1.81

devices of the same type (e.g., there can be a TV in the bedroom, a TV in the living room, and a TV in the guest room). Our corpus contains dialogues where the system and the user need to make sure that they are both referring to the same device, which in turn leads to very realistic and complex system-user interactions.

For the experiments presented in this paper we selected 232 dialogues from 6 categories (washer, speaker, bulb, TV, air-conditioner, multiple devices in the same dialogue), taking care to include as many realistic dialogue phenomena as possible. For each category we have 2 sub-categories: dialogues without any misunderstandings and dialogues with misunderstandings. Thus in total we have 12 dialogue categories with about 20 dialogues per category.

Our corpus of 232 dialogues has been annotated manually, automatically, and semi-automatically with a variety of features both on the utterance level and the full dialogue level. More specifically, we have automatically calculated the following features: number of system and user turns per dialogue, number of total words from system and user per dialogue, average number of words per system and user utterance in a dialogue, and number of occurrences of specific words and expressions, e.g., "yes/yeah/yep/yup", "no/nope", "ok/okay", "alright/all right", "done", "system", "thanks/thank you", "good/great", "not at all", "sure", "sure thing", "got it", "no problem", "sorry/apologize/apologies", "naturally", "obviously", etc. Table 1 shows statistics of our corpus of 232 dialogues.

We have also developed a novel annotation scheme and performed the following annotations on the utterance level for both system and user utterances:

- System utterances:
  - Assess action:
    A-something (system does something: "I'm connecting the speaker.")
    A-nothing (system does nothing: "Which speaker?")
    A-valid (system does requested thing: "U: Turn on the kitchen light. S: I'm turning on the kitchen light.")

A-invalid (system does not do requested thing: "U: Turn on the kitchen light. S: I'm turning on the porch light.")
- Describe current understanding:
  CU-confirm (confirm request before doing: "Shall I turn on the light?")
  CU-lack (describe lack of understanding: "Sorry I don't understand.")
- Action acknowledge:
  AA-past (action specified in the past: "The light has been turned on.")
  AA-present (action specified in the present: "I'm turning on the light.")
  AA-future (action specified in the future: "I'll turn on the light in 5 min.")
  AA-ANS (action not specified: "U: Turn on the light. S: Done.")
  AA-AI (action impossible: "I can't open the door while the cycle is running.")
  AA-null (action is done but not acknowledged: "U: Turn on the light. S: Anything else?")
- Specify state:
  SS-done (explicit action, done: "The light is now on.")
  SS-NA (explicit action, not applicable: "The light is already on.")
  SS-unclear (explicit action, unclear: "The light is on."—it is not clear whether the light was already on or the system performed the action)
- Requests:
  Req-location (missing parameter, location: "Which light?")
  Req-dev (missing parameter, device: "What should I connect to Wifi?")
  Req-time (missing parameter, time: "When should I do that?")
  Req-temp (missing parameter, temperature: "What temperature?")
  Req-other (missing parameter, other: "What should I connect it to?")
  Req-action (request more actions: "Anything else?")
  Req-repeat (request repeat: "Could you repeat?")
- Other response:
  O-null (equivalent to silence)
  O-pleasant (system pleasantry: "You are welcome.")
- Level of specificity:
  explicit (parameters explicit: "U: Turn on the light. S: The light has been turned on.")
  implicit (parameters implicit: "U: Turn on the light. S: It's been turned on.")
- Register:
  Reg-direct (direct: "U: Turn on the light. S: I'm turning on the light.")
  Reg-conv (conversational: "U: Turn on the light. S: Sure thing, the light is now on.")
- Grammaticality:
  gram (grammatical responses: "Which light shall I turn on?")
  ungram (ungrammatical responses: "Which light shall I open?")

- User utterances:
  - Request action:
    RA-dev (RA-dev-wash, RA-dev-speaker, etc., depending on the device: "Turn on the speaker.")

RA-location (specified location: "Turn on the speaker in the bedroom.")
RA-time (specified time: "Turn on the TV in 10 min.")
RA-temp (specified temperature: "Decrease temperature to 40°.")
RA-end-state (specified end state: "I feel like listening to music.")
RA-other (specified other: "Connect the speaker to Bluetooth.")
RA-action (specified action: "Turn it off.")

– Response to system:
RS-yes (yes)
RS-no (no)
RS-null (silence)
RS-restate (restate request: "S: I don't understand. Which light should I turn on? U: The one in the bedroom.")
RS-decline (decline further action: "S: Anything else? U: No, thanks.")
RS-param (provide parameters: "S: When? U: Today at 4 pm.")

– Pleasantries:
P-greet (greeting: "Hello system!")
P-thank (thanks: "Thanks.")

– Level of specificity:
explicit (parameters explicit: "The washer in the bathroom.")
implicit (parameters implicit: "The one in the bathroom.")

Our corpus of 232 dialogues was annotated for the features presented above by our principal annotator. It was these annotations that were used to generate some of the feature correlations which ultimately informed our dialogue quality evaluation functions (see Sect. 5). To measure inter-annotator agreement, a smaller subset of 30 dialogues was annotated by a second annotator, and used for comparison with the annotations performed by the principal annotator. Overall there was raw agreement of 97% and a Krippendorff's alpha value of 0.867, including cases where neither annotator annotated anything for a category. If we look only at cases where at least one annotator entered a tag, we still have 83.3% agreement. In those cases raw agreement was above 80% for most categories, ranging from 58.6% for "Response to system" to 100% for "Other response".

Other manual or semi-automatic annotations that we performed at the whole dialogue level are as follows. Note that semi-automatic annotations were based on manual annotations, and were manually checked afterwards.

- Number of all tasks and successful tasks in the dialogue.
- Number and list of misunderstandings in the dialogue.
- Number of system confirmation requests in the dialogue. (This can be derived from "CU-confirm" above.)
- Number of system requests for repetition in the dialogue. (This can be derived from "Req-repeat" above.)
- Number of system requests for more information in the dialogue. (This can be derived from "Req-dev", "Req-location", "Req-temp", "Req-time", and "Req-other" above.)
- Number and list of immediate and scheduling tasks in the dialogue.

- Number and list of devices in the dialogue.
- Number of cases in which the system says that it cannot perform an action. (This can be derived from AA-AI above.)

Note that for each of the annotations above we compute two features. One feature gives us the exact count of e.g., misunderstandings, system confirmation requests, scheduling tasks, etc. The other feature has a binary value (yes/no) and keeps track of whether e.g., misunderstandings, system confirmation requests, scheduling tasks, etc., are present in the dialogue or not regardless of their frequency.

4 Collection of Human Ratings Via Crowdsourcing

We grouped our 232 dialogues in sets of 5 dialogues and asked human raters on Amazon Mechanical Turk (MTurk) to rank them. Each rater had to perform 4 tasks. In each task, raters were presented with 8 sets of 5 dialogues and asked to rank them (best dialogue to worst dialogue in the set) based on which system they would most like to interact with. The types of dialogues and contexts of the individual tasks are described below:

- Task 1: All dialogues in a set had the same task goal and device (i.e., turn on TV, connect speaker to Bluetooth, etc.).
- Task 2: Dialogues in a set represented a mix of task goals and devices.
- Task 3: Dialogues in a set represented a mix of task goals and devices, and raters were presented with a description of the pre-conversation system status, i.e., the state and location of each device before the dialogue starts.
- Task 4: Dialogues in a set represented a mix of task goals and devices, and raters were presented with a description of the pre-conversation and post-conversation system status, i.e., the state and location of each device before the dialogue starts and after the dialogue ends.

We collected rankings from 199 people on MTurk. From these rankings we generated pairwise comparisons for all dialogues. For example, assuming that we have 3 dialogues D1, D2 and D3, we can generate the pairs D1-D2, D1-D3, D2-D3 and calculate for each pair how many times the first dialogue of the pair is ranked higher than the second dialogue of the pair and vice versa. Thus for each dialogue we can generate a score by dividing the number of times this dialogue wins in all pairwise comparisons by the number of all comparisons of this dialogue. So in the example above, the score for dialogue D1 would be ("number of times D1 beats D2" + "number of times D1 beats D3")/("number of times D1 competes with D2" + "number of times D1 competes with D3"). This score (from now on referred to as "Score") is a real number between 0 and 1 and will be used in the calculations of correlations and evaluation functions in Sect. 5.

5 Dialogue Quality Evaluation Functions

We calculated pairwise Pearson correlations between all features in our annotations (automatic, semi-automatic, and manual) as well as Pearson correlations between the "Score" and each feature in our annotations. Note that "***" means that the correlation is very significant ($p < 0.001$), "**" means that the correlation is significant ($p < 0.01$), and "*" means that the correlation is borderline significant ($p < 0.05$). Table 2 shows a list of some of the most indicative (higher and/or more significant) correlations that we found between the "Score" and dialogue features.

For each feature we experimented both with binary values (the feature exists or not) and counts (frequency of occurrence). Counts may be affected by the number of tasks and multiple devices in the interaction and we wanted to account for that. From

Table 2 Correlations of "Score" with features (***: $p < 0.001$, **: $p < 0.01$, *: $p < 0.05$)

Dialogue feature	Pearson's r
Number of misunderstandings	−0.76***
Misunderstandings exist or not? (Binary)	−0.77***
Number of system confirmation requests	−0.50***
System confirmation requests exist or not? (Binary)	−0.50***
Number of system requests for more information	0.27***
System requests for more information exist or not? (Binary)	0.28***
Number of silence occurrences	−0.67***
Silence exists or not? (Binary)	−0.68***
Number of times the user says "I mean/I meant"	−0.17**
User says "I mean/I meant" or not? (Binary)	−0.20**
Number of times the user says "I said"	−0.23***
User says "I said" or not? (Binary)	−0.23***
Number of times the user says "no/nope"	−0.72***
User says "no/nope" or not? (Binary)	−0.73***
Number of system turns per dialogue	−0.61***
Number of user turns per dialogue	−0.61***
Number of all turns per dialogue	−0.61***
Number of system words per dialogue	−0.44***
Number of user words per dialogue	−0.45***
Number of all words per dialogue	−0.51***
Number of times the system does nothing (A-nothing)	−0.54***
System does nothing (A-nothing) or not? (Binary)	−0.36***
Number of times the system does something invalid (A-invalid)	−0.33***
System does something invalid (A-invalid) or not? (Binary)	−0.33***
System has a conversational style or not? (Binary)	0.17**
Number of times the user specifies the location of a device (RA-location)	−0.22**
User specifies the location of a device (RA-location) or not? (Binary)	−0.26***

Table 3 Correlations of misunderstandings with features (***: $p < 0.001$, **: $p < 0.01$, *: $p < 0.05$)

Dialogue feature	Pearson's r (for counts)	Pearson's r (for binary values)
Occurrences of silence	0.42***	0.44***
Occurrences of "I mean/I meant"	0.22***	0.24***
Occurrences of "I said"	0.25***	0.29***
Occurrences of "no/nope"	0.66***	0.64***
Number of system turns per dialogue	0.58***	–
Number of user turns per dialogue	0.58***	–
Number of all turns per dialogue	0.58***	–
Number of system words per dialogue	0.55***	–
Number of user words per dialogue	0.47***	–
Number of all words per dialogue	0.60***	–
System does nothing (A-nothing)	0.47***	0.33***
System does something invalid (A-invalid)	0.41***	0.40***
User specifies the location of a device (RA-location)	0.17**	0.24***

the above correlations we can see that misunderstandings and system confirmation requests are indicative of poor quality dialogues. On the other hand when the system asks for more information (e.g., "U: Can you connect the speaker? S: What should I connect it to?") this is indicative of having understood what the user wants and it is something that the human raters liked. Also, human raters seem to like a more conversational style for the system (i.e., when the system says "sure thing", etc.). The user action of specifying the location of a device usually occurred together with misunderstandings and thus negatively correlated with the "Score".

Table 3 shows a list of some of the most indicative (higher and/or more significant) correlations that we found between misunderstandings and dialogue features. Note that the correlations for counts are derived when we compare the number of misunderstandings with the counts of the dialogue features. In the same way, the correlations for binary values are derived when we compare whether misunderstandings exist or not with the binary values of the dialogue features. The high (or relatively high) correlation between misunderstandings and each one of these features entails that once we include misunderstandings in our evaluation function we do not also need to consider all these features.

The next step was to perform regression experiments to come up with evaluation functions that are predictive of human ratings. We excluded as redundant the aforementioned features that were highly correlated with misunderstandings, and we experimented with variations of the following features that as we saw in Table 2 were highly correlated with the "Score":

- Number of misunderstandings (Mu)
- Misunderstandings exist or not? (Binary-Yes/No) (Mu_b)

Table 4 Evaluation functions and corresponding root mean square error (RMSE) values (the best, i.e., lowest, RMSE values are shown in bold); reward-based function is normalized; Mu: number of misunderstandings, Mu_b: misunderstandings exist or not (binary), Conf: number of system confirmation requests, Conf_b: system confirmation requests exist or not (binary), Inf: number of system requests for more information, Inf_b: system requests for more information exist or not (binary), Conv_b: system has a conversational style or not (binary)

Description	Evaluation function	RMSE
Reward-based function	100*Task_success−5*Num_system_turns	0.5224
Mu	−0.21*Mu+0.62	**0.0902**
Mu_b	−0.23*Mu_b+0.62	**0.0920**
Conf	−0.16*Conf+0.56	0.1318
Conf_b	−0.17*Conf_b+0.56	0.1325
Inf	0.09*Inf+0.49	0.1281
Inf_b	0.09*Inf_b+0.48	0.1270
Conv_b	0.05*Conv_b+0.49	0.1320
Mu+Conf	−0.18*Mu−0.05*Conf+0.62	**0.0919**
Mu_b+Conf_b	−0.2*Mu_b−0.05*Conf_b+0.63	0.0929
Mu+Inf	−0.2*Mu+0.02*Inf+0.61	**0.0899**
Mu_b+Inf_b	−0.23*Mu_b+0.01*Inf_b+0.62	**0.0919**
Conf+Inf	−0.15*Conf+0.04*Inf+0.55	0.1281
Conf_b+Inf_b	−0.16*Conf_b+0.04*Inf_b+0.55	0.1283
Mu+Conf+Conv_b	−0.19*Mu−0.05*Conf−0.02*Conv_b+0.63	0.0944
Mu_b+Conf_b+Conv_b	−0.2*Mu_b−0.05*Conf_b−0.01*Conv_b+0.63	0.0944
Mu+Inf+Conv_b	−0.21*Mu+0.02*Inf−0.01*Conv_b+0.62	**0.0911**
Mu_b+Inf_b+Conv_b	−0.23*Mu_b+0.01*Inf_b−0.01*Conv_b+0.62	**0.0924**
Conf+Inf+Conv_b	−0.15*Conf+0.04*Inf+0.01*Conv_b+0.55	0.1269
Conf_b+Inf_b+Conv_b	−0.16*Conf_b+0.04*Inf_b+0.01*Conv_b+0.55	0.1271
Mu+Conf+Inf	−0.18*Mu−0.05*Conf+0.01*Inf+0.62	**0.0915**
Mu_b+Conf_b+Inf_b	−0.2*Mu_b−0.05*Conf_b+0.01*Inf_b+0.63	0.0928
Mu+Conf+Inf+Conv_b	−0.18*Mu−0.05*Conf+0.01*Inf−0.02*Conv_b+0.63	0.0939
Mu_b+Conf_b+Inf_b+Conv_b	−0.2*Mu_b−0.05*Conf_b+0.005*Inf_b−0.01*Conv_b+0.63	0.0943

- Number of system confirmation requests (Conf)
- System confirmation requests exist or not? (Binary-Yes/No) (Conf_b)
- Number of system requests for more information (Inf)
- System requests for more information exist or not? (Binary-Yes/No) (Inf_b)
- System has a conversational style (i.e., at least half of system responses are annotated as "Reg-conv") or not? (Binary-Yes/No) (Conv_b)

We randomly split our corpus in a training set and a test set (75% for training and 25% for testing). We applied linear regression to the training set, calculated our evaluation functions, and then measured how these evaluation functions performed on the test set (i.e., how predictive they were of the actual human ratings). To do that we calculated the root mean square error (RMSE) as shown in Eq. (1) where n

is the number of dialogues, $Score_{iPredicted}$ is the predicted "Score" for dialogue i (calculated by our evaluation function), and $Score_{iActual}$ is the actual "Score" for dialogue i (derived from the human ratings).

$$RMSE = \sqrt{\frac{1}{n}\sum_{i=1}^{n}(Score_{iPredicted} - Score_{iActual})^2} \quad (1)$$

Obviously the lower the RMSE the better. We also constructed a reward-based evaluation function based on the literature of reinforcement learning of dialogue policies. In this case a typical reward function is to give 100 points for a fully successful dialogue minus a penalty (−5) per system turn. We normalized so that the resulting scores of this function range from 0 to 1, and thus they are comparable to the rest of the "Scores" derived from the human ratings and the evaluation functions.

Table 4 shows the derived evaluation functions and the corresponding RMSE values. As we can see, the evaluation functions that include "misunderstandings", and to a lesser extent "system confirmation requests", "system requests for more information", and "conversational style" are all good predictors of the real "Scores" (derived from the human ratings). This is not true for the reward-based evaluation function which results in a much higher RMSE.

Note that we did not find any statistically significant correlation between the "Score" and the "level of specificity" in the system's responses (explicit vs. implicit). As part of our human ratings' data collection process, we asked raters on MTurk to provide qualitative feedback about which features of dialogues they liked or disliked. Some raters consistently mentioned the ability of the system to explicitly state the action that it was about to perform (grounding) and request more information (as a clarification request) as positive dialogue features. However, other people appreciated brevity and preferred more implicit system responses. This means that it is hard to come up with one evaluation function that captures the preferences of all users, and that we may need to develop dialogue quality evaluation functions tailored to specific types of users.

6 Conclusion and Future Work

We analyzed a corpus of system-user dialogues in the IoT domain. Our corpus was automatically, semi-automatically, and manually annotated with a variety of features both on the utterance level and the full dialogue level. The corpus also includes human ratings of dialogue quality collected via crowdsourcing. We calculated correlations between annotated features in our corpus and human ratings, and developed dialogue quality evaluation functions that were shown to be highly predictive of human ratings when tested on a held-out portion of our corpus.

For future work, we plan to develop models that can recreate rankings within a set of dialogues and see whether these derived rankings agree with the actual human rankings in our data set. Furthermore, we would like to collect more realistic system-user dialogues in a Wizard of Oz setting and/or with a real dialogue system. We also plan to develop evaluation functions that are tailored to specific users or groups of users (user modeling). User modeling will also include studying how to make the system-user interaction more engaging, which in turn will facilitate establishing rapport between the system and the user.

Acknowledgements This work was funded by Samsung Electronics Co., Ltd. Some of the authors were partly supported by the U.S. Army Research Laboratory. Any statements or opinions expressed in this material are those of the authors and do not necessarily reflect the policy of the U.S. Government, and no official endorsement should be inferred.

References

1. Artstein R, Gandhe S, Gerten J, Leuski A, Traum D (2009) Semi-formal evaluation of conversational characters. In: Grumberg O, Kaminski, M, Katz, S, Wintner S (eds) Languages: from formal to natural. Essays dedicated to Nissim Francez on the occasion of his 65th birthday. Lecture Notes in Computer Science 5533. Springer, pp 22–35
2. Foster ME, Giuliani M, Knoll A (2009) Comparing objective and subjective measures of usability in a human-robot dialogue system. In: Proceedings of ACL, pp 879–887. Suntec, Singapore
3. Galley M, Brockett C, Sordoni A, Ji Y, Auli M, Quirk C, Mitchell M, Gao J, Dolan B (2015) DeltaBLEU: a discriminative metric for generation tasks with intrinsically diverse targets. In: Proceedings of ACL (short papers), pp 445–450. Beijing, China
4. Gandhe S, Traum D (2008) Evaluation understudy for dialogue coherence models. In: Proceedings of SIGDIAL, pp 172–181. Columbus, Ohio, USA
5. Georgila K, Henderson J, Lemon O (2005) Learning user simulations for information state update dialogue systems. In: Proceedings of Interspeech, pp 893–896. Lisbon, Portugal
6. Georgila K, Henderson J, Lemon O (2006) User simulation for spoken dialogue systems: learning and evaluation. In: Proceedings of Interspeech, pp 1065–1068. Pittsburgh, Pennsylvania, USA
7. Guo F, Metallinou A, Khatri C, Raju A, Venkatesh A, Ram A (2017) Topic-based evaluation for conversational bots. In: Proceedings of NIPS Workshop on Conversational AI: Today's Practice and Tomorrow's Potential. Long Beach, California, USA
8. Hastie H (2012) Metrics and evaluation of spoken dialogue systems. In: Lemon O, Pietquin O (eds) Data-driven methods for adaptive spoken dialogue systems. Springer, pp 131–150
9. Henderson J, Lemon O, Georgila K (2008) Hybrid reinforcement/supervised learning of dialogue policies from fixed datasets. Comput Linguist 34(4):487–511
10. Hone KS, Graham R (2000) Towards a tool for the subjective assessment of speech system interfaces (SASSI). J Nat Lang Eng 6(3–4):287–303
11. Jeon H, Oh HR, Hwang I, Kim J (2016) An intelligent dialogue agent for the IoT home. In: Proceedings of the AAAI Workshop on Artificial Intelligence Applied to Assistive Technologies and Smart Environments, pp 35–40. Phoenix, Arizona, USA
12. Jung S, Lee C, Kim K, Jeong M, Lee GG (2009) Data-driven user simulation for automated evaluation of spoken dialog systems. Comput Speech Lang 23(4):479–509
13. Liu CW, Lowe R, Serban IV, Noseworthy M, Charlin L, Pineau J (2016) How NOT to evaluate your dialogue system: an empirical study of unsupervised evaluation metrics for dialogue response generation. In: Proceedings of EMNLP, pp 2122–2132. Austin, Texas, USA

14. Lowe R, Serban IV, Noseworthy M, Charlin L, Pineau J (2016) On the evaluation of dialogue systems with next utterance classification. In: Proceedings of SIGDIAL, pp 264–269. Los Angeles, California, USA
15. Paksima T, Georgila K, Moore JD (2009) Evaluating the effectiveness of information presentation in a full end-to-end dialogue system. In: Proceedings of SIGDIAL, pp 1–10. London, UK
16. Purandare A, Litman D (2008) Analyzing dialog coherence using transition patterns in lexical and semantic features. In: Proceedings of FLAIRS, pp 195–200. Coconut Grove, Florida, USA
17. Robinson S, Roque A, Traum D (2010) Dialogues in context: an objective user-oriented evaluation approach for virtual human dialogue. In: Proceedings of LREC, pp 64–71. Valletta, Malta
18. Schatzmann J, Georgila K, Young S (2005) Quantitative evaluation of user simulation techniques for spoken dialogue systems. In: Proceedings of SIGDIAL, pp 45–54. Lisbon, Portugal
19. Traum DR, Robinson S, Stephan J (2004) Evaluation of multi-party virtual reality dialogue interaction. In: Proceedings of LREC, pp 1699–1702. Lisbon, Portugal
20. Walker M, Kamm C, Litman D (2000) Towards developing general models of usability with PARADISE. J Nat Lang Eng 6(3–4):363–377

Estimating User Satisfaction Impact in Cities Using Physical Reaction Sensing and Multimodal Dialogue System

Yuki Matsuda, Dmitrii Fedotov, Yuta Takahashi, Yutaka Arakawa, Keiichi Yasumoto and Wolfgang Minker

Abstract Following the increase in use of smart devices, various real-time environmental information becomes available everywhere. To provide more context-aware information, we also need to know emotion and a satisfaction level in a viewpoint of users. In this paper, we define it as "a user satisfaction impact (USI)" and propose a method to estimate USI by combining dialogue features and physical reaction features. As dialogue features, facial expression and acoustic feature are extracted from multimodal dialogue system on a smartphone. As physical reactions, head motion, eye motion, and heartbeat are collected by wearable devices. We conducted the preliminary experiments in the real-world to confirm the feasibility of this study in the

Y. Matsuda (✉) · Y. Takahashi · K. Yasumoto
Graduate School of Science and Technology, Nara Institute of Science and Technology, 8916-5 Takayama-cho, Ikoma City, Nara 630-0101, Japan
e-mail: yukimat@is.naist.jp

Y. Takahashi
e-mail: takahashi.yuta.to2@is.naist.jp

Y. Matsuda
Research Fellow of Japan Society for the Promotion of Science, Tokyo 102-0083, Japan

Y. Matsuda · K. Yasumoto
Center for Advanced Intelligence Project AIP, RIKEN, Tokyo 103-0027, Japan
e-mail: yasumoto@is.naist.jp

D. Fedotov · W. Minker
Institute of Communications Engineering, Ulm University, Albert-Einstein-Allee 43, 89081 Ulm, Germany
e-mail: wolfgang.minker@uni-ulm.de

D. Fedotov
ITMO University, Kronverksky Pr. 49, 197101 St. Petersburg, Russia
e-mail: dmitrii.fedotov@uni-ulm.de

Y. Arakawa
Graduate School and Faculty of Information Science and Electrical Engineering, Kyushu University, 744 Motooka, Nishi-ku, Fukuoka 819-0395, Japan

JST Presto, Tokyo 102-0076, Japan
e-mail: arakawa@ait.kyushu-u.ac.jp

L. F. D'Haro et al. (eds.), *9th International Workshop on Spoken Dialogue System Technology*, Lecture Notes in Electrical Engineering 579,
https://doi.org/10.1007/978-981-13-9443-0_15

tourism domain. Among various features, we confirmed that eye motion correlates with satisfaction level up to 0.36.

1 Introduction

With the spread of smart devices including smartphones and wearable devices, various environmental information becomes available everywhere. To provide more context-aware information, the user status needs to be taken into account as well, since the emotional status or satisfaction level differs across different users and even for the same user during a certain period of time. For example in an urban environment, the situation of "congested" at the event venue can be regarded as "exciting." On the other hand, it is nothing other than a situation of "hindering the passage" on the road. The aim of our research is to combine context with inter- and intra-user status information and to integrate this information in an end to end prototype for intelligent user companion and smart environment.

Actually, there are many related projects that try to estimate emotion/satisfaction level of the people with various methods [1, 7, 8]. However, many of those have restrictions (e.g., data comprehensiveness, accuracy) for applying to the real-world, and did not describe predicting future emotion/satisfaction level.

In this study, we define "a user satisfaction impact (USI)," and propose the method to estimate USI. This method uses physical reaction features (head motion, eye motion, and heartbeat) collected by wearable devices, in addition to dialogue features (facial expressions and acoustic features) extracted from the conversation with multimodal dialogue system on a smartphone. Moreover, it builds the USI model using the urban environmental data simultaneously gathered by sensors embedded in the users' device.

We conducted preliminary experiments in the real-world to confirm the feasibility of this study in the tourism domain. As a result, we found the correlation up to 0.36 between satisfaction level and eye gaze data. Accordingly, we confirmed there is a possibility that physical reaction features can be used for estimating USI. In future research, we will derive new features based on raw features and deploy system for time-continuous estimation of user satisfaction utilizing deep recurrent neural models.

2 Related Work

Resch et al. proposed an emotion collecting system, called "Urban Emotions," for urban planning [8]. The paper describes that wrist-type wearable device and social media were used for emotion measurements. However, this approach relied on assuming that posts on the social media are written in-situ.

An emotion recognition system based on acoustic features via a dialogue system on a mobile device has been proposed in [7]. Actually, the method is based only on the audio features from mobile devices and has not yet achieved a realistic accuracy.

In the tourism domain, the significant part of the research adopts a questionnaire-based survey for measuring the tourist satisfaction [1]. However, methods relying on questionnaires have problems in sustainability and spatial coverage of the survey.

Furthermore, most of the related work did not describe predicting future emotion/satisfaction level. However, we need to make the contents based on not only the estimation but also on the prediction of them.

3 Concept of Estimating User Satisfaction Impact (USI)

We define the emotion/satisfaction level affected by the urban environment as "a user satisfaction impact (USI)." Figure 1 shows the concept of the method for estimating USI. This method yields the following three important steps and challenges:

1. Estimating the USI (emotions, satisfaction level) from users.
2. Building the USI model with urban environmental data and USI.
3. Predicting the future USI using the built model and observed environmental data.

To estimate the USI from users (the first challenge), we focused on the fusion of features including "physical reaction features" in addition to dialogue features (facial expressions and acoustic features) extracted from the conversation with a multimodal dialogue system. Physical reaction features include head motion, eye motion, heartbeat and others that can be implicitly collected by using wearable devices. Collected

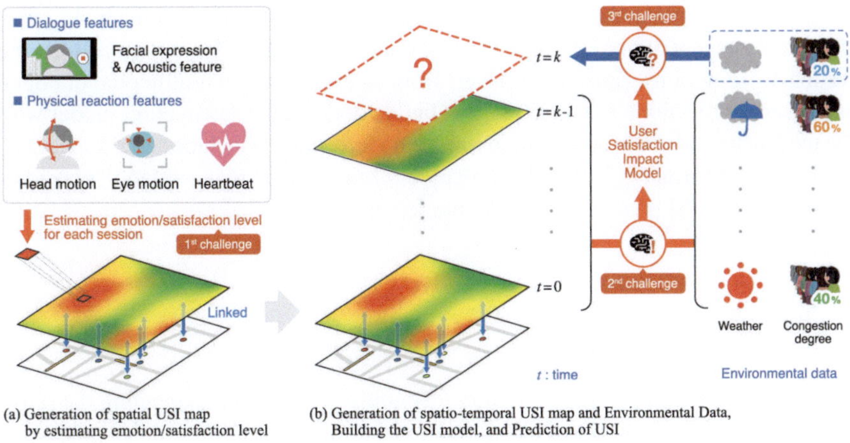

Fig. 1 Concept of user satisfaction impact (USI) estimating method

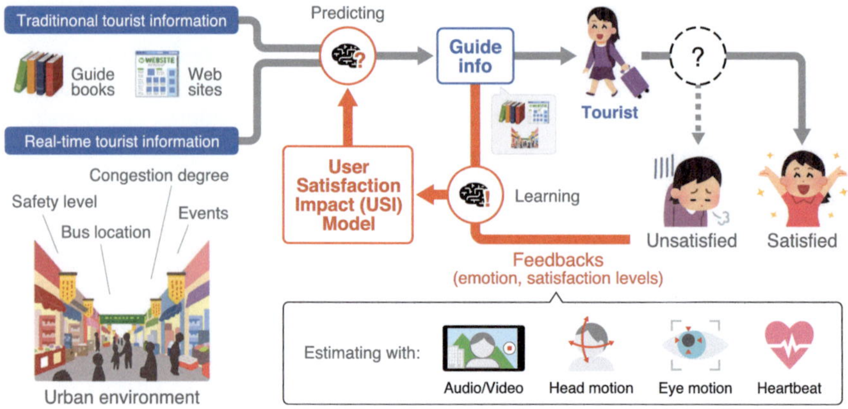

Fig. 2 USI-based tourist guidance system

data then divided into the small periods those include fusion of features, called "session," and USI is estimated for each session. This USI data of the city is generated as the spatial map shown in Fig. 1a. In the second and third challenge (Fig. 1b), it continuously collects the spatial USI map and urban environmental data, and builds the USI model using the collected spatio-temporal data. These urban environmental data can be obtained by participatory sensing approach using sensors embedded in user devices [4]. Finally, it predicts the USI status of next period with USI model using the current urban environmental data as input.

As the use case, we especially focused on the "tourism" area. There is an increasing interest on "smart tourism" domain that utilizes traditional tourist information and real-time tourist information such as congestion degree and event holding situation, thanks to the sensor network, participatory sensing and others [3, 5].

In such kind of use cases, the USI estimating method works effectively. Figure 2 shows the concept of our USI-based tourist guidance system. The system collects the feedbacks (emotions, satisfaction level) from tourists through multimodal dialogue and physical reactions sensing. Then, it builds/updates the satisfaction model with the feedbacks and guide information (the guide information includes traditional/real-time tourist information). Finally, the guide information is regenerated in consideration with the USI predicted by real-time features.

4 Features

Recent research in the field of emotion recognition focused on expanding the range of used modalities from traditional (audio, visual) to new, more complex ones. One of the most popular direction in this area relates to physiological features [2, 9, 10]. They can be separated into several groups: heart-related (electrocardiogram, heartbeat), skin- and blood-related (electro dermal activity, blood-pressure), brain-related (electroencephalography), eye-related (eyes gaze, pupil size) and movement-related (gestures, gyroscopic data). Some of the features are easily covered in the real-life conditions, e.g., heartbeat and skin response can be collected by smart watches and other wearable devices; other can be measured only in the laboratory environment, e.g., electroencephalography; and some of them can be hard to use in real-life scenario at the moment, but it may become much easier in the nearest future, e.g., eye-movement with wider usage of smart glasses.

In the context of our study, we used four devices to record the features in real-time: smartphone Asus Zenfone 3 Max ZC553KL (GPS-data, accelerometer data, gyroscope, magnetic field, short videos from frontal camera and integrated microphone), smart band Xiaomi MiBand 2 (heartbeat), mobile eye tracking headset Pupil with two 120 Hz eye cameras (eyes gaze, pupil features) and sensor board SenStick [6] mounted on an ear of eye tracking device (accelerometer, gyroscope, magnetic field, brightness, UV level, air humidity, temperature, air pressure).

5 Experiment

We conducted preliminary experiments in real-world conditions to confirm the correlation between the data obtained from wearable devices and the user's emotion. Six participants were asked to make a short (approximately 1.5 km) sightseeing tour in the city center of Ulm, Germany. Figure 3 shows the touristic route of this study with satisfaction level and emotion labels. Participants should have visited 8 sightseeing spots and rate each one afterwards using two scales: satisfaction level from

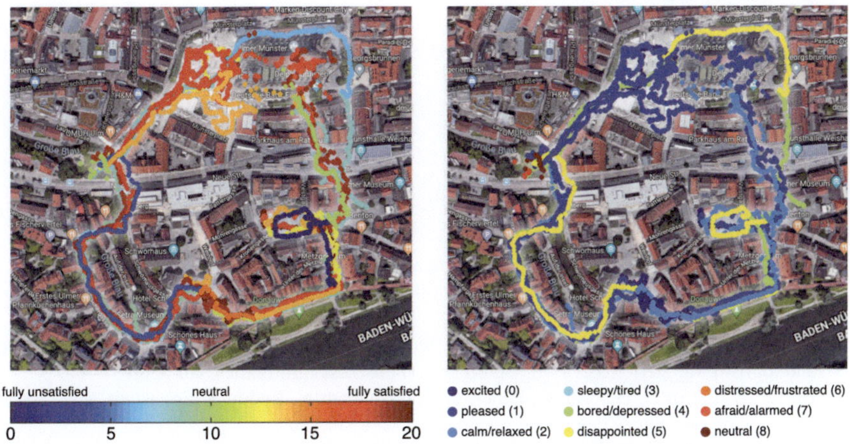

(a) Satisfaction level

(b) Emotion labels

Fig. 3 Maps of tourist satisfaction (**a**) and emotions (**b**) during the city tour

0 (fully unsatisfied) to 20 (fully satisfied) and the most relevant emotion from the following list: excited, pleased, calm/relaxed, sleepy/tired, bored/depressed, disappointed, distressed/frustrated, afraid/alarmed or neutral. They also recorded a short video for each sightseeing spot describing their impression using their native language. Each recording contains 8 sessions and has a duration of about 1 hour. For some participants, one or several sets of features can be missing due to technical problems.

Using raw features we have found correlations between some of the eyes gaze and pupil features and both satisfaction and emotion labels. The most correlated features are: pupil diameter (correlation up to 0.21), projection of pupil sphere (up to 0.26) and eye center in 3-dimensional representation (up to 0.36). Raw features can be used in further research as a basis for deriving new features, e.g., gaze behavior in a context of several seconds can be obtained from raw eyes gaze features. Additionally, raw features can be used in deep recurrent neural model to build a hierarchy of feature maps, whose will be used for further analysis.

6 Conclusion

In this study, we proposed a user satisfaction impact (USI) estimating method based on dialogue features and physical reaction features. A typical use case of such an approach is gathering tourist satisfaction during a city tour. In our preliminary experiments on this topic, we used several sensors and have found a correlation between raw eyes-related features (eyes gaze, pupil size) and tourist satisfaction and emotions. This proves the potential feasibility of building such system. In future research, we will derive new features based on raw features. This will be helpful to build a powerful recognition model.

Acknowledgements This work was supported by JSPS KAKENHI Grant Number 16J09670, 16H01721.

References

1. Alegre J, Garau J (2010) Tourist satisfaction and dissatisfaction. Ann Tour Res 37(1):52–73
2. AlHanai TW, Ghassemi MM (2017) Predicting latent narrative mood using audio and physiologic data. In: AAAI, pp 948–954
3. Balandina E, Balandin S, Koucheryavy Y, Mouromtsev D (2015) Iot use cases in healthcare and tourism. In: 2015 IEEE 17th conference on business informatics, vol 2, pp 37–44
4. Burke JA, Estrin D, Hansen M, Parker A, Ramanathan N, Reddy S, Srivastava MB (2006) Participatory sensing. Center for embedded network sensing
5. Morishita S, Maenaka S, Daichi N, Tamai M, Yasumoto K, Fukukura T, Sato K (2015) Sakurasensor: quasi-realtime cherry-lined roads detection through participatory video sensing by cars. In: Proceedings of the UBICOMP '15, pp 695–705

6. Nakamura Y, Arakawa Y, Kanehira T, Fujiwara M, Yasumoto K (2017) Senstick: comprehensive sensing platform with an ultra tiny all-in-one sensor board for IOT research. J Sens
7. Quck WY, Huang DY, Lin W, Li H, Dong M (2016) Mobile acoustic emotion recognition. In: 2016 IEEE region 10 conference (TENCON), pp 170–174
8. Resch B, Summa A, Sagl G, Zeile P, Exner JP (2014) Urban emotions – geo-semantic emotion extraction from technical sensors, human sensors and crowdsourced data. pp 199–212
9. Ringeval F, Eyben F, Kroupi E, Yuce A, Thiran JP, Ebrahimi T, Lalanne D, Schuller B (2015) Prediction of asynchronous dimensional emotion ratings from audiovisual and physiological data. Pattern Recognit Lett 66:22–30
10. Soleymani M, Pantic M, Pun T (2012) Multimodal emotion recognition in response to videos. IEEE Trans Affect Comput 3(2):211–223

Automated Lexical Analysis of Interviews with Individuals with Schizophrenia

Shihao Xu, Zixu Yang, Debsubhra Chakraborty, Yasir Tahir, Tomasz Maszczyk, Yi Han Victoria Chua, Justin Dauwels, Daniel Thalmann, Nadia Magnenat Thalmann, Bhing-Leet Tan and Jimmy Lee Chee Keong

Abstract Schizophrenia is a chronic mental disorder that contributes to poor function and quality of life. We are aiming to design objective assessment tools of schizophrenia. In earlier work, we investigated non-verbal quantitative cues for this purpose. In this paper, we explore linguistic cues, extracted from interviews with patients with schizophrenia and healthy control subjects, conducted by trained psychologists. Specifically, we analyzed the interviews of 47 patients and 24 healthy age-matched control subjects. We applied automated speech recognition and linguistic tools to capture the linguistic categories of emotional and psychological states. Based on those linguistic categories, we applied a binary classifier to distinguish patients from matched control subjects, leading to a classification accuracy of about 86% (by leave-one-out cross-validation); this result seems to suggest that patients with schizophrenia tend to talk about different topics and use different words. We provided an in-depth discussion of the most salient lexical features, which may provide some insights into the linguistic alterations in patients.

S. Xu (✉) · T. Maszczyk · Y. H. V. Chua · J. Dauwels
School of Electrical and Electronic Engineering, Nanyang Technological University, Singapore, Singapore
e-mail: XUSH0019@e.ntu.edu.sg

D. Chakraborty · Y. Tahir · D. Thalmann · N. M. Thalmann
Institute for Media Innovation, Nanyang Technological University, Singapore, Singapore

Z. Yang · B.-L. Tan · J. L. C. Keong
Institute of Mental Health, Singapore, Singapore

B.-L. Tan
Health and Social Sciences, Singapore Institute of Technology, Singapore, Singapore

J. L. C. Keong
Lee Kong Chian School of Medicine, Nanyang Technological University, Singapore, Singapore

L. F. D'Haro et al. (eds.), *9th International Workshop on Spoken Dialogue System Technology*, Lecture Notes in Electrical Engineering 579,
https://doi.org/10.1007/978-981-13-9443-0_16

1 Introduction

Schizophrenia is a chronic mental disorder affecting millions of people globally. Its symptoms are typically classified into three broad groups: positive (hallucinations and delusions), negative (apathy, blunting of affect, alogia), and cognitive (problems with attention, memory, and executive functions) [1]. The heterogeneous linguistic alterations associated with schizophrenia have been extensively studied since the last century [2, 3]. As words and expressions contain a wealth of information regarding emotions and behavior [4, 5], linguistic analysis is becoming a common tool for research in psychiatry [6, 7]. In recent years, technological advances have made it increasingly faster and more cost-effective to collect a variety of data, a data-driven approach may lead to novel discoveries and treatments in the realm of mental health [8]. In this paper, we follow such a data-driven approach.

A typical lexical based tool for analyzing language is the Linguistic Inquiry and Word Count (LIWC) [9]. The latest version, LIWC 2015 [10], provides several subsets of words representing different emotional states or characteristics. Some studies have applied LIWC to essays written by patients with schizophrenia. A linguistic analysis of written self-descriptive tasks completed by patients with schizophrenia and controls found no differences in the use of words related to emotions (both positive and negative) between patients and control subjects when describing themselves [11]. On the other hand, analysis of standardized written assignments by psychiatric outpatients and control subjects revealed that patients used fewer words pertaining to optimism/energy, basic cognitive mechanisms, exclusion, and bodily functions [12]. In another study by Deutsch-Link et al. [13], essays written by patients with schizophrenia included more external referential language and fewer mentions of the word *I* than control subjects.

LIWC has also been applied to transcripts of interviews with patients with schizophrenia. A study by Minor et al. [14] consisted of analyzing structured interviews with patients with schizophrenia. Interviews were recorded, manually transcribed, and then processed for lexical analysis by LIWC, where *anger* words significantly predicted greater symptoms. Similarly, Hong et al. [15] extracted lexical features to distinguish between patients and controls from the manually transcribed speech of patients and control subjects. In their study, the speech of patients with schizophrenia featured less usage of *I* and *adverb*, and more frequently included words from the categories of *friend* and *relative* instead, when compared to control subjects.

In all above-mentioned studies, the texts were either written by patients or manually transcribed from audio recordings. In our study, we explored the feasibility of applying linguistic analysis to automated transcriptions of interviews with patients and control subjects, conducted by psychologists. The audio recordings of 71 participants (47 patients and 24 control subjects) were automatically transcribed through speech recognition software. We then apply LIWC to the automatically transcribed text to explore linguistic differences between patients and controls. This research is in alignment with our previous works to develop automated, objective methods to examine behavioral deficits in schizophrenia, in which we studied non-verbal cues related to speech [16] and movement [17].

Speech recognition technology has greatly improved in recent years as a result of breakthroughs in the domain of artificial intelligence. Many existing APIs support transcription of microphone streams and audio files directly to text files [18]. Dozens of languages are supported by these platforms, but the performance varies as they employ different language models and machine learning algorithms. Generally, it has been observed that Deep Neural Network (DNN) based tools outperform those based on Gaussian mixture models (GMM) [19].

In our analysis, Google Cloud Speech API was utilized to convert the interview recordings to texts. By means of lexical cues, we were able to classify patients and healthy control subjects at an accuracy of 86% (by leave-one-out cross-validation). We discovered that the speech produced by patients featured fewer *informal* words (e.g., okay, coz, oh) and speech from female patients more frequently contained *female family* words (e.g., sister, mother). Similar to the findings in [15, 20], we also observed that patients were more likely to use *feelings* words (especially feeling related to themselves), and less likely to use adverbs in their speech compared to control subjects.

This paper is organized as follows. In Sect. 2, we describe the experimental design and demographics of the participants. We then elaborate on the steps of our analysis in Sect. 3, and present the numerical results for speech recognition and linguistic analysis in Sect. 4. In Sect. 5, we investigate the most salient linguistic features and offer our concluding remarks in Sect. 6.

2 Experiment Design

This experiment is in collaboration with the Institute of Mental Health Singapore (IMH). 71 individuals participated in this experiment. There are two groups of participants: 47 *Patients* who are diagnosed with schizophrenia, and 24 *Controls*, who do not have any pre-existing mental disorders. Other inclusion criteria of the study include aged 16 to 65, English speaking and fit to provide informed consent. The exclusion criteria include the history of strokes, traumatic brain injuries and neurological disorders such as epilepsy. The participants have provided written informed consent and receive monetary compensation for their participation in the study. The study protocol has been approved by the National Healthcare Group's Domain-specific Review Board in Singapore. The demographic information of participants is displayed in Table 1.

In this experiment, each participant underwent an interview conducted by a psychometrician from IMH. The interview is a semi-structured one where the participants are asked questions follow from the Negative Symptoms Assessment (NSA-16) rating instrument, which is 16-point scale specifically designed to reflect on the emotions and activities in the life of a patient suffering from negative symptoms of schizophrenia. Based on their replies, the psychometrician rates the behavior of the

Table 1 Demographics of participants

		Patients (N = 47)	Healthy Controls (N = 24)
Age	Mean (years)	30.4	29.8
	Range (years)	20–49	19–47
Gender	Male	22	10
	Female	25	14
Ethnicity	Chinese	39	20
	Malay	5	3
	Indian	3	1
Education	University	6	4
	Diploma/JC/ITE	26	14
	High School	15	6

participants on the NSA-16 instrument on a scale of 1–6, where a rating of 1 denotes no recognizable symptoms and a rating of 6 denotes severe symptoms. There is no pre-determined time limit for the interview, nor role-playing during the interview. On average, the interviews last for around 25 min. We analyzed the entire length of the interview recordings. The total length of audio analyzed is about 34 h.

3 System Overview

Figure 1 depicts the different steps in our analysis.

We employed portable equipment to record audio during the interview. Both the psychometrician and patient wore a lapel microphone, and the two audio streams were recorded by an H4n recorder. The two-channel audio recordings were saved in a single .wav file on a laptop. The psychometrician and the participant were seated about 2 meters apart, and this setting minimizes cross-talk from the psychometrician channel onto the participant channel.

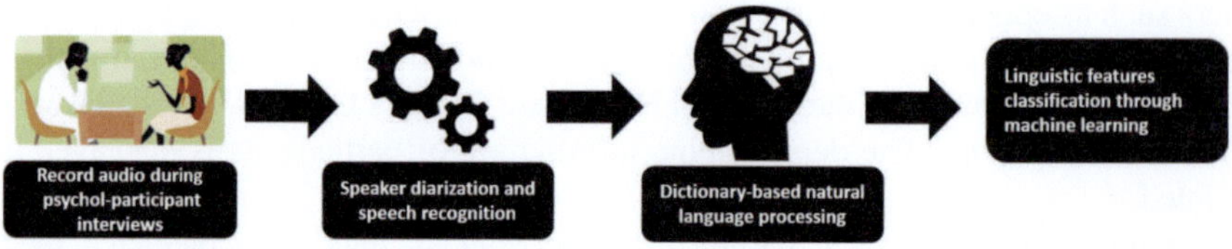

Fig. 1 Diagram of the analysis pipeline

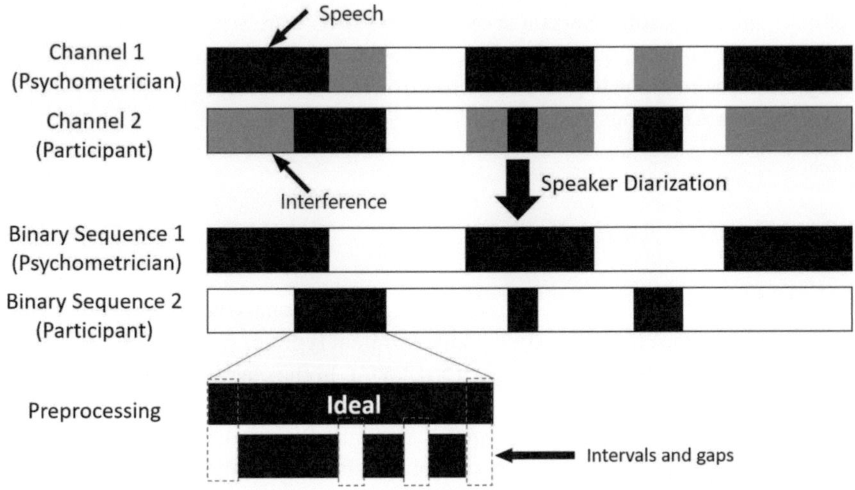

Fig. 2 Illustration of speech preprocessing. We first reduce the interference between both audio channels. Channel 1 and 2 are the original signals, from which we derive binary sequences 1 and 2, indicating when the psychometrician and participant respectively are speaking. Next, we apply one-dimensional erosion and dilation to the binary vectors (shown at the bottom of the figure)

3.1 Speaker Diarization and Speech Recognition

The speech of participant and the psychometrician were recorded on separate channels, as mentioned earlier. Nevertheless, there is still some interference from the psychometrician channel onto the participant channel. We apply automated speaker diarization to remove the psychometrician voice from the participant channel. As illustrate in Fig. 2, we extract binary sequences from both speech signals, indicating when the participant and the psychometrician respectively were speaking (0: not speaking; 1: speaking).

To improve the effectiveness of speech recognition, preprocessing is necessary. In order to obtain cohesive speech segments, we apply one-dimensional erosion and dilation to the binary sequence of the participant. Firstly, we dilate the binary sequence 2 by a one-second structuring element, which filled up small gaps (less than 1s) in one speech segment without filling up the adjacent two sentences. Next, we erode and dilate the binary sequence by a two-second structuring element. These steps reduce noise and incorrect automated transcriptions. At last, we obtain the filtered speech signal, containing mostly speech from the participant, by multiplying the participant audio channel (channel 2) with the binary sequence associated with the participant (sequence 2).

The filtered audio was then subjected to automated speech-to-text API. Currently, many groups and companies provide speech-to-text services, like Google,[1] CMU

[1]https://cloud.google.com/speech/.

Sphinx,[2] and Microsoft.[3] Several speech-to-text APIs were assessed under different audio resources in [21], where Google Cloud Speech API outperformed other speech recognition tools and achieved 9% word error rate. Therefore, we applied Google Cloud Speech API in our analysis.

3.2 Linguistic Analysis

Following speech-to-text conversion, we applied a dictionary-based method of Natural Language Processing. We counted the different types of words or phrases using Linguistic Inquiry and Word Count 2015 (LIWC). In the classification task, the linguistic extracted from the text were used as attributes in several supervised machine learning algorithms to classify the participants into *Patient* or *Control* groups. We used the Weka tool in JAVA [22] to perform classification with leave-one-out cross-validation. We tested the following classifiers: Support Vector Machine (SVM), Multilayer Perceptron (MLP), Logistic Regression (LR), Multinomial Nave Bayes and 1-Nearest Neighbor (1NN) [23]. Because each feature has specific semantic meaning, and not all features may be relevant to our task, feature selection becomes essential during the classification task. We applied three feature ranking methods: SVM-Recursive Feature Elimination (RFE), Relief Attribute Evaluator (Relief), and chi-Squared Attribute Evaluator [24]. For all combinations of the classifier, ranking method, and the number of top features, we selected the top features of the training data and classified the test data for each fold in cross-validation. We present the classification results for the three ranking methods in Sect. 4.2. In Sect. 5, we elaborate on the most salient linguistic features, since they give insight into characteristic behaviors of schizophrenia patients.

4 Results

4.1 Speech Recognition

We measured the performance of Google Cloud Speech API on our dataset. We randomly chose 9 different audio files from our dataset, then extracted the first 5 min of each audio to test the accuracy of speech recognition. We determined the number of correct words in every transcription manually. We calculated the accuracies of translation of different APIs by dividing the number of correctly transcribed words by the number of total words (ground truth). The average transcription accuracy and its standard deviation are listed in Table 2, where we transcribed on average 170

[2]https://cmusphinx.github.io/.

[3]https://azure.microsoft.com/en-us/services/cognitive-services/speech/.

Table 2 Transcription accuracy of google cloud speech API, tested on 9 random recordings in our dataset

Participant ID											
Speech API	1	2	3	4	5	6	7	8	9	Mean	STD
Google cloud	0.87	0.79	0.84	0.84	0.66	0.82	0.78	0.83	0.93	**0.82**	0.07

Table 3 Patient (P) versus controls (C) classification with LIWC features

Classifier	Confusion matrix			Precision	Recall	F-score	AUC	Accuracy (%)	Baseline (%)	Ranking method
		P	C							
SVM	P	45	2	0.85	0.96	0.90	0.81	85.9	66.2	Relief
	C	8	16	0.89	0.67	0.76	0.81			
LR	P	41	6	0.91	0.87	0.89	0.87	85.9	66.2	SVM-RFE
	C	4	20	0.77	0.83	0.80	0.87			
1NN	P	44	3	0.85	0.94	0.89	0.80	84.5	66.2	chi-square
	C	8	16	0.84	0.74	0.74	0.80			

words in each file. These results indicate that Google Cloud speech API an average accuracy of about 82% and a standard deviation of around 7% (total transcribed words per recording: M = 709).

4.2 *Classification*

In this paper, we tested several classifiers and feature ranking methods with the normalized LIWC features as attributes and class labels as targets. We present results for the three best-performing classifiers and their ranking method in Table 3. Both SVM and LR yielded the highest accuracy of 86%, which indicated the schizophrenia patients and controls in our dataset can be well separated by lexical features. We discuss the most salient linguistic features in Sect. 5.

5 Discussion

In Fig. 3, we show how the classification accuracy varies with the number of linguistic features, for the best classifiers obtained by the three feature selection procedures. The SVM-RFE ranking method yields higher classification accuracy than the chi-square ranking method and leads to the same level of accuracy with fewer features compared to relief ranking method. Therefore, we selected the top 20 features by means of the SVM-RFE. Next, we apply the Kruskal-Wallis Test to these 20 linguistic features

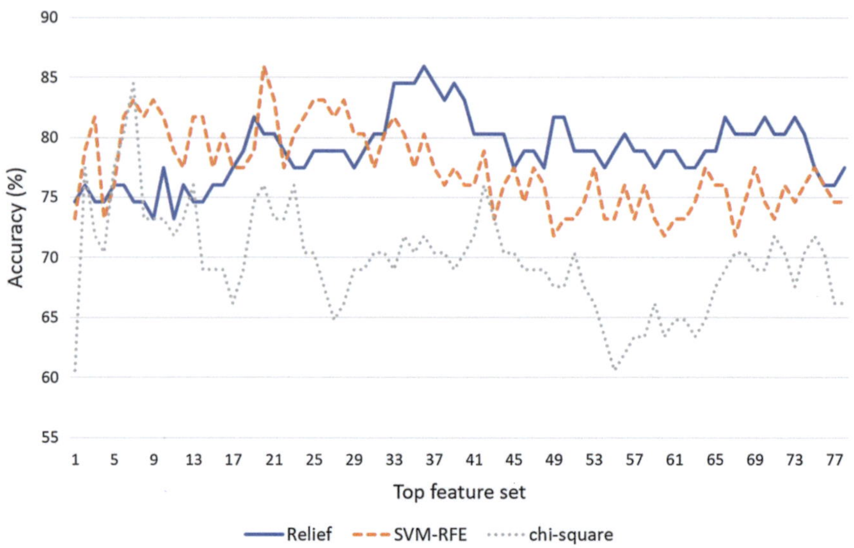

Fig. 3 Classification accuracy of three feature ranking methods as a function of the number of features. The results for the best performing classifiers for each feature selection approach are shown here

Table 4 Word categories showing with different average counts for patients and healthy controls

Words patients used *less*			Words patients used *more*		
Category	Examples	p-value	Category	Examples	p-value
Informal	netspeak, swear words	0.0029	Female family	lady, sister, mom	0.0024
Netspeak	btw, lol, thx	0.0046	Feel	feels, touch, anger	0.0602
Assent	agree, ok, yes	0.0056	Family	mom, father, wife	0.0935
Adverbs	very, really, quickly	0.0090			

and computed the corresponding p-values. The resulting most salient categories and their p-values are shown in Table 4.

We separated the categories into two classes: the type of words patients used *less* and *more* respectively compared to control subjects. The box plots of the four most salient categories are shown in Fig. 4. We found that patients were less likely to use *adverb* words and *informal* words, which include the *assent* category and *netspeak* category. On the other hand, we found that *female family* words and *feel* words appeared more frequently in the speech produced by patients compared to controls.

We computed the correlation coefficient for *I* and *feeling*, and for *family* and *feeling*; we also conducted the Pearson correlation test. The results of this analysis are summarized in Tabel 5. We can see from this table that patients with schizophrenia tend to use both the word I and words related to feelings simultaneously more often

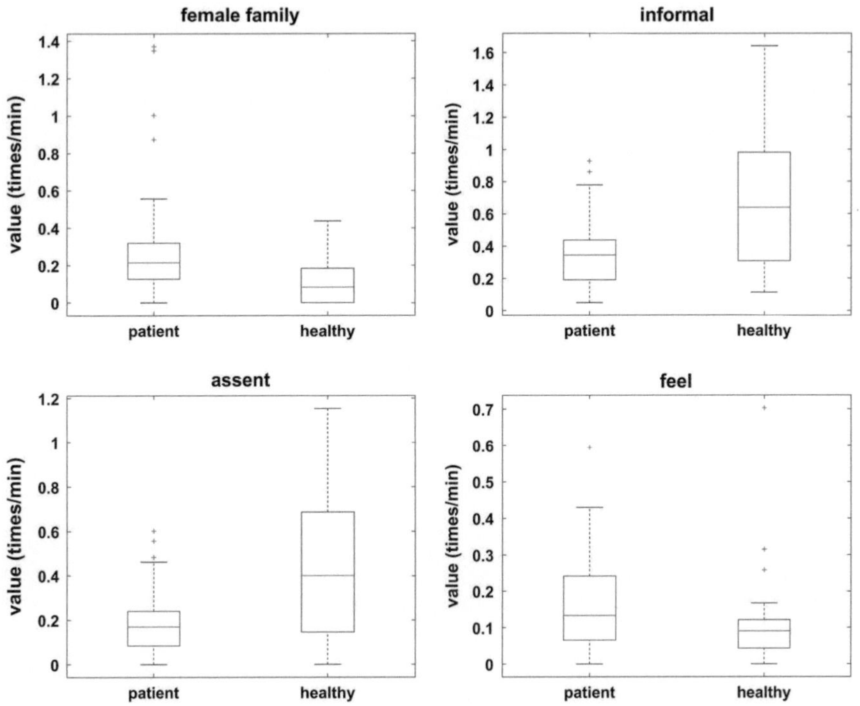

Fig. 4 Box plots of *female family*, *informal*, *assent*, and *feel* categories for patients and healthy controls

Table 5 Correlation coefficients and p-values of the Pearson correlation test for *feeling* versus *I* and *feeling* versus *family*

	Feeling & I		Feeling & Family	
	Correlation	p-value	Correlation	p-value
Patients	0.648	8.66E-07	0.287	0.05
Healthy	0.5	0.0129	0.026	0.905

compared to control subjects. This effect is also present, yet less pronounced, for the simultaneous usage of family and feeling words. These speech pattern results appear to be in line with the research on social cognitive impairs in schizophrenia [25]. Studies have found that due to impairment in social cognition [26, 27], patients tend to focus on themselves, their feelings and thoughts in a conversation, reflecting an inability to perceive, interpret and generate responses to the intentions, dispositions, and behaviors of others [20]. In [15], it was observed that patients were more likely to focus on their own feelings when writing autobiographic narratives as compared to healthy controls.

Table 6 Word frequency of *female family* and *male family* category on patient and healthy people

Female family words (per min)				*Male family* words (per min)			
Patient		Mean	STD	Patient		Mean	STD
	Female	**0.39**	0.37		Female	0.22	0.14
	Male	0.17	0.11		Male	0.15	0.08
Healthy		Mean	STD	Healthy		Mean	STD
	Female	**0.10**	0.13		Female	0.22	0.32
	Male	0.16	0.16		Male	0.21	0.10

Table 7 Patients versus Controls p-values of *female family* words and *male family* words

	Female family words	*Male family* words
Female patients	**0.00055**	0.250
Male patients	0.515	0.046

We also found in our study that the *family* words were more frequently mentioned by patients. It is plausible that for patients, their family members are often the ones they interact with the most within their social circles, outside of therapeutic environments. Thus, the frequent mention of *family* is plausible when referencing themselves in the conversation.

Studies that investigate differences in LIWC categories in patients and healthy controls often make use of the LIWC 2007 version. *Female* and *informal* are a new category in the 2015 version that we applied here. Therefore, it would be hard to make direct comparisons of the results of this study with earlier studies. In our study, we observed that female patients used far more *female family* vocabulary (mean = 0.39 per min) than healthy female controls (mean=0.1 per min) as shown in Table 6. This trend did not show up between male patients and healthy male controls. In Table 7, we further confirm that only female patients used more *female family* words ($p = 0.00055$).

The existing literature suggests that female patients often report higher levels of social support, social functioning, and social withdrawal than male patients [28, 29]. Traditionally, caregiving roles often fall to female members of the family [30], and female caregiving of patients perceive caregiving of their family members with schizophrenia more rewarding as well [31]. Consequently, female patients may have greater access and higher proximity to a social support network of female caretakers (aunt, mother, sister), which could account for the higher occurrence of *female family* words in their speech.

6 Conclusion

In this paper, we analyzed linguistic features extracted from interviews of patients with schizophrenia and healthy control subjects. More precisely, we applied LIWC2015 to provide a dictionary-based words counting method for extracting linguistic features from the audio recordings. We applied several classification and feature ranking methods to distinguish patient and healthy subjects. We obtained an accuracy of 86% for distinguishing patients from healthy control subjects on our dataset of 71 participants. We observed that patients used *informal* and *adverb* words less frequently, but instead used more words correlated to *female family* and *feel*. However, the results are limited by the accuracy of speech recognition, since the accuracy of converting Singapore English to text is significantly lower than for native English (US and UK). Moreover, the sample size (47 patients and 24 control subjects) is relatively small. More research is warranted to further explore and demonstrate the results of this study. In future work, we will also explore how combining non-verbal and verbal features may provide a more comprehensive characterization of patients with schizophrenia.

Acknowledgements This study was funded by the Singapore Ministry of Health National Medical Research Council Center Grant awarded to the Institute of Mental Health Singapore (NMRC/CG /004/2013) and by NITHM grant M4081187.E30. This research is also supported in part by the Being Together Centre, a collaboration between Nanyang Technological University (NTU) Singapore and University of North Carolina (UNC) at Chapel Hill. The Being Together Centre is supported by the National Research Foundation, Prime Minister's Office, Singapore under its International Research Centres in Singapore Funding Initiative. Besides, this project is funded in part by the RRIS Rehabilitation Research Grant RRG2/16009. The authors also acknowledge support from the Interdisciplinary Graduate School at NTU.

References

1. Demily C, Franck N (2008) Cognitive remediation: a promising tool for the treatment of schizophrenia. Expert Rev Neurother 8(7):1029–1036
2. Covington MA, He C, Brown C, Nai L, McClain JT, Fjordbak BS, Semple J, Brown J (2005) Schizophrenia and the structure of language: the linguist's view. Schizophr Res 77(1):85–98
3. Andreasen NC (1979) Thought, language, and communication disorders: I. Clinical assessment, definition of terms, and evaluation of their reliability. Arch Gen Psychiatry 36(12):1315–1321
4. Slatcher RB, Trentacosta CJ (2012) Influences of parent and child negative emotionality on young children's everyday behaviors. Emotion 12(5):932
5. Rude S, Gortner EM, Pennebaker J (2004) Language use of depressed and depression-vulnerable college students. Cogn & Emot 18(8):1121–33
6. Losh M, Gordon PC (2014) Quantifying narrative ability in autism spectrum disorder: a computational linguistic analysis of narrative coherence. J Autism Dev Disord 44(12):3016–25
7. Jackson RG, Patel R, Jayatilleke N, Kolliakou A, Ball M, Gorrell G, Roberts A, Dobson RJ, Stewart R (2017) Natural language processing to extract symptoms of severe mental illness from clinical text: the clinical record interactive search comprehensive data extraction (CRIS-CODE) project. BMJ Open 7(1):e012012
8. Nature Publishing Group (2018) Thinking big in mental health. Nat Med 24(1):1

9. Tausczik YR, Pennebaker JW (2010) The psychological meaning of words: LIWC and computerized text analysis methods. J Lang Soc Psychol 29(1):24–54
10. Pennebaker JW, Boyd RL, Jordan K, Blackburn K (2015) The development and psychometric properties of LIWC2015
11. St-Hilaire A, Cohen AS, Docherty NM (2008) Emotion word use in the conversational speech of schizophrenia patients. Cogn Neuropsychiatry 13(4):343–56
12. Junghaenel DU, Smyth JM, Santner L (2008) Linguistic dimensions of psychopathology: a quantitative analysis. J Soc Clin Psychol 27(1):36–55
13. Deutsch-Link S (2016) Language in schizophrenia: what we can learn from quantitative text analysis
14. Minor KS, Bonfils KA, Luther L, Firmin RL, Kukla M, MacLain VR, Buck B, Lysaker PH, Salyers MP (2015) Lexical analysis in schizophrenia: how emotion and social word use informs our understanding of clinical presentation. J Psychiatr Res 31(64):74–8
15. Hong K, Kohler CG, March ME, Parker AA, Nenkova A (2012) Lexical differences in autobiographical narratives from schizophrenic patients and healthy controls. In: Proceedings of the 2012 Joint conference on empirical methods in natural language processing and computational natural language learning. Association for Computational Linguistics pp 37-47
16. Tahir Y, Chakraborty D, Dauwels J, Thalmann N, Thalmann D, Lee J (2016) Non-verbal speech analysis of interviews with schizophrenic patients. In: 2016 IEEE international conference on acoustics, speech and signal processing (ICASSP) 2016 Mar 20. IEEE pp 5810–5814
17. Chakraborty D, Tahir Y, Yang Z, Maszczyk T, Dauwels J, Thalmann D, Thalmann NM, Tan BL, Lee J (2017) Assessment and prediction of negative symptoms of schizophrenia from RGB+ D movement signals. In: 2017 IEEE 19th international workshop on multimedia signal processing (MMSP). IEEE pp 1-6
18. Lange P, Suendermann-Oeft D (2014) Tuning sphinx to outperform googles speech recognition API. In: Proceedings of the ESSV 2014, conference on electronic speech signal processing, pp 1-10
19. Hinton G, Deng L, Yu D, Dahl GE, Mohamed AR, Jaitly N, Senior A, Vanhoucke V, Nguyen P, Sainath TN, Kingsbury B (2012) Deep neural networks for acoustic modeling in speech recognition: The shared views of four research groups. IEEE Signal Process Mag 29(6):82–97
20. Green MF, Penn DL, Bentall R, Carpenter WT, Gaebel W, Gur RC, Kring AM, Park S, Silverstein SM, Heinssen R (2008) Social cognition in schizophrenia: an NIMH workshop on definitions, assessment, and research opportunities. Schizophr Bull 34(6):1211–20
21. Kpuska V, Bohouta G (2017) Comparing speech recognition systems (Microsoft API, Google API and CMU Sphinx). Int J Eng Res Appl 7:20–24
22. Witten IH, Frank E, Trigg LE, Hall MA, Holmes G, Cunningham SJ (1999) Weka: practical machine learning tools and techniques with Java implementations
23. Hall M, Frank E, Holmes G, Pfahringer B, Reutemann P, Witten IH (2009) The WEKA data mining software: an update. ACM SIGKDD Explor Newsl 11(1):10–8
24. Frank E, Hall M, Holmes G, Kirkby R, Pfahringer B, Witten IH, Trigg L (2009) Weka-a machine learning workbench for data mining. In: data mining and knowledge discovery handbook 2009. Springer, Boston, MA pp 1269–1277
25. Howes C, Purver M, McCabe R (2013) Using conversation topics for predicting therapy outcomes in schizophrenia. Biomed Inform Insights 6(Suppl 1):39
26. Penn DL, Corrigan PW, Bentall RP, Racenstein J, Newman L (1997) Social cognition in schizophrenia. Psychol Bull 121(1):114
27. Savla GN, Vella L, Armstrong CC, Penn DL, Twamley EW (2012) Deficits in domains of social cognition in schizophrenia: a meta-analysis of the empirical evidence. Schizophr Bull 39(5):979–92
28. Hfner H (2003) Gender differences in schizophrenia. Psychoneuroendocrinology 30(28):17–54
29. Ochoa S, Usall J, Cobo J, Labad X, Kulkarni J (2012) Gender differences in schizophrenia and first-episode psychosis: a comprehensive literature review. Schizophr Res Treat 8:2012
30. Diekman AB, Goodfriend W (2006) Rolling with the changes: A role congruity perspective on gender norms. Psychol Women Q 30(4):369–83

31. Chen FP, Greenberg JS (2004) A positive aspect of caregiving: The influence of social support on caregiving gains for family members of relatives with schizophrenia. Community Ment Health J 40(5):423–35

Impact of Deception Information on Negotiation Dialog Management: A Case Study on Doctor-Patient Conversations

Nguyen The Tung, Koichiro Yoshino, Sakriani Sakti and Satoshi Nakamura

Abstract Almost all of existing negotiation systems assume that their interlocutors (the user) are telling the truth. However, in negotiations, participants can tell lies to earn a profit. In this research, we proposed a negotiation dialog management system that detects user's lies and designed a dialog behavior on how should the system react with. As a typical case, we built a dialog model of doctor-patient conversation on living habits domain. We showed that we can use partially observable Markov decision process (POMDP) to model this conversation and use reinforcement learning to train the system's policy.

1 Introduction

Recently, the focus of studies about dialog system have switched from passive role systems (e.g. restaurant, tourist information provider) to become more actively systems that can influence the user's decision (persuasive technology) [8, 10]. Most of current researches mainly deal with cooperative dialog, in which the system and the user work together to reach a balanced point. However, there are situations that the mutual goal cannot be reached (non-cooperative dialog) [3, 11]. There are a number of works on non-cooperative negotiation using trading game scenario [3, 12]. A drawback of these studies is that they do not cover the situation in which participants use "false promises" tactics. For example, someone pretends to give up unhealthy

N. The Tung (✉) · K. Yoshino · S. Sakti · S. Nakamura
Nara Institute of Science and Technology, Takayama-cho, Ikoma, Nara 6300192, Japan
e-mail: nguyen.tung.np5@is.naist.jp

K. Yoshino
e-mail: koichiro@is.naist.jp

S. Sakti
e-mail: ssakti@is.naist.jp

S. Nakamura
e-mail: s-nakamura@is.naist.jp

L. F. D'Haro et al. (eds.), *9th International Workshop on Spoken Dialogue System Technology*, Lecture Notes in Electrical Engineering 579,
https://doi.org/10.1007/978-981-13-9443-0_17

custom (e.g. smoking/drinking) after receiving advices from friends or family but he actually does not have intention to follow the advice.

In this paper, we focus on a typical and interesting type of negotiation, the doctor-patient conversation where they discuss to find the best treatment for the patient. In this dialog, the patient has their own perspective and opinions about their health condition. The doctor needs to consider these opinions when making a treatment plan (recommend a new habit) and negotiates with users to reach a plan that satisfy both user's demand and requirements of the treatment. Patients sometimes tell lies because they do not want to change their habit, thus considering deceptive information will improve the negotiation strategy of the doctor.

2 Scenario and Modeling

2.1 Dialog Scenario

This work considers a dialog scenario between a system (doctor) and a user (patient). They discuss about user's living habit, which can be about: sleeping, food, working/studying, exercise, social media usage and leisure activities. The system tries to convince the user that they need to change to a more healthy living habit. The system persuades the user by giving them information about the new habit (system's recommendation), health benefits of the new habit and negative effects of user's current habit. This action is denoted as Framing in this research. On the other hand, the user wants to continue the current habit and gives reasons to show that it is too difficult for him to change. The system behaves cooperatively; if user's reason is honest, system will give an easier recommendation.

To make the conversation simpler, only the system can propose recommendations, the user cannot suggest what habit they should change to. However, the users are allowed to use dishonest reasons to make the system offer an easier recommendation. The user can also pretend to accept the system's offer while they actually do not intend to change their current habit.

Figure 1 describes the proposed dialog behavior, which considers deceptions of the user. In this flowchart, rectangles indicate system actions. The set of dialog acts for the system includes:

Offer: the system suggests the user should change to a new habit.
Framing: the system provides arguments to persuade the user.
End: the system ends the conversation.

Similar to the work in [8], we also use framing as one of system's dialog acts. The user can react with different actions as described below:

Accept: the user agrees to change habit.
Reject: the user gives reason why they cannot change their habit.

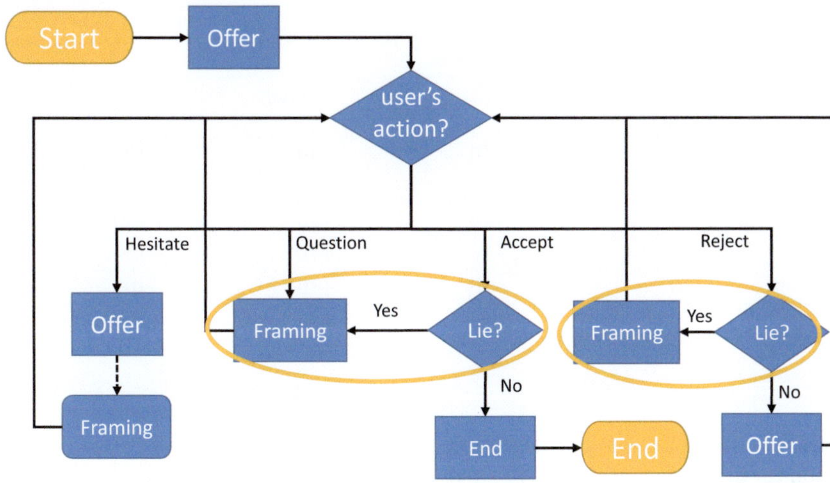

Fig. 1 Flow chart of the conversation

Hesitate: the user says he/she is unsure about whether to accept the offer or not.
Question: the user asks the system about the new habit.

According to a study by [6], when the patient is lying, the doctor should tell the patient about the necessity and benefits of the treatment plan and consequences if patient refuses to follow it. Applying to the "living habit" scenario, the most logical reaction when the user is telling lie (uses fake reasons or pretend to agree) is Framing. In contrast, a conventional negotiation system that does not consider user's deception always offer a new recommendation when the user rejects and end the conversation when the user agrees regardless whether they are telling the truth or not.

2.2 *Policy Management Using POMDP*

To find the best strategy of dialog system against the user deception, considering errors of dialog act classification and deception detection for user utterance is necessary, because these models do not have 100% accuracy. Partially observable Markov decision process (POMDP) is widely used to learn the best strategy of dialog systems for such error containing cases [14].

The equation for updating belief state of a POMDP can be written as:

$$b^{t+1} \propto P(o^{t+1}|s^{t+1}) \sum_{s_i} P(s^{t+1}|s^t, \hat{a}^t) b^t \tag{1}$$

Table 1 Rewards in each turn

Dialog state		Rewards		
User DA (s)	User deception (d)	Offer	Framing	End
Accept	0	−10	−10	+100
	1	−10	+10	−100
Reject	0	+10	+10	−100
	1	−10	+10	−100
Question	0	−10	+10	−100
Hesitate	0	+10	+10	−100

Apart from user's action s^t, the proposed dialog system also uses deception information for dialog management. To solve this problem, we used method in a similar work for user focus by [14]. By extending Eq. (1) with deception information of the current turn d^t and next turn d^{t+1}, we have the belief update of the proposed system:

$$b_{s,d}^{t+1} \propto \quad P(o_s^{t+1}, o_d^{t+1}|s^{t+1}, d^{t+1}) \sum_s \sum_d P(s^{t+1}, d^{t+1}|s^t, d^t, \hat{a}^t) b_{s,d}^t \tag{2}$$

With the observation result come from SLU and Deception Detection modules being denoted as o_s and o_d respectively.

In this research, we use Q-learning [13], a popular method to train the optimal policy π^*. We utilize Grid-based Value Iteration method proposed by [2]. The belief is calculated by:

$$b_{s_i} = \begin{cases} \mu & \text{if } s = o \\ \frac{1-\mu}{|S|-1} & \text{otherwise} \end{cases} \tag{3}$$

μ represent the rounded probability for every 0.1 that the observation comes from Spoken Language Understanding and Deception Detection equal to actual user's dialog act and deception information. Table 1 show the reward received for each turn.

2.3 Deception Detection Using Multi-modal Approach

There are various clues that can help us to detect lies, which include lexical modal, acoustic modal, gestures and facial expressions. As a result, multi-modal approach that combines those modalities was proved to be very efficient in detecting deception [7]. Thus, we also utilized multi-modal approach with acoustic and facial features to build the deception detection module.

To extract facial features, we used OpenFace toolkit developed by [1]. All features provided by this are used for deception detection. From this tool, we were able to

Table 2 Deception detection accuracy

Models	Audio (%)	Video (%)	Features combine (%)	Decision combine (%)	Hierarchical (%)
Accuracy	55.89	52.95	61.75	58.83	64.71
F-measure	51.61	42.85	55.17	56.25	60.00

extract 14 face AU (Action Unit) regression and 6 AU classification values as well as head position, head direction parameters. All these values are then normalized and discretized into 5 different levels of intensity.

Acoustic features are extracted from audio files using OpenSMILE tool [4]. The acoustic feature template was taken from the work by [5]. From the pitch and loudness values, we calculated maximum (max), minimum (min), mean ($mean$) and standard deviation (std). The duration-related features include: percentage of frame with voice, percentage of frame with lower pitch than previous frame (falling pitch), percentage of frame with higher pitch than previous frame (raising pitch).

The classification model that we used for deception detection is Multi Layer Perceptron with hierarchical structure to combine acoustic and facial features. This method was proposed by [9] for emotion detection task. In particular, for our research, we put facial features directly into the first layer of the network while acoustic features are incorporated at the second hidden layer.

To test the effectiveness of the deception detection module, we used data from recorded conversations between 2 participants. One of them plays the role of a doctor and the other one's role is a patient. In this conversation, the "patient" will try to get a prescription from the "doctor" by telling lies about his health condition. Deception label for each utterance was manually annotated by themselves. The total number of utterances in this data is 146. We took 34 of them (17 honest, 17 deceptive) to use as test data. Results of the experiment are shown in Table 2.

3 Experiments of Dialog Management System

3.1 Data

Due to the difficulties of recruiting actual doctors and patients to collect the data, the participants who took part in data collection are students who have good level of English fluency. All participants are working at the same academic environment at the time of data collection. There are total of 7 participants who take part in the recordings. 4 of them played the role of the system and 6 played the role of the user.

The data used for training are recorded using the "living habits" dialog scenario in Wizard-of-Oz (WoZ) setup. Each recording session is carried out by 2 participants who play the role of the system and the user respectively. Each session consists of 6

Table 3 Deception and dialog acts statistic

Data	System DA			User DA				% lie in user's
	End (%)	Framing (%)	Offer (%)	Hesitate (%)	Question (%)	Accept (%)	Reject (%)	Utterances (%)
Train	14.43	43.30	42.27	21.69	3.61	51.81	22.89	18.07
Test	17.54	36.26	46.20	17.64	9.86	19.72	52.82	19.01

dialogs for each of the living habit topic. The participants who play the role of the user (patient) are given payment as reward for the outcome of the conversation. If they pretend to agree with the system's offer, they will receive lower payment. On the other hand, if they choose to truly agree with the system's offer they will get higher payment with a condition that they will need to change to the new habit for one week. The payment is to create the situation where the user has to choose between an easy activity (continue current habit) with low reward and an activity that is difficult but has higher reward in return (change to new habit) to observe more lies. In the end, the recorded training data are about 3 h 20 min long and contains 29 dialogs with average of 5.72 turns per dialog. Labels of DA are annotated by one expert and labels of deception are provided by the participant who made the deception.

With test corpus, recordings are done as direct conversations between participants. The recording setup is similar to WoZ scenario but without the help of TTS since the participants are now talking directly with each other. The test data set is about 2 h 35 min in length and contains 30 dialogs with average of 4.73 turns per dialog. Table 3 shows statistics of the collected data.

3.2 Results

First, we test the negotiation efficiency of the learned policy by comparing it with a baseline negotiation policy that does not concern about user's deception. We let the system interacted with a simulated user created from test data for 100,000 dialogs. The simulator is created with the same method as described in [14]. In particular, user's dialog acts and deceptions are generated using **intention model** and **deception model** as below:

$$\begin{aligned} \textbf{intention model}: P(s^{t+1}|d^{t+1}, s^t, d^t, \hat{a}^t) \\ \textbf{deception model}: P(d^{t+1}|s^t, d^t, \hat{a}^t) \end{aligned} \tag{4}$$

For evaluation, these probabilities are calculated from test data using maximum likelihood.

Performance is evaluated with success rate and average offer per succeeded dialog. Success rate is the percentage of dialogs in which user truly accepts the system's offer.

Table 4 Success rate and average offer per succeeded dialogs

Dialogue policy	Success rate (%)	Avg. offer
Conventional	21.83	2.472
Proposed policy	**29.82**	2.447

Table 5 Dialog selection acts accuracy

Dialog system	DA accuracy (%)	Deception handling (%)
Conventional	68.15	35.00
Proposed policy + gold-label deception	**80.45**	**80.00**
Proposed policy + predicted deception	**79.32**	**55.00**

For the second evaluation metric, average offer; since every time the system makes an Offer action, the new habit will be easier but gives less health benefit, thus it is less favorable for the system. Therefore, using fewer offers to successfully persuade the user is better. From the results shown in Table 4, it is clear that our proposed system outperforms the baseline.

Next, we evaluate the performance of the system's dialog acts decision. The accuracy are measured with 2 metrics. DA accuracy refers to the accuracy of the system's chosen dialog acts against reference actions that were chosen by human. Deception handling indicates the accuracy of dialog acts decision when user is lying. From the results of Table 5, we can see that our proposed system outperformed the baseline again. Especially, for deception handling, the proposed system achieved higher accuracy in both situations: using gold-label and using predicted results.

4 Conclusions

In this paper, we present a negotiation strategy that counter deception in doctor-patient conversation. We also propose a dialog system that utilize this strategy and showed the construction and performance of each module. Experiment results proved that the proposed strategy outperformed normal negotiation strategy significantly, beating it by more than 8% in chance of successful persuasion. DA accuracy experiments indicated that the learned policy achieved a good level of naturalness when compared to human behavior in the same scenario. In future, we would like to conduct experiments of the system with human.

Acknowledgements Part of this work was supported by JSPS KAKENHI Grant Number JP17H06101.

References

1. Baltrušaitis T, Robinson P, Morency L-P (2016) Openface: an open source facial behavior analysis toolkit. In: 2016 IEEE winter conference on applications of computer vision (WACV). IEEE, pp 1–10
2. Bonet B (2002) An e-optimal grid-based algorithm for partially observable markov decision processes. In: Proceedings of the 19th international conference on machine learning (ICML-02)
3. Efstathiou I, Lemon O (2014) Learning non-cooperative dialogue behaviours. In: Proceedings of the 15th annual meeting of the special interest group on discourse and dialogue (SIGDIAL). Association for Computational Linguistics, Philadelphia, PA, USA, pp 60–68. http://www.aclweb.org/anthology/W14-4308
4. Eyben F, Wollmer M, Schuller B (2010) Opensmile: the munich versatile and fast open-source audio feature extractor. In: Proceedings of the 18th ACM international conference on Multimedia. ACM, New York, pp 1459–1462
5. Hirschberg J, Benus S, Brenier JM, Enos F, Friedman S, Gilman S, Girand C, Graciarena M, Kathol A, Michaelis L et al (2005) Distinguishing deceptive from non-deceptive speech. In: Interspeech, pp 1833–1836
6. Kjellgren KI, Ahlner J, Säljö R (1995) Taking antihypertensive medication controlling or co-operating with patients? Int J Cardiol 47(3):257–268
7. Pérez-Rosas V, Abouelenien M, Mihalcea R, Burzo M (2015) Deception detection using real-life trial data. In: Proceedings of the 2015 ACM on international conference on multimodal interaction. ACM, New York, pp 59–66
8. Takuya H, Graham N, Sakriani S, Tomoki T, Satoshi N (2014) Reinforcement learning of cooperative persuasive dialogue policies using framing. In: COLING, pp 1706–1717
9. Tian L, Moore J, Lai C (2016) Recognizing emotions in spoken dialogue with hierarchically fused acoustic and lexical features. In: 2016 IEEE spoken language technology workshop (SLT). IEEE, pp 565–572
10. Torning K, Oinas-Kukkonen H (2009) Persuasive system design: state of the art and future directions. In: Proceedings of the 4th international conference on persuasive technology. ACM, New York, p 30
11. Traum DR (2008) Computational models of noncooperative dialogue
12. Vourliotakis A, Efstathiou I, Rieser V (2014) Detecting deception in noncooperative dialogue: a smarter adversary cannot be fooled that easily. In: Proceedings of the 18th workshop on the semantics and pragmatics of dialogue (SemDIAL). Edinburgh, Scotland, UK, pp 252–254
13. Watkins CJCH, Dayan P (1992) Q-learning. Mach Learn 8(3–4):279–292
14. Yoshino K, Kawahara T (2015) Conversational system for information navigation based on pomdp with user focus tracking. Comput Speech Lang 34(1):275–291

Part IV
End-to-End Systems

An End-to-End Goal-Oriented Dialog System with a Generative Natural Language Response Generation

Stefan Constantin, Jan Niehues and Alex Waibel

Abstract Recently advancements in deep learning allowed the development of end-to-end trained goal-oriented dialog systems. Although these systems already achieve good performance, some simplifications limit their usage in real-life scenarios. In this work, we address two of these limitations: ignoring positional information and a fixed number of possible response candidates. We propose to use positional encodings in the input to model the word order of the user utterances. Furthermore, by using a feedforward neural network, we are able to generate the output word by word and are no longer restricted to a fixed number of possible response candidates. Using the positional encoding, we were able to achieve better accuracies in the Dialog bAbI Tasks and using the feedforward neural network for generating the response, we were able to save computation time and space consumption.

1 Introduction

A goal-oriented dialog system should fulfill the user's request. If there are ambiguities or missing information in the request, the system should ask for a clarification or for the missing information. In such a dialog system, there can be a priori knowledge about the domain. However, dialog systems without any a priori knowledge about the domain and which are only trained with a training dataset make for a very interesting research topic. They allow extending or changing the domain only by training them with an extended or a changed training dataset, respectively. Recently advancements in deep learning allow the development of such end-to-end trained goal-oriented

S. Constantin (✉) · J. Niehues · A. Waibel
Karlsruhe Institute of Technology, Institute for Anthropomatics and Robotics, Karlsruhe, Germany
e-mail: stefan.constantin@kit.edu

J. Niehues
e-mail: jan.niehues@kit.edu

A. Waibel
e-mail: waibel@kit.edu

L. F. D'Haro et al. (eds.), *9th International Workshop on Spoken Dialogue System Technology*, Lecture Notes in Electrical Engineering 579,
https://doi.org/10.1007/978-981-13-9443-0_18

dialog systems. No human written rules are necessary. Bordes et al. [1] describe an end-to-end goal-oriented dialog system without a priori knowledge based on Memory Networks, which achieved good results.

This dialog system ignores the position of the words in an utterance. That means, the encoded utterance of the utterance "I'm in the office and not in the production hall" is equal to the encoded utterance of the utterance "I'm in the production hall and not in the office". This work presents an encoding that encodes the position of the words of an utterance.

The dialog system described in [1] generates a response by predicting the best response from a fixed set of response candidates. This approach has some performance drawbacks. The computation time and the space consumption increase linearly with the number of response candidates. This work presents an approach that eliminates these drawbacks by generating a response word by word.

Furthermore, this work investigates how sensitive the dialog system to user input that differ from the fictional user input in the training dataset is.

2 Related Work

In [1], an end-to-end goal-oriented dialog system based on Memory Networks with good results is presented. Memory Networks are described in general in [12] and the end-to-end trainable Memory Network variant is described in [9].

To train a dialog system without a priori knowledge for a domain, a good training dataset is needed. In [8], more than 50 datasets are presented. Only four datasets are presented in [2], but the information on which skill of a dialog system the dataset is testing is added to every dataset. Datasets can be created by humans in a fashion in which how they would be using the system or synthetically from patterns. Synthetic datasets often do not have so many variances than real dialogs, but they are easier to create. In [11], the intention for synthetic datasets is given. If a dialog system cannot handle synthetic datasets, it will not be able to handle real datasets. The opposite, in general, is false.

For the Natural Language Understanding (NLU) component and the Dialog Manager (DM) component there are also other appropriate neural network architectures besides Memory Networks: the Recurrent Entity Networks [5], the Differentiable Neural Computer [4], the Neural Turing Machine [3], and the Dynamic Memory Networks [7].

This work uses a feedforward neural network (FNN) for Natural Language Generation (NLG). A promising approach for an end-to-end trainable NLG is the semantically controlled long short-term memory (SC-LSTM), which is presented in [10]. It is a further development of the long short-term memory (LSTM).

3 Dialog System

The structure of the dialog system is based on [1]. Section 3.1 briefly describes the structure of the dialog system described in [1]. All the equations are from [1]. Section 3.2 presents the improvements: position encoding and temporal encoding. The NLG word by word approach is presented in Sect. 3.3.

3.1 Foundation

A dialog D is split into subdialogs D_i. The first subdialog D_1 includes the first utterance of the user c_1^u and the first utterance of the system c_1^r. Every subdialog D_i includes its predecessor D_{i-1} and the i-th utterance of the user and of the system. The subdialogs are split into three parts: the last utterance of the user c_i^u, the last utterance of the system c_i^r and the other utterances compose the history.

Each utterance of the history is encoded in a V-dimensional vector. V is the size of the vocabulary. Every element of the vector corresponds to a different word in the vocabulary. An element has the value 1 if the corresponding word occurs in the utterance; otherwise it has the value 0. This encoding is called bag-of-words (BOW) encoding. The mapping from an utterance to such a V-dimensional vector is called Φ.

The vectors are multiplied by a $d \times V$-dimensional matrix A and saved as d-dimensional utterance-embeddings in the memory:

$$m = (A\Phi(c_1^u), A\Phi(c_1^r)..., A\Phi(c_{t-1}^u), A\Phi(c_{t-1}^r))$$

In the same way, the last utterance of the user c_i^u is calculated to an utterance-embedding q.

The relevance p_i of a memory entry m_i to q is calculated in the following manner:

$$p_i = Softmax(q^T m_i)$$

Every memory entry is multiplied by its relevance and then these results are summed up. The sum is multiplied by a matrix R and this result is the vector o:

$$o = R \sum_i p_i m_i$$

The output of the Memory Network is the sum of q and o.

The described Memory Networks can be stacked together. The sum of q_h and o_h is used as the new input q_{h+1} for the next Memory Network in the stack. The number of the Memory Networks in the stack is called N and the output of the last Memory Network is q_{N+1}.

The temporal order of the utterances is directly encoded in the conversation. Therefore, t keywords are defined, which are only allowed to be used for the temporal encoding. The keywords are added to the vocabulary. To every utterance, the keyword that encodes how many utterances were made before this utterance is added. The keyword for the t-th dialog before this utterance is used for the utterances before as well.

For the NLG, a candidates' selection approach is used. The result of the multiplication of the output of the Memory Network q_{N+1} by a trainable matrix W is multiplied by all the C encoded response candidates y_i:

$$\hat{a} = Softmax({q_{N+1}}^T W\Phi(y_1), ..., {q_{N+1}}^T W\Phi(y_C))$$

The result $\hat{a}$ is a C-dimensional vector. The corresponding candidate to the element of $\hat{a}$ that has the highest value is the predicted response.

3.2 Position and Temporal Encoding

To retain the order of the words in the utterances, a position encoding is described in [9]. A memory entry m_i is calculated with this encoding in the following manner:

$$m_i = \sum_j l_j \cdot Ax_{ij}, \; l_{kj} = (1 - j/J) - (k/d)(1 - 2j/J)$$

J is the maximum utterances length, j the index of the regarded word, k the index of the regarded element of the embedding vector, and x_{ij} is the j-th word of the i-th utterance.

Sukhbaatar et al. [9] present an alternative to the temporal encoding with keywords. A trainable matrix T_A is used to calculate the utterance-embeddings of the memory entries. The calculation of m_i with the position encoding and the temporal encoding is:

$$m_i = \sum_j l_j \cdot Ax_{ij} + T_A(i)$$

$T_A(i)$ is the i-th row of T_A. In this paragraph, the number of rows of T_A is called t. With the temporal encoding only dialogs with a maximum length of t are supported. This alternative approach is not used for the dialog system in this work, because the dialog system in this work should not have this constraint. A workaround for this constraint is to use T_A for the last t utterances of the history and $T_A(0)$ for all the utterances before or to remove old utterances from the history.

3.3 NLG Word by Word Approach

For every response prediction the result of the multiplication output of the Memory Network by the matrix W must be multiplied by every encoded response candidate. Therefore, the computation time of the NLG is linear in the number of the response candidates. Since all response candidates must be saved, the space consumption is also linear in the number of response candidates. To eliminate these drawbacks, a word by word approach is developed for this work. This approach uses an FNN.

The input for the FNN in timestep t is the output of the Memory Network q_{N+1} and the word-embeddings of the last m words $w_t, w_{t-1}, ..., w_{t-m+1}$. For words with an index smaller than 1, the word-embedding of a keyword that must only be employed for this case is used. The variable m is a hyperparamter. The size of m is especially relevant in the utterances in which words occur multiple times. In the case where m has the size 1, there is, for example, a problem in the utterance "Do you prefer a water or a cola?". The FNN after the output of one of the two "a" in both cases has the same input. The output of the Memory Network is constant and the last outputted word is "a". This leads to a result in which at least one output of the FNN is false. Therefore, m should at least have the size of the longest repeating n-gram that can occur in a possible response.

The input of the FNN is fully connected with a hidden layer. The number of neurons is variable. The output layer has as many neurons as there are different words in the vocabulary. Every word is clearly allocated to a neuron. The word that belongs to the neuron with the highest value is the word that is outputted. The output layer is fully connected with the hidden layer and every neuron of the output layer has a bias. The softmax function is applied to the output values. To the outputted word w_{t+1}, a word-embedding w'_{t+1} is calculated. This word-embedding is used in the next time step $t + 1$ as the input for the FNN. The described structure of the FNN is depicted in Fig. 1.

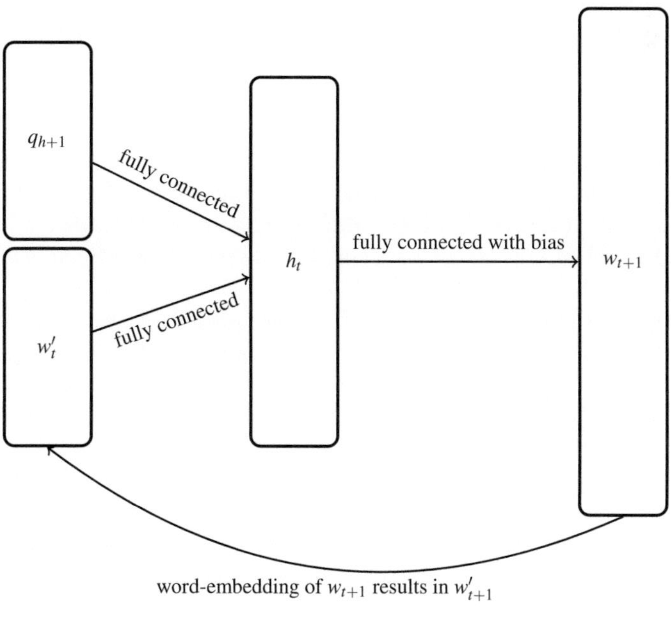

Fig. 1 NLG word by word approach with an FNN ($m = 1$)

The weights of the FNN are randomly and equally initialized in the interval -1 to 1. Without this large span, the FNN delivers bad results. The FNN is used as many times as the many words that the longest outputted utterance should have. In the case that a response is shorter, after the last word of this response, a keyword that must be only used in this case is outputted. The FNN builds together with the Memory Network, which calculates q_{N+1}, a neural network. The whole neural network can be trained end-to-end. The weights of the Memory Network are normally distributed with an expected value of 1 and a standard derivation of 0.1. The error function is cross entropy and the Adam optimizer is used. The batch size is 32.

4 Results

The dialog system described in [1] was evaluated with the Dialog bAbI Tasks. These tasks are also described in [1] and contain six tasks. Task 1 tests if a dialog system can issue an API call with four parameters. The return of the API call is the return of a database query. The user gives 0, 1, 2, 3, or 4 parameters and the dialog system must ask for the missing parameters. Task 2 tests if after issuing the API call a new API call with updated parameters can be issued. The ability to select the best element from the returned elements of the database query is tested by task 3 and the ability to give information about a certain element of the return of the database query is tested by task 4. Task 5 combines the first four tasks. Real dialogs from the the Second Dialog State Tracking Challenge [6] was formatted in the format of the Dialog bAbI Tasks and constitute task 6.

To compare the performance of the dialog system described in this work with the dialog system described in [1], the Dialog bAbI Tasks are used as the evaluation dataset.

The accuracies of the presented dialog system in the Dialog bAbI Tasks are depicted in Table 1. All accuracies are the accuracies of the test dataset, evaluated at the state after the epoch of the training with the highest accuracy in the validation dataset regarded the subdialogs. All the tasks were trained six times with 100 epochs and there was an evaluation after every fifth epoch for the NLG candidates' selection approach. For the NLG word by word approach there was an evaluation after every epoch.

The accuracies given in parenthesis are the accuracies for the complete dialogs, and the other value is the accuracy for the subdialogs. For the dialog system with the NLG candidates' selection approach the following hyperparameter values were chosen for the evaluation: 0.0058 for the learning rate of the Adam optimizer, 44 for the embedding-size, and 1 for the number of stacked Memory Networks. The NLG word by word approach has the following hyperparameter values: 0.0022 for the learning rate of the Adam optimizer, 59 for the embedding-size, 3 for the number of stacked Memory Networks, and 50 for the number of neurons of the hidden layer of the FNN. The FNN got only the last outputted word as the input. The batches were shuffled before every epoch.

Table 1 Accuracies in percent in the Dialog bAbI Tasks

	Candidates, evaluated in [1]	Candidates with position encoding	Word by word with position encoding	Candidates with BOW encoding	Word by word with BOW encoding
Task 1	99.6 (99.6)	99.95 (99.70)	99.98 (99.90)	99.24 (95.70)	100.00 (100.00)
Task 2	100 (100)	99.92 (99.20)	99.23 (92.70)	98.55 (86.30)	100.00 (100.00)
Task 3	74.9 (2)	74.98 (0.00)	74.90 (0.00)	74.59 (0.00)	74.90 (0.00)
Task 4	59.5 (3)	57.26 (0.10)	57.18 (0.00)	57.20 (0.00)	57.18 (0.00)
Task 5	96.1 (49.4)	95.71 (46.90)	89.07 (10.50)	92.31 (20.50)	86.84 (6.20)
Task 6	41.1 (0)	40.72 (0.81)	22.26 (0.00)	37.54 (0.18)	21.65 (0.00)
Task 1[a]	– (–)	99.78 (98.70)	100.00 (100.00)	99.19 (95.20)	99.98 (99.90)

[a]With 30 788 dummy candidates

It is important that a dialog system can generalize the trained dialogs. The inputs of a real user can very likely differ from the inputs of the fictional user in the training dataset. Here, it is investigated whether the ability to generalize can be evaluated with the Dialog bAbI Tasks. In task 1, the training dataset does not include any of the API calls of the test dataset. This fact and the good results of the dialog system in task 1 lead to the fact that the dialog system can build API calls autonomously.

All the utterance patterns that occur in the test dataset also occur in the training dataset. Therefore, for this work, a dialog from task 1 of the Dialog bAbI Tasks was selected. In this dialog the user gives 0 parameters and the dialog system asks for the missing 4 parameters. 5 user utterances were modified. The utterances were shortened without to change the sense of them and after that the word "please" was added. The dialog, the possible parameter values and the modified utterances are depicted in Fig. 2. All combinations of parameter values and one modified utterance per dialog generate 13 500 dialogs. In these dialogs, the NLG candidates' selection approach had an accuracy of 60.00% with the BOW encoding and 60.00% with the position encoding and the NLG word by word approach had an accuracy of 98.60% with the BOW encoding and 94.20% with the position encoding. The accuracies are the accuracies for the complete dialogs.

Figure 3 depicts the accuracies of task 1 from the six runs of both NLG approaches with the position encoding evaluated every fifth epoch (totaling 100 epochs). This figure should show how good the results can be reproduced. The accuracy is the accuracy of the test dataset. In summary, the result of the NLG candidates' selection approach has a good reproducibility. Whereas the NLG word by word approach often needs multiple runs for good results. These good results are often not stable but rather decrease with higher epochs. Therefore, it is important to evaluate after every epoch to get good results and buffer the state of the neural network. The large span in the initialization of the weights in the FNN, as described in Sect. 3.3, is probable the reason for this bad reproducibility.

original dialog:

1. hello↹hello what can i help you with today
2. can you book a table↹i'm on it
3. <SILENCE>↹any preference on a type of cuisine
4. with *cuisine* food↹where should it be
5. in *location*↹how many people would be in your party
6. we will be *number* people↹which price range are looking for
7. in a *price* price range please↹ok let me look into some options for you
8. <SILENCE>↹api_call italian rome four cheap

parameter values:
cuisine: british, french, indian, italian, spanish
number: two, four, six, eight
location: bombay, london, madrid, paris, rome
price: cheap, moderate, expensive

modified user utterances:

2. book a table please↹i'm on it
4. *cuisine* food please↹where should it be
5. *location* please↹how many people would be in your party
6. *number* please↹which price range are looking for
7. please *price* price range↹ok let me look into some options for you

Fig. 2 Modified user utterances

In the tested datasets with a maximum of 4212 candidates the dialog system with the NLG word by word approach needed longer computation times. By increasing the number of candidates of task 1, the dialog system with the NLG candidates' selection approach showed a worse computation time and the computation time of the dialog system with the NLG word by word approach was independent of the number of candidates. With 4212 candidates, the NLG word by word approach needed 2.80 more computation time than the NLG candidates' selection approach. However, with 35 000 candidates the NLG candidates' selection approach needed 1.08 more computation time than the other NLG approach. For all the computation times, it was used 100 epochs training of task 1, position encoding as encoding, and the training state was evaluated after every fifth epoch. The computation time of the NLG word by word approach did not get worse by adding more candidates. These computation times were measured by a computer with an Intel Core i7 7700 CPU, an Nvidia GeForce 1070 GTX GPU and 16 GB main memory (the swap was not used).

For the NLG candidates' selection approach, all the possible response candidates have to be multiplied by the output of the Memory Network. The main memory consumption depends on the implementation. More memory is needed if all the vector representations of the candidates are in the memory instead of computing the vector representation on the fly, which increases the computation time. However, the candidates must be on the disk. For the computation times in this work, the vector representations of all the candidates were held in the memory.

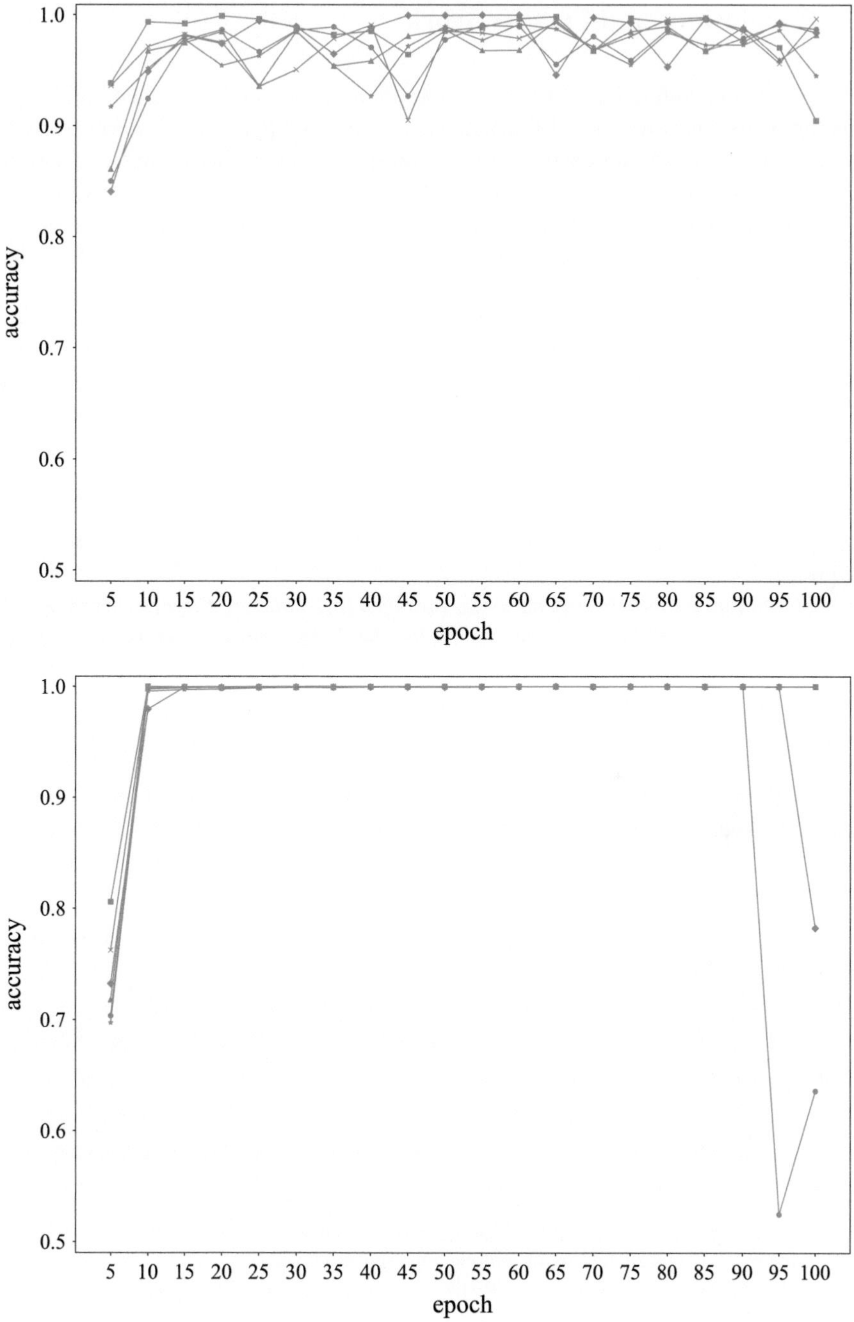

Fig. 3 Task 1 of the Dialog bAbI Tasks with the NLG word by word approach (above) and the candidates' selection approach (below), position encoding is used for both approaches

5 Conclusions and Further Work

The position encoding had better accuracies than the bag-of-words encoding in the Dialog bAbI Tasks for the NLG candidates' selection approach. However, for the NLG word by word approach with the position encoding, the accuracies were in some tasks worse. For the experiment with utterances that was modified from the user utterances in the training dataset, the bag-of-words encoding produced a better accuracy for the NLG word by word approach and both encodings produced the equal accuracy for the NLG candidates' selection approach.

The presented NLG word by word approach showed in task 1 of the Dialog bAbI Tasks a better accuracy than the NLG candidates' selection approach. Furthermore, the accuracy of the NLG word by word approach was higher in the experiment with the modified utterances. The space consumption was lower with the NLG word by word approach and the computation time was lower with large numbers of candidates (more than 35 000 response candidates).

In future work, there must be deeper research to find whether more sophisticated neural network architectures can improve the performance of the word by word approach.

The utterance-embeddings have the problem that small differences can lead to other utterance-embeddings. This means that spelling errors of words can lead to other utterance-embeddings. To avoid learning all the possible spelling errors, a spelling corrector component before the Memory Network could mitigate this problem. A normalization component can also help in reducing the necessary size of the datasets. Future works might investigate if these proposed upstream components can lead to a better performance or if these components themselves introduce errors and lead to a lower accuracy.

Acknowledgements This work has been conducted in the SecondHands project which has received funding from the European Union's Horizon 2020 Research and Innovation programme (call:H2020- ICT-2014-1, RIA) under grant agreement No 643950. This work was supported by the Carl-Zeiss-Stiftung.

References

1. Bordes A, Boureau YL, Weston J (2017) Learning end-to-end goal-oriented dialog. In: Proceedings of the fifth international conference on learning representations (ICLR)
2. Dodge J, Gane A, Zhang X, Bordes A, Chopra S, Miller AH, Szlam A, Weston J (2016) Evaluating prerequisite qualities for learning end-to-end dialog systems. In: Proceedings of the fourth international conference on learning representations (ICLR)
3. Graves A, Wayne G, Danihelka I (2014) Neural turing machines. CoRR http://arxiv.org/abs/1410.5401v2
4. Graves A, Wayne G, Reynolds M, Harley T, Danihelka I, Grabska-Barwinska A, Colmenarejo SG, Grefenstette E, Ramalho T, Agapiou J, Badia AP, Hermann KM, Zwols Y, Ostrovski G, Cain A, King H, Summerfield C, Blunsom P, Kavukcuoglu K, Hassabis D (2016) Hybrid computing using a neural network with dynamic external memory. Nature 538(7626):471–476

5. Henaff M, Weston J, Szlam A, Bordes A, LeCun Y (2017) Tracking the world state with recurrent entity networks. In: Proceedings of the fifth international conference on learning representations (ICLR)
6. Henderson M, Thomson B, Williams JD (2014) The second dialog state tracking challenge. In: Proceedings of the 15th annual meeting of the special interest group on discourse and dialogue (SIGDIAL)
7. Kumar A, Irsoy O, Ondruska P, Iyyer M, Bradbury J, Gulrajani I, Zhong V, Paulus R, Socher R (2016) Ask me anything: Dynamic memory networks for natural language processing. In: Proceedings of the 33rd international conference on machine learning (ICML)
8. Serban I, Lowe R, Henderson P, Charlin L, Pineau J (2018) A survey of available corpora for building data-driven dialogue systems. D&D 9:1–49
9. Sukhbaatar S, Szlam A, Weston J, Fergus R (2015) End-to-end memory networks. In: Proceedings of the 28th international conference on neural information processing systems (NIPS)
10. Wen TH, Gasic M, Mrkšić N, Su PH, Vandyke D, Young S (2015) Semantically conditioned lstm-based natural language generation for spoken dialogue systems. In: Proceedings of the 2015 conference on empirical methods in natural language processing (EMNLP)
11. Weston J, Bordes A, Chopra S, Mikolov T (2016) Towards ai-complete question answering: A set of prerequisite toy tasks. In: Proceedings of the fourth international conference on learning representations (ICLR)
12. Weston J, Chopra S, Bordes A (2015) Memory networks. In: Proceedings of the third international conference on learning representations (ICLR)

Enabling Spoken Dialogue Systems for Low-Resourced Languages—End-to-End Dialect Recognition for North Sami

Trung Ngo Trong, Kristiina Jokinen and Ville Hautamäki

Abstract In this paper, we tackle the challenge of identifying dialects using deep learning for under-resourced languages. Recent advances in spoken dialogue technology have been strongly influenced by the availability of big corpora, while our goal is to work on the spoken interactive application for the North Sami language, which is classified as one of the less-resourced languages spoken in Northern Europe. North Sami has various variations and dialects which are influenced by the majority languages of the areas in which it is spoken: Finnish and Norwegian. To provide reliable and accurate speech components for an interactive system, it is important to recognize the speakers with their Finnish or Norwegian accent. Conventional approaches compute universal statistical models which require a large amount of data to form reliable statistics, and thus they are vulnerable to small data where there is only a limited number of utterances and speakers available. In this paper we will discuss dialect and accent recognition in under-resourced context, and focus on training an attentive network for leveraging unlabeled data in a semi-supervised scenario for robust feature learning. Validation of our approach is done via two DigiSami datasets: conversational and read corpus.

1 Introduction

Recent advances in dialect and accent recognition have been strongly influenced by the availability of big corpora for popular languages. Dialogue technology applications for major languages are widely available, but for many languages this is not the case: there is no commercial interest in developing speech technology nor corpora or

T. N. Trong (✉) · V. Hautamäki
University of Eastern Finland, Joensuu, Finland
e-mail: trungngo.trong@uef.fi

K. Jokinen
AI Research Center, AIST Tokyo Waterfront, Tokyo, Japan
e-mail: kristiina.jokinen@aist.go.jp

L. F. D'Haro et al. (eds.), *9th International Workshop on Spoken Dialogue System Technology*, Lecture Notes in Electrical Engineering 579,
https://doi.org/10.1007/978-981-13-9443-0_19

resources to enable further development in interactive applications. It is thus important to study how to create such technology for languages which are under-resourced in that the available data is not large or it is not in the digital format ready to be used for technical applications. The lack of digital "presence" can be threatening to such languages as the speakers are forced to compromise and trade off their native language for languages which provide better and wider communication with the world and the society. The recent interest in revitalizing such languages has initiated both general and scientific effort to collect and develop tools and applications for less-resourced languages [5, 6, 13]. From the research and development point of view, the focus has been on enabling technology which allows creating applications and technology given low data resources and shortage of staff, bringing in questions of how to best address basic speech technology needs with minimum effort available.

Robot systems are getting popular as communicating devices, and there is a wide range of applications from small sociable robotic devices to advanced interactive applications such as WikiTalk [36] and ERICA [10] which enable natural language human-robot dialogues. Much research goes into development of speech technology, but applications in social robotics with the human-centered view as the core concept to support natural interaction capabilities have gradually been brought into the center of dialogue system development. The main hypothesis is that the more engaging the interaction is in terms of communicative competence, the better results are obtained by the system in terms of enjoyment and reliability. Concerning under-resourced languages, new technology can play a pivotal role in boosting revitalization of threatened languages, as the language users can see the language as a meaningful and useful part in the globalized and technologized world, cf. Ó Laoire [43].

Multilingual aspects have been addressed by Laxström et al. [38], who point out the need for software localization and internationalization to produce systems that can be used in different language and cultural contexts. Multilingualism relates to the need to cater for speakers with different dialects and speech accents so as to allow the spoken interaction to proceed smoothly. In real world applications, dialect and accent recognition contributes to the performance of speech processing systems, and the task has attracted increased attention in the speech community. Spoken dialogue systems are typical applications which can benefit from enhanced multilingualism: if they can recognize the user's language preferences, they can customize the interface to digital services accordingly [8]. The models can enhance the performance of processing tasks such as SR and ASR, and contribute positively to the performance of the whole system, and consequently, to the user experience and evaluation.

In our previous work, we studied North Sami spoken variation in Finland and in the Finnmark area in Norway [14]. We chose North Sami as our target language, since it is the object of research in the DigiSami project [15], which concerns speech and language technology to support small Finno-Ugric language communities. North Sami is an official language in the six northernmost counties in Norway, and legally recognized in Finland and Sweden. The Sami speakers are at least bilingual and can also speak the majority language of the country they live in (Norwegian, Swedish or Finnish), while North Sami is used as a lingua franca among the Sami people

[15]. The speaker's country of origin can be fairly easily distinguished based on their speaking manner. In Jokinen et al. [14], we hypothesized that the variation in North Sami dialects is due to the majority language, rather than individual variation, i.e. that there is more variation among the speakers of North Sami who live in the different majority language locations in Norway and in Finland, than among the speakers who live in different location within the same majority language context.

In this paper we continue the work on language change, assimilation, and dialect variation based on the North Sami data. We address the challenge of identifying minority dialects in a restricted data context using deep learning. Our approach augments neural architecture to form a robust and consistent dialect representation from a small corpus, while keeping an end-to-end design to maximize the potential application to similar problems. Specifically, the experiments establish two crucial situations recognizing minority languages. First, the set of samples is available only for development and consists of a small number of samples from different sources which were gathered from distinct contexts and speakers. Second, a partial set of dialects is presented during the training process, and the algorithm performs semi-supervised learning to efficiently recognize new dialects during the test time. We validate our approach through systematic experiments on two DigiSami datasets: conversational and read corpus. Our experimental findings are corroborated by outperforming the recent state-of-the-art i-vector approach.

The paper is structured as follows. We discuss the DigiSami datasets and visualise their properties in Sect. 2. We describe the Deep Learning method focusing on its use for small corpus dialect recognition in Sect. 3. We present our experiments and results in Sect. 4, and conclude with future views in Sect. 5.

2 DigiSami Datasets

Availability of large corpora in speech processing has been one of the major driving forces advancing speech technologies. This has allowed recent state-of-the-art systems to obtain impressive performance in recognizing spoken languages [2, 21, 23]. As for under-resourced languages, research is carried out in several projects and initiatives which focus on data collection and speech technology development. Most of the development for those languages have been concentrated on two directions: bootstrapping the system using adaptation of pre-trained model [33], and introducing closely related "out-of-languages" data [5]. These approaches require additional corpus, which is problematic for under-resourced languages.

The DigiSami project [15] aims to study the effect of digitalisation on small endangered languages, and to support visibility and revitalisation of Finno-Ugric language communities by creating digital content as well as developing language and speech technology tools, resources, and applications that can be used for automatic speech and language processing. The project focuses on the North Sami language, the largest of the Sami languages with about 20,000 speakers, and explores various spoken language issues (speaker identification, multimodal conversation analysis, laughing), with the challenging goal of demonstrating viability of an interactive dialogue system in the North Sami language, SamiTalk, following the multilingual open-domain robot application WikiTalk [15]. WikiTalk is an interactive robot application that enables users to find out more about subjects that interest them by discussing with the humanoid robot. They can navigate through the Wikipedia articles, ask for more information on interesting subjects, and get the robot to read the related Wikipedia article for them [16, 39].

The project organized Sami language data collection and Wikipedia article writing through series of community events in the central Sami speaking areas. The participants took part in three different tasks: discussion and writing Wikipedia articles, reading aloud of existing Wikipedia texts, and taking part in a free conversation which was video recorded. Locations were selected to represent different North Sami dialects, and consisted of three villages in Finland: Utsjoki (Ohcejohka), Inari (Anár) and Ivalo (Avvil), and two villages in Norway: Kautokeino (Guovdageaid) and Karasjoki (Kárásjohka). See more of the DigiSami data and data collection in [13, 17].

There were 28 participants, 10 men and 18 women with age range 16–65 years. They (or their parents in case of under-aged participants) gave explicit agreement to allow the data to be used for research purposes. All the participants were native speakers of North Sami, and almost all (26) reported using North Sami daily. All participants were bilingual and spoke either Finnish (Utsjoki, Ivalo, Inari), or Norwegian (Kautokeino and Karasjoki). Most participants had lived their life in the Sápmi area, although not in the same place. Ten participants had also lived in bigger cities in the southern part of the area for a short period of time.

The read speech part of the corpus contains speech samples from 28 participants who read Wikipedia articles written in North Sami. The conversational corpus has eight casual conversations with two or three participants, and the topics vary from everyday life (next vacation, driving school, cars) to translation between Sami and other languages and to technological tools that have been made to help writing North Sami more correctly. Annotation of the corpus was done with Praat and consists of 5 time-aligned tiers: a phonological/phonetic transcription, the words, the sentence in orthographic form, a Finnish translation, and remarks on things like dialectal variation. Conversations are also annotated with respect to topics and laughter [12]. The corpus has been made available for general use through the CSC website.

Overview of the datasets is given in Table 1, and Fig. 1 visualizes the closeness of the dialects in the read and conversational corpora. There is a separation between read speech and conversational speech. We observe that conversational speech is more separated than read speech in the majority languages (Finnish and Norwegian), which can be explained by the formal mode of speaking in the read speech versus informality of the conversational speech. We also noticed occasional code-switching in conversational speech to majority language. The samples group into isolated regions, but dialects do not form clearly separated clusters. The same dialect can form multiple clusters, which possibly represent the effect of unwanted variation (i.e. speaker variation, recording variation, etc.).

Table 1 DigiSami data overview

Dialects	Read corpus		Conversational corpus	
	#Speakers	Duration (hour)	#Speakers	Duration (hour)
Kautokeino	4	1.03	–	–
Karasjoki	6	0.72	6	1.5
Ivalo	6	0.72	7	0.72
Utsjoki	5	1.07	6	1.03
Inari	4	0.73	–	–
Total	**25**	**3.26**	**19**	**4.28**

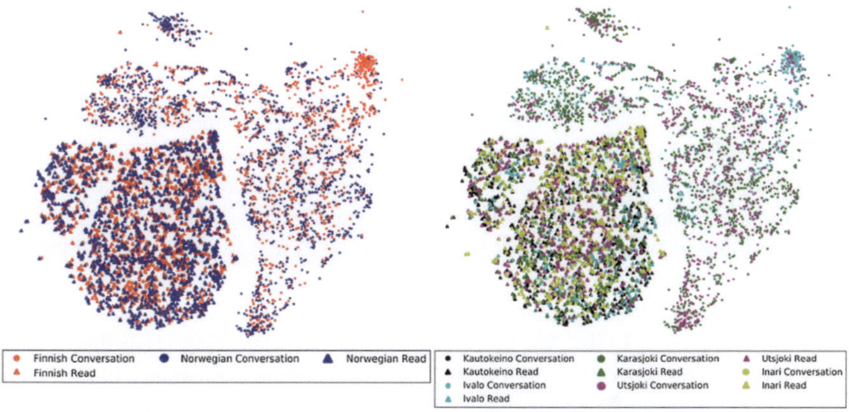

Fig. 1 t-SNE visualization of MFCCs features with 10 frames for left and right context for the conversational corpus (dot) and read corpus (triangle). Left figure highlights the majority languages, right one illustrates five different North Sami dialects

3 Deep Learning for Small Corpus Dialect Recognition

We tackle the problem discriminative learning for a small corpus in dialect recognition by addressing certain challenges. First, the complexity of speech signal is emphasized in the small dataset, with a limited number of utterances per speaker. Conventional approaches to dialect recognition compute universal background models from closely related languages [2, 21, 30], and as a result, the system requires big data to compute reliable statistics. For minority languages, collecting additional data is difficult, and as the language is typically distinguishable from the prominent language groups, introducing external corpus might lead to an unpredictable bias [33]. Moreover, separating feature learning and discriminative learning can lead to an unoptimized representation for the classification objective. We thus simplified our approach by constructing an end-to-end trained deep neural network that takes into account the channel variation to learn robust dialect representation.

Second, deep networks confront two crucial issues as non-linear models. Theoretically, neural networks can approximate Bayesian posterior probabilities under the assumptions which concern accurate estimation: the number of parameters is enough, there are enough training examples, and the priori class distribution of training set must be representative for the whole data set [18, 27, 41]. In recent years, much effort has been directed to advance neural architectures and optimization algorithms for deep learning [7, 21, 28, 42]. However, the issue concerning small, imbalanced datasets remains open. Inspired by the new perspective in understanding generalizability of deep network [31], we employ the implicit regularization techniques to directly combat overfitting within the network architectures.

3.1 Supervised Attention for Language Identification

We improve the language identification accuracy by forcing the model attended to speech segments within the utterances. Since the speech was recorded in various conditions, from conversations with ambient noise to formal reading sessions. There are differences among the dialect distribution of training and evaluation data, and the limited amount of training requires that the training process extracts a more precise representation. On the other hand, conventional attention-based approaches require large datasets since the attended weights are automatically learnt together with the main task objective [3, 40]. We introduce a supervised attention algorithm for the speech processing task as inspired from [26]. We compute the distance between the machine attentions and the "true" alignments of speech segments, and integrate the cost to the LID objective. The energy-based Voice Activities Detection (VAD) is used to generate the labels to supervise the attention network. Also, soft attention mechanism [40] is implemented to handle uncertainty of the VAD labels.

3.2 Semi-supervised End-to-End Learning

Figure 2 (next page) shows the proposed architecture for semi-supervised end-to-end (SSEE) learning. The design is composed of three deep learning architectures: convolutional neural network (CNN), long short-term memory network (LSTM), and fully connected network (FNN) [19]. The semi-supervised function is introduced by the convolutional decoder which learns to reconstruct the original signal. We do not apply decoding process after LSTM, since the recurrent neural network learns temporal patterns, and decoding of temporal signals involves alignment of long sequences which requires additional constraints and weakens our main objective to learn a robust features representation. Moreover, the weights of the decoder and encoder are tied, and random Gaussian noise is presented in the encoder during the training process. The learning process is a joint optimization of discriminative objective and reconstruction cost balanced by the hyperparameter α

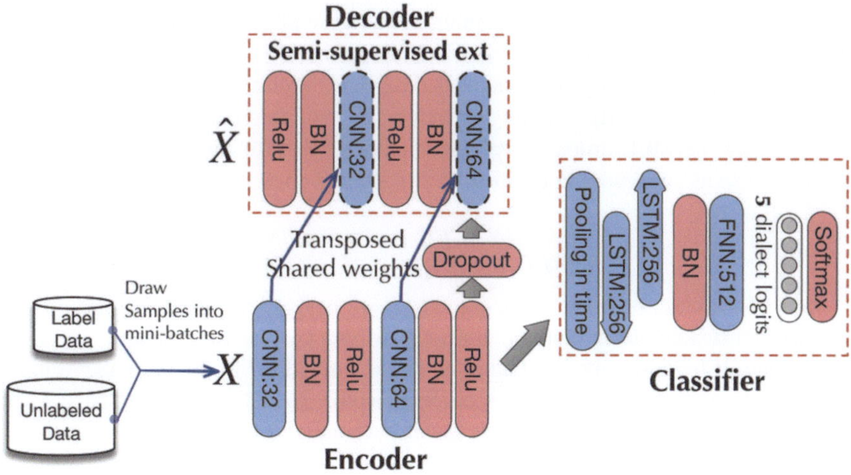

Fig. 2 Architecture of a semi-supervised end-to-end dialect recognition system with the different algorithms. A fully-supervised network can be achieved by removing the decoder

$$\alpha * \sum_{i}^{n} \log(p_\theta(y_i|X_i, z_i)) + (1-\alpha)E\left[\left\|X_i - \hat{X}_i\right\|_2\right]. \tag{1}$$

By including the second term, and the weight of encoder and decoder are tied, optimizing the objective also denoises the corrupted encoder. In order to stabilize the optimization process and enforce the network learning robust statistics from training set, we introduce batch normalization after each layer using Eq. 2 where batch normalization $BN_{\gamma,\beta}$ is given by Eq. 3. As a result, the network seeks for the optimal weights and layers statistics that preserve the most representative features within its convolutional structures.

$$fw\big(BN_{\gamma,\beta}(X, \epsilon)\big) = X \tag{2}$$

3.3 *Compensate Channel Variances Using Implicit Regularization*

Regularization is used to prevent the learning algorithm from overfitting the training and thus boost the model's generalizability. The two main strategies for regularizing a neural network are explicit regularization and implicit regularization [42]. Explicit regularization applies prior to constraint network parameters (e.g. norm constraint, dropout) and concentrates on penalizing the algorithm from overfitting; the resulted model is neither necessary nor sufficiently generalized. Implicit regularization considers the mismatch between the training set and the population, and integrates its policies into the optimization process. Early stopping and batch normalization (BN) have shown to be effective approaches to implicit regularization. We use generalization loss (GL) as early stopping criterion [28] and decrease learning rate by 2 whenever the network drops its validating score. We also modify BN to include internal noise as suggested in Radford et al. [29] and shown in formula (3)

$$y = f\left(\frac{(X - E[h])}{\sqrt{\mathrm{Var}(X)}}\gamma + \epsilon + \beta\right), \tag{3}$$

where ε is a residual term that explaines the differences between training and evaluation data. It is represented as Gaussian noise, and added to the normalized input before applying activation to force BN to learn a more robust normalized space. Conversely, the approach in Eq. 4 creates instable statistics and decelerates convergent process.

$$y = f\left(\mathrm{BN}_{\gamma,\beta}(X)\right) + \epsilon \tag{4}$$

3.4 Bayesian Cross-Entropy Objective

The modified cross-entropy [7] takes into account prior distribution of training set, and scales the loss value appropriately for each class

$$L(\theta|(X, y)) = -\frac{1}{Kn}\sum_{i=1}^{n} y_i * \frac{\log(f(x_i, \theta))}{p(y_i)}, \tag{5}$$

n is the number of training examples, K is the number of classes, and $p(y_i)$ is the probability of class y_i given our training set. This objective heavily relies on the assumption that the training set encapsulates the same distribution as the population. Since the assumption is unlikely to be sound for small datasets, we use mini batch statistics to aggregate the prior probability of each class, i.e.

$$p(y_c) = \sum_i^n \mathbb{1}(y_i = c)/n$$

This approach has been proved to stabilize the gradients and lead to better results in our experiments. We also found out that softmax activation outperforms other activation functions (i.e. sigmoid, rectifier, tanh), and is more stable for learning imbalanced data due to its normalization term, the gradients are equally distributed to anti-model neurons. Consequently, the deep network becomes a probabilistic inference model, since the parameters define a probability distribution of discrete random variable for each class conditional on the training data.

4 Experiments in Dialect Recognition

All the experiments were repeated three times to minimize the effect of random initialisation, and the final reported numbers are the mean and the standard deviation of the experiments. All the audio files were down-sampled to 16 kHz and partitioned into 30 s chunks. Our experiments have showed that log mel-filter banks features with Δ and Δ Δ are more suitable for deep network, and the same observation has been found in Ossama et al. [1], Trong et al. [12]. Using RNN enables us to leverage longer temporal windows, we segment each utterance into chunks of 200 consecutive frames, and each chunk is shifted forward 100 frames to form the next sample. One may finds augmenting the data by decreasing the shift distance, however, our experiments had provided no improvement with smaller distance. As suggested in [20], we normalize our speech frames using global mean and variance calculated from given training set.

4.1 Evaluation Metrics

Results are reported in terms of average detection cost (C_{avg}) which is the mean of all the binary detection cost (C_{DET}) for each language. C_{DET} is defined as in [23]:

$$C_{DET} = C_{miss} P_{tar} P_{miss}(L_a) + C_{fa}(1 - P_{tar}) \frac{1}{J-1} \sum_{k \neq j} P_{fa}(L_j, L_k) \tag{6}$$

where P_{miss} denotes the miss probability (or false rejection rate), i.e., a test segment of dialect L_i is rejected as being in that dialect. $P_{fa}(L_i; L_k)$ is the probability of a test segment of dialect L_k accepted as being of dialect L_i. The costs, C_{miss} and C_{fa} are both set to 1 and P_{tar}, the prior probability of a target accent, is set to 0.5 as in [23].

4.2 Baseline System

We impose the state-of-the-art i-vector approach to LID as our baseline [30, 37], the system has been implemented for our task in [14]. An utterance is represented using the fixed length and low-dimensional latent variable vector in the total variability space [8]. This is commonly called an i-vector, and it contains the variability in the utterance, such as dialect, speaker and the recording session. The Gaussian mixture model (GMM) supervector, M, of an utterance is represented as

$$M = m + Tw, \tag{7}$$

where m is the utterance independent component (the universal background model or UBM supervector), T is a rectangular low rank matrix and w is an independent random vector of distribution N(0; I). T represents the captured variabilities in the supervector space. It is estimated by the expectation maximization (EM) algorithm similar to estimating the V matrix in joint factor analysis (JFA) [25], with the exception that every training utterance of a given model is treated as belonging to a different class. The extracted i-vector is then the mean of the posterior distribution of w.

As the extracted i-vectors contain both intra- and inter-dialect variability, we use heteroscedastic linear discriminant analysis (HLDA) to project i-vectors onto a space where inter-dialect variability is maximized and intra-dialect variability is minimized. In standard HLDA technique, the vectors of size n are projected into subspace $p < n$, using HLDA matrix

$$\mathbf{A} \in \Re^{n \times n}$$

Within-class covariance normalization (WCCN) is used to compensate unwanted intra-class variations in the variability space [4]. Given two i-vectors w_{test} and $w_{dtarget}$ for dialect d, cosine similarity score t is computed:

$$t = \frac{\hat{\mathbf{w}}_{\text{test}}^{T} \hat{\mathbf{w}}_{\text{target}}^{d}}{\left\| \hat{\mathbf{w}}_{\text{test}} \right\| \left\| \hat{\mathbf{w}}_{\text{target}}^{d} \right\|} \tag{8}$$

where

$$\hat{\mathbf{w}}_{\text{test}} = \mathbf{A}^{T} \mathbf{w}_{\text{test}} \tag{9}$$

Further, w_{target}^{d} is the average i-vector over all the training utterances in dialect d. This score a calculated for all target languages, and the dialect is identified by the highest degree of similarity. Only the dialect labels are involved in computing HLDA, hence, the system doesn't know which utterances belong to Norwegian or Finnish, and provides unbiased results concerned the effect of majority languages.

4.3 End-to-End Deep Learning Systems

Following [32, 35, 34], we design our network using multiple neural architectures which are complementary in their modelling capabilities to capture different patterns.

While FNN using multiple processing layers is able to extract hierarchical representations that benefit the discriminative objective, CNN has ability to extract local invariant features in both time and frequency domain [9]. RNN combines the input vector x_t (i.e. t-th frames of utterances) with their internal state vector to exhibit dynamic temporal pattern in signal. As sequence-training is critical for speech processing, conventional FNN approaches have been proven inefficient in both lan-

Table 2 Different end-to-end neural architectures. (1) = Lopez-Moreno et al. [24], (2) = Ganapathy et al. [9], (3) = Gonzalez-Dominguez et al. [11]

Network	Design	# of parameters
FNN (1)	FNN(2560-2560-1024)	2.8×10^6
CNN (2)	CNN(32-64-128-256)	2.4×10^5
LSTM (3)	RNN(512)	6.1×10^6
Our system	CNN(32-64); RNN(256-256); FNN(512)	3.2×10^6

guage and speaker identification task [9, 24]. Our observation shows that DigiSami datasets contains long conversation with continual silence between each talk, hence, the frames-level features extracted by FNN can introduce extra biases and noises to the network. As a result, our algorithm focuses on adapting CNN and RNN architectures to address the difference between distribution of training and evaluation data. We further compare our networks to the approaches in [9, 11, 24], where the networks' hyperparameters (i.e. number of layers, number of hidden units, activation function, and parameters initialization) are fine-tuned for our task. The designs of the networks are shown in Table 2.

4.4 Supervised Language Identification

The algorithms are developed on the read speech corpus using a restricted closed training set, and their performance is verified on both read and conversational speech corpus. For training, we randomly split the read speech corpus into three datasets: training set (50% of the corpus), validation set and test set (25% each). The segmentation process ignores speaker information, and so the three sets contain three disjoint sets of utterances from the shared speaker space. Validation set is used for early-stopping. The test set from the read speech corpus is used for testing, together with the conversational speech corpus which is not used for training. Table 3 shows the results of the four different architectures. All the systems seem to generaliz well, however, they are overfitting to individual speakers which is indicated by poor performance on conversational data.

However, the result in Table 3 show the effect of channel variation as the data splitting used a shared speaker set. We thus split the development data so that one

Table 3 Performance of the different network designs on a shared speaker set. (1) = Lopez-Moreno et al. [24], (2) = Ganapathy et al. [9], (3) = Gonzalez-Dominguez et al. [11]

	$C_{avg} \times 100$	
Networks	Read speech (test set)	Conversational speech
FNN (1)	2.56 ± 0.25	22.60 ± 2.42
CNN (2)	0	17.76 ± 0.84
LSTM (3)	0	15.60 ± 1.1
Our system	0	21.05 ± 0.25

Table 4 Performance of different network designs using LOSO. (1) = Lopez-Moreno et al. [24], (2) = Ganapathy et al. [9], (3) = Gonzalez-Dominguez et al. [11]

	$C_{avg} \times 100$	
Networks	Read speech (test set)	Conversational speech
FNN (1)	32.30 ± 1.60	24.67 ± 2.36
CNN (2)	26.06 ± 1.97	23.82 ± 2.52
LSTM (3)	25.77 ± 2.69	21.03 ± 1.85
Our (Eq. 3)	14.68 ± 0.42	19.78 ± 3.48
Our (Eq. 4)	18.98 ± 1.22	22.08 ± 2.32
Our (crossentropy)	18.49 ± 1.75	19.26 ± 1.34
i-vector (4)	17.79	–

speaker from each dialect is randomly selected for validation and another speaker for testing, cf. [4, 14]. The remaining speakers are used for training the classifier. The results of the LOSO (leave-one-speaker-out) method are shown in Table 4, and they are comparable to our previous i-vector approach [14]. Table 4 emphasizes the importance of implicit regularization and the multiple-architecture design for end-to-end learning. Moreover, the best network outperforms i-vector system in both datasets.

4.5 *Semi-supervised Scenario*

We also compared the performance of a semi-supervised and full-supervised system. We used the same configuration as described in the LOSO experiment except that we removed the label from the validation set and feed it into the unsupervised system as unsupervised samples. Both the labelled and unlabelled data are shuffled and mixed into mini-batches for training.

Table 5 shows the results on both data sets and emphasizes the role of α in balancing the supervised and unsupervised objectives for the final performance. It should be noted that the value of α also variates depending on the ratio between the amount of supervised and unsupervised data available during training process.

We chose the optimized $\alpha = 0.5$ for our SSEE system. The results in Table 6 indicate that semi-supervised system outperforms the supervised one in both datasets.

Table 5 Semi-supervised learning with different α values

α	Read speech (test set)	Conversational speech
0.1	24.68 ± 0.22	22.42 ± 1.91
0.2	22.12 ± 3.26	18.83 ± 2.13
0.5	12.42 ± 1.79	16.52 ± 1.33
0.8	29.51 ± 1.21	25.12 ± 1.84

Table 6 Performance of different network designs on LOSO configurations

	Fully supervised	Semi-supervised
Read speech (test set)	14.68 ± 0.42	12.42 ± 1.79
Conversational speech	19.78 ± 3.48	16.52 ± 1.33

As a result, we conclude that SSEE has learnt an internal structure within the unsupervised samples to support the discriminative task of dialect recognition.

5 Discussion and Conclusion

The goal of the study is to enable automatic spoken interaction in the less-resourced North Sami language. We have focused on the dialect recognition task, which is one of the main issues in speech technology in general. In this paper we presented the first profound study concerning end-to-end learning on a small corpus for dialect recognition. Our results indicate the potential of end-to-end deep learning approach, and also validate the possibility of applying semi-supervised learning for auditory signal to improve the performance in restricted data context.

The results also support localization of speech applications to endangered languages. Such applications can be beneficial to these languages in multiple ways. They can provide motivation to use the language and have a favourable effect on the prestige of the language. For instance, there exists a growing number of Wikipedia articles in North Sami, and the SamiTalk application, based on the existing WikiTalk technology [16], allows the user to use North Sami Wikipedia by conducting a conversation with a humanoid robot in North Sami. The speech components need to be integrated and tested on the robot software. They are being developed separately: the DigiSami project worked on speech recognition [22], and recently a commercial company has started a project on North Sami speech recognition. The dialect recognition component described in this paper is one of the enabling technologies that can be used in the development of interactive applications for North Sami.

Acknowledgements The paper is partially based on results obtained from the Academy of Finland project *Fenno-Ugric Digital Citizens* (grant n°270082) and the *Future AI and Robot Technology Research and Development* project commissioned by the New Energy and Industrial Technology Development Organization (NEDO) in Japan.
The research was partially funded by the Academy of Finland (grant no. 313970) and Finnish Scientific Advisory Board for Defense (MATINE) project no. 2500 M-0106. We gratefully acknowledge the support of NVIDIA Corporation with the donation of the Titan Xp GPU used for this research.

References

1. Abdel-Hamid O, Mohamed A-R (2014) Convolutional neural networks for speech recognition. IEEE/ACM Transactions on Audio, Speech, and Language Processing 22:1533–1545
2. Amodei D, Anubhai R, Battenberg E et al (2015) Deep speech 2: End-to-end speech recognition in English and Mandarin. CoRR, vol. abs/1512.02595
3. Bahdanau D, Chorowski J, Serdyuk D, Brakel P, Bengio Y (2015) End-to-end attention-based large vocabulary speech recognition. CoRR, vol. abs/1508.04395
4. Behravan H, Hautamäki V, Siniscalchi SM, Kinnunen T, Lee C-H (2016) I-vector modeling of speech attributes for automatic foreign accent recognition. Audio, Speech, Lang Process, IEEE/ACM Trans 24(1):29–41
5. Besacier L, Barnard E, Karpov A, Schultz T (2014) Automatic speech recognition for under-resourced languages: a survey. Speech Commun 56:85–100
6. Crystal D (2000). English as a global language. Cambridge
7. Dalyac A, Shanahan M, Kelly J (2014). Tackling class imbalance with deep convolutional neural networks. Thesis, Imperial College London
8. Dehak N, Kenny P, Dehak R, Dumouchel P, Ouellet P (2011) Front-end factor analysis for speaker verification. IEEE Trans Audio Speech Lang Process 19(4):788–798
9. Ganapathy S, Han K, Thomas S et al (2014) Robust language identification using convolutional neural network features. In: Proceedings of the fifteenth annual conference of the international speech communication association
10. Glas DF, Minato T, Ishi CT, Kawahara T, Ishiguro H (2016) Erica: the erato intelligent conversational android. In: 2016 25th IEEE international symposium on robot and human interactive communication (RO-MAN). IEEE, pp 22–29
11. Gonzalez-Dominguez J, Lopez-Moreno I, Sak H (2014) Automatic language identification using long short-term memory recurrent neural networks. Interspeech
12. Hiovain K, Jokinen K (2016) Acoustic features of different types of laughter in north sami conversational speech. In: Proceedings of the LREC Workshop Just talking—casual talk among humans and machines, Portorož, Slovenia
13. Jokinen K (2014) Open-domain interaction and online content in the sami language. In: Proceedings of the language resources and evaluation conference (LREC 2014)
14. Jokinen K, Trong TN, Hautamäki V (2016) Variation in Spoken North Sami Language. Interspeech-2016, pp. 3299–3303
15. Jokinen K, Hiovain K, Laxström N, Rauhala I, Wilcock G (2017) DigiSami and digital natives: Interaction technology for the north sami language. In: Jokinen K, Wilcock G (eds) Dialogues with social robots. Springer, pp 3–19
16. Jokinen K, Wilcock G (2013) Multimodal open-domain conversations with the Nao robot. In: Natural interaction with robots, knowbots and smartphones: putting spoken dialogue systems into practice. Springer, pp 213–224
17. Jokinen K, Wilcock G (2014) Community-based resource building and data collection. In: Proceedings of the 4th international workshop on spoken language technologies for under-resourced languages (SLTU'14). St Petersburg, Russia, pp 201–206
18. Kirkpatrick K, Pascanu R, Rabinowitz NC et al (2017) Overcoming catastrophic forgetting in neural networks. Proc Natl Acad Sci USA 114(13):3521–3526
19. LeCun Y, Bengio Y, Hinton G (2015) Deep learning. Nature 521(7553):436–444
20. LeCun Y, Bottou L, Orr GB, Müller KR (1998) Efficient Back-Prop. Springer, Berlin Heidelberg, Berlin, Heidelberg, pp 9–50
21. Lee KA, Li H, Deng L, Hautamäki V et al (2016) The 2015 NIST language recognition evaluation: The shared view of i2r, fantastic4 and singams. Interspeech
22. Leinonen J (2015) Automatic speech recognition for human-robot interaction using an under-resourced language. Master's thesis, Aalto University, School of Electrical Engineering, Department of Signal Processing and Acoustics, Espoo
23. Li H, Ma B, Lee KA (2013) Spoken language recognition: From fundamentals to practice. Proc IEEE 101(5):1136–1159

24. Lopez-Moreno I, Gonzalez-Dominguez J, Plchot O (2014) Automatic language identification using deep neural networks. ICASSP
25. Matrouf D, Scheffer N, Fauve BGB, Bonastre J-F (2007) A straightforward and efficient implementation of the factor analysis model for speaker verification. Interspeech, pp 1242–1245
26. Mi H, Wang Z, Ittycheriah A (2016) Supervised attentions for neural machine translation, CoRR, vol. abs/1608.00112
27. Morin F, Bengio Y (2005) Hierarchical probabilistic neural network language model. AISTATS05, pp 246–252
28. Prechelt L (2012) Neural Networks: Tricks of the Trade, 2nd edn. Chapter "Early Stopping—But When?". Springer, Berlin, Heidelberg, pp 53–67
29. Radford A, Metz L, Chintala S (2015) Unsupervised representation learning with deep convolutional generative adversarial networks. CoRR, vol. abs/1511.06434
30. Richardson F, Reynolds DA, Dehak N (2015) A unified deep neural network for speaker and language recognition. CoRR, vol. abs/1504.00923
31. Sainath TN, Kingsbury B, Saon G, Soltau H, Mohamed A, Dahl G, Ramabhadran B (2014) Deep convolutional neural networks for large-scale speech tasks. Neural Netw, pp 1–10
32. Sainath T, Vinyals O, Senior A, Sak H (2015) Convolutional, long short-term memory, fully connected deep neural networks. ICASSP, pp 4580–4584
33. Thomas S, Seltzer ML, Church K, Hermansky H (2013) Deep neural network features and semi-supervised training for low resource speech recognition. In: IEEE international conference on acoustics, speech and signal processing, pp 6704–6708
34. Trong TN, Hautamäki V, Lee KA (2016) Deep Language: a comprehensive deep learning approach to end-to-end language recognition. Speaker Odyssey, Bilbao, Spain
35. Trong TN, Hiovain K, Jokinen K (2016) Laughing and co-construction of common ground in human conversations. The 4th European and 7th Nordic symposium on multimodal communication, Copenhagen, Denmark
36. Wilcock G, Jokinen K (2014) Advances in Wikipedia-based Interaction with Robots. In: Proceedings of the ICMI workshop on multi-modal, multi-party, real-world human-robot interaction, pp 13–18
37. Wilcock G, Jokinen K (2015) Multilingual WikiTalk: Wikipedia-based talking robots that switch languages. In: Proceedings of the SIGDial conference, pp 162–164
38. Wilcock G, Laxström N, Leinonen J, Smit P, Kurimo M, Jokinen K (2016) Towards SamiTalk: A sami-speaking robot linked to sami wikipedia. In: Jokinen K, Wilcock G (eds) Dialogues with Social Robots. Springer, pp 343–351
39. Wilcock G, Jokinen K (2013) WikiTalk human-robot interactions. In: Proceedings of the 15th ACM international conference on multimodal interaction (ICMI), pp 73–74
40. Xu K, Ba J, Kiros R, Cho K et al (2015) Show, Attend and tell: Neural image caption generation with visual attention. In: Proceedings of the 32nd international conference on machine learning, pp 2048–2057
41. Zhang S, Qin Y (2013) Semi-supervised accent detection and modelling. In: IEEE international conference on acoustics, speech and signal processing, pp 7175–7179
42. Zhang Z, Bengio S, Hardt M, Recht B, Vinyals O (2016) Understanding deep learning requires rethinking generalization
43. Ó Laoire M (2008) Indigenous language revitalization and globalization. Te Kaharoa 1

Part V
Empathic Dialogue Systems

Human-Robot Dialogues for Explaining Activities

Kristiina Jokinen, Satoshi Nishimura, Kentaro Watanabe and Takuichi Nishimura

Abstract The paper focuses on dialogue modelling that enables communication between users and social robots, on topics that concern experience and knowledge of people in service industries, especially in elder-care services. We discuss different knowledge types and explication of knowledge using dialogues based on goal-oriented ontologies. We present an architecture and an example of a dialogue system that allows humans to communicate with a robot partner and be engaged in interactions concerning activities related to basic care-giving tasks.

1 Introduction

In present-day societies with aging populations, changes in family structures, social as well as physical environments, and digitalization reaching all realms of society, there is a need for innovative solutions that can tackle problems in elder care and medical care services focusing on increasing medical and long-term care costs and diminishing number of carers, while accommodating independent and healthy living. Large-scale plans and frameworks have been initiated to support elder care and medical care services, and active discussions concern the use of AI research and novel technologies such as robotics and cloud services, to support greater operational efficiency. Studies have focused on the adaptation and integration of technologies into various service systems. For instance, Watanabe and Mochimaru [20] discuss

K. Jokinen (✉) · S. Nishimura · K. Watanabe · T. Nishimura
Artificial Intelligence Research Center, National Institute of Advanced Industrial Science and Technology (AIST), Tokyo, Japan
e-mail: kristiina.jokinen@aist.go.jp

S. Nishimura
e-mail: satoshi.nishimura@aist.go.jp

K. Watanabe
e-mail: kentaro.watanabe@aist.go.jp

T. Nishimura
e-mail: takuichi.nishimura@aist.go.jp

L. F. D'Haro et al. (eds.), *9th International Workshop on Spoken Dialogue System Technology*, Lecture Notes in Electrical Engineering 579,
https://doi.org/10.1007/978-981-13-9443-0_20

the impact of technology-assisted service systems in the industrial and societal levels, and how to generalise such technologies for multiple service systems. Also, the digitalized IoT (Internet of Things) environment calls for interactive solutions and applications that implement advance AI technology, including dialogue models, robotics, speech and language technology, and cognitive science. Projects such as the EU funded *Empathic* (http://cordis.europa.eu/project/rcn/212371_en.html), the Japanese *Robotic Devices for Nursing Care* (http://robotcare.jp/?lang=en), or *Metese* (http://www.vtt.fi/sites/METESE) are examples of activities that aim at developing and exploring innovate new paradigms, platforms and services especially for senior citizens.

Robotic devices have been developed to assist care-givers and elder people in daily tasks: walking, eating, lifting, and reminding of important events and tasks, as well as providing distance monitoring or support for safe independent living. Various types of communication robots are available ranging from humanoid type robot agents to sociable small robots for entertainment and companionship. Much research goes into social robotics with the human-centred design view as the core concept, powered by advanced speech and dialogue technology which enables natural language dialogues with the user. Such robot applications can provide useful information or just talk about interesting topics like WikiTalk [21] and ERIKA [14], or they are built on QA systems aiming at home assistance like commercial intelligent speakers.

Our work focuses on exploring the design and architecture for human robot dialogues that deploy ontological information about actions and activities to explicate the correct sequence of actions to the user [11]. In practical care-work, such communicating robot agents can be of valuable help in providing useful information to the users, assisting novice care-givers in their tasks or instructing them how to do a particular task, or just to provide companionship and interesting information to talk about. Moreover, activities that promote health and create community through hobbies such as dance and music can benefit from interactive robot applications which offer new ways to learn and communicate.

Our domain concerns service industries such as elder-care, nursing, and education, for which social robots can initiate novel services and work procedures with the help of their natural interaction capability. By engaging the user in interaction, the robot can provide instructions for the correct procedure to perform a task, help the user in searching for certain information on a task, or to confirm the user's own knowledge of how to do a task. The user can also access the knowledge, shared by the other care workers, of the relevant actions and activities for certain tasks. Accessing the knowledge via a spoken dialogue system has a further advantage of allowing instruction and interaction even though the user is busy acting and using his or her hands for a particular task.

The challenge for social robot agents is that they need to be furnished with knowledge that enables them to communicate in a flexible and natural manner: they require rich knowledge of the tasks and activities in various kinds of care-giving and everyday situations, and of the participants and the context in which the dialogues take place. For this, it is essential to gather and analyse information on the ways people act and interact with fellow humans and with intelligent devices, and on the knowledge they possess of the facts, events, and objects in care-giving tasks. By systematizing human knowledge into ontologies and creating dialogues based on these structured knowledge bases, it is possible to reason about appropriate actions and create natural interactions to instruct and guide the user through the task. By observing human users interacting with a social robot, it is also possible to gain more information about the dialogue strategies and functionalities that affect usefulness of such systems in practice. However, an ultimate goal for knowledge acquisition is to enable an intelligent robot to expand its knowledge by learning through interaction with the users: by imitating the partner's actions, by reinforcement learning, or by explicit teaching of the objects and actions in the environment and to learn to ground knowledge to the physical world [17].

Our research goals are summarised as:

1. Gathering and modelling of knowledge related to activities that underlie human social behaviour, especially in service scenarios,
2. Modelling interaction that enables humans and social robots to interact with each other in a natural and multimodal manner,
3. Exploring novel solutions for social robots to improve the quality of life of citizens and the quality of services in various industries, create a community through shared tasks, and offer opportunities to learn and communicate.

In this paper, we discuss the first step of these challenging aims, a framework for a knowledge-based dialogue model that enables interaction between users and services. We start with a discussion on knowledge and knowledge explication, then continue with a short survey of the research on dialogue systems, and finally present our AIRI dialogue system on a Nao robot to explain elder-care knowledge.

2 Knowledge and Interaction Concerning Actions and Activities

2.1 Task-Ontologies for Elder-Care

Various work situations in elder care, education, etc. include tacit knowledge of how to deal with various activities. In order to improve services, it is essential to gather and analyze knowledge that people possess when working in these kinds of workplaces and make implicit knowledge explicit. For this purpose, we have constructed task hierarchies that describe knowledge about the common tasks and actions in care-giving. Instead of hand-crafted ontologies, the knowledge-base was built with the help of the care-givers themselves, using a community-sourcing method of co-creating knowledge. The focus was on the procedural knowledge of how particular care supporting tasks are performed, and what are the risks involved. The knowledge was collected in workshops where the people who work in the area could describe the work and provide information about risks and proper ways to conduct the tasks.

The explication process was goal-oriented so that the knowledge could be structured into a hierarchy which shows how the various tasks are interrelated and dependent on each other [15]. In this way we were able to use the knowledge within the community and capture knowledge which is often implicit in the cooperation of the workers.

The task graphs currently consist of eight of the so-called *direct care support tasks* in elder care services. Each of the eight tasks can be regarded as the main goal for which the graph shows subtasks and action sequences that are needed in order to reach the goal, i.e. to perform the task successfully. All subtasks are obligatory to reach the goal, and the graph edges specify if they need to be performed in the given order or if they are commutable. Each subtask can also function as a goal and have subtasks of its own which describe the actions relevant to reaching that goal. Altogether there are 1746 tasks and subtasks, and they are stored in a json-type database. As the graph structure represents the plan for performing a certain task, there is no need for a reasoner. The ordering and subtask information is already included in the graph, and the knowledge can be stored in a more lightweight way.

Left side of Fig. 1 lists the eight main tasks, i.e. the upper-most goals for the system to execute, in the order of descending number of sub-tasks included in the task, and the right side visualizes a part of the task hierarchy related to the task of transfer assistance. The hierarchy shows action sequences for changing a person's position if the person is lying in the bed (there are separate action sequences if the person is sitting in a chair or wanting to stand up—these are not shown in the figure). The options for changing a person's position include a choice of moving the person horizontally or to a side-lying or seating position, to sit on a chair, or stand up, and each of them has its own subtasks. For instance, sub-tasks of the option of moving the person horizontally (not shown in the graph) include tasks such as moving the pillow, reducing friction, moving the person's upper body by supporting head and shoulders, and moving lower body by moving legs.

Care support	# of actions
Feeding	140
Oral care	170
Position change	197
Dressing	226
Moving	236
Transferring	247
Bathing	253
Excretion	277

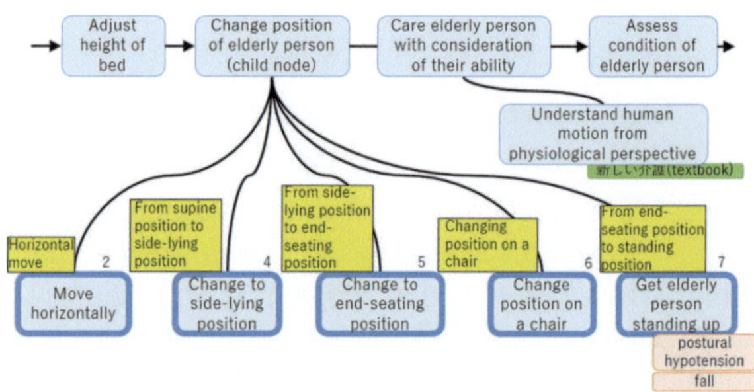

Fig. 1 On the left: a list of support tasks and the number of actions and subtasks for each type. On the right: a subtree of the task hierarchy depicting a care-giver's knowledge of how to change a position of an elderly person (translated into English). Each of the five subtasks (circled nodes) have their own subtasks (not shown in the figure), and the last one is also associated with a risk (reddish box). The yellow boxes describe the action involved

Some of the tasks are regarded as preconditions and post-actions for the main task. They are marked with an arrow pointing to the dependency relationship. For instance, for the task of changing a person's position there is a need to check the context and the person's situation by assessing if the person is conscious and able to move (these actions are not shown in Fig. 1) as well as adjusting the height of the bed (shown to the left of the "change position" node). The post-actions are needed to finish off the task properly, and in the case of changing a person's position, these include taking care of the person by considering their psycho-physiological ability and assessing their condition (shown to the right of the "change position" node), as well as adjusting the height of the bed back to the original (not included in the figure). Figure 1 also shows that the task to consider the person's (psycho-physiological) ability requires task hierarchy also includes risks associated with the tasks, and e.g. the task of getting a person stand up brings in a risk of the person falling due to postural hypotension.

It should be noticed that the knowledge graphs are not action or task hierarchies as such but represent best practice knowledge of the community to perform basic care support tasks. Accordingly, each node is interpreted as referring to steps in a procedure which explains how to reach a particular care support goal. The nodes in the graph in Fig. 1 are thus not basic actions in the same sense as those represented in common ontologies (e.g. in OWL) in terms of Subject-Predicate-Object triplets, but rather, they are abbreviations of human knowledge concerning specific actions such as lifting the person standing up or adjusting the height of the bed.

2.2 *Knowledge and Knowledge Explication*

Knowledge is often associated with explicitness, the agent's conscious deliberation and ability to exert voluntary control on actions and communicate about what they know of. The ability to produce conceptual representations of the actions and events in the physical world actuates symbolic reasoning and enables agents to talk about their experiences: actions and events are referred to by concepts which can be mapped to natural language expressions. Much of the knowledge is also implicit, i.e. used without conscious effort or deliberation, and not always verbalizable. For instance, many everyday activities such as recognizing faces or using language, or learnt skills like riding a bike or playing guitar, are based on knowledge which is not possible (or even desirable) to be described in detail.

The agent's conscious actions are aimed at changing the current state of the world into another state, the goal with certain desirable properties, and they presuppose intention and deliberate planning. However, the agent need not be conscious of all the details of the action plan or of the correct procedure to reach the intended goal. The agent may not know some pertinent facts or be able to express verbally how to perform an action, yet be able to act smoothly and accurately, or the agent may be able to verbalize an action, yet not be able to perform it in a sophisticated manner. The agent's experience of the world is also mediated by processes that the agent may not be aware of due to sensor-level limitations in perceiving the world and cognitive inability to find suitable interpretations for the perceived events.

The agent's knowledge and understanding of the world can thus be partitioned into different types [3]. We distinguish different types of knowledge by drawing distinctions along two separate dimensions: the *representation* of knowledge, and the agent's *awareness* of the knowledge. The first dimension is based upon the fact that representational issues are important for the deliberation of actions: following [5] we say that representations are tools for inquiry. The second dimension refers to the agent's consciousness of its own actions, of the events and objects in the world, and self-reflexion of knowing that it knows. These two dimensions are consistent with the recent views of memory-based architectures [2] and that all knowledge is embodied [19]. They facilitate talking about the different types of knowledge available when building computational systems for knowledge explication.

We use the term *explicit* to refer to the knowledge that the agent can explicate: the agent is aware of possessing the knowledge and has a representation for its verbalisation. The term *implicit* is used as a general term to refer to all non-explicit knowledge that the agent cannot explicate either because of not being aware of it or having no appropriate conceptual representation for it. Implicit knowledge can be further divided into three subtypes: *tacit, action-centred*, and *unknown.*

The term *tacit* is used to refer to the knowledge that is not fully known to the agent but for which there is an appropriate representation that would allow the agent to talk about it. Tacit knowledge is implicit in the sense that it can be made explicit by explication, i.e. it has an implicit conceptual representation which can be taught to the agent so that the agent is aware of the knowledge and can talk about it. For instance, a novice care-giver can be said to have tacit knowledge of the care support tasks, while a senior care-giver or a robot can help the apprentice to become aware of this by explaining how to do the care-giving tasks, thus turning tacit knowledge into explicit knowledge.

The term *action-centred* is opposite to tacit knowledge and refers to knowledge which the agent is aware of, but for which there is no direct representation available that would allow the agent to explicate it. Automatic actions without conscious effort like bike riding or talking are typical examples of action-centred knowledge: they can be learnt by imitation and practice, but they avoid clear conceptualization and are beyond the effort of explication. *Unknown* knowledge is the exact opposite of explicit: the knowledge cannot be explicated since the agent is unaware of it and accordingly, has no representation for its verbalisation. It can be made known (explicit, tacit, or action-centred) by learning or discovery. It is worth pointing out that explication presupposes conceptual representation that enables the agents to verbalize their knowledge: the agent already knows (explicit knowledge) or can be taught to know (tacit knowledge). However, if no conceptual representation exist, verbal explication is impossible: the agent either knows the actions intuitively (action-centered knowledge) or is totally ignorant of them (unknown knowledge). The different knowledge types are shown in Table 1.

Considering the knowledge graph of Fig. 1, the nodes with child nodes are explainable by referring to the children, while the terminal nodes are inexplicable actions

Table 1 Knowledge types. All non-explicit knowledge is implicit knowledge, while unknown knowledge can be made known by learning or discovery

		Conceptual representation	
		Yes	No
Agent's awareness	Yes	Explicit knowledge	Action-centred knowledge
	No	Tacit knowledge	Unknown knowledge

within the given knowledge space. Consequently, a robot agent can explain a task with the help of the given subtasks (including preconditions, post-actions, and risks), and a human agent, if not already aware of the subtasks, can be made aware of them when learning the task by verbal explanations: e.g., knowledge of the different types of position changes can be made explicit to a novice care-giver. However, a proper way of lifting the person up is action-centred knowledge as a novice care-giver must learn to detect the person's body posture and react to body movements during the lifting action itself, the verbal characterisation of which is difficult. In fact, this example shows the limitations of verbal knowledge transfer: implicit action-centred knowledge (using the terminology above) is not possible to be explicated, since it presupposes experience of the actual execution of the action, not just conceptual representations of the action.

Dialogues among the various types of communicating agents (humans, robots, and intelligent devices) require understanding of how knowledge about smart objects and relevant activities in such ambient environments affect multi-agent interactions, and what kind of consequences such dynamic IoT networks impose on the shared contexts in human interactions. Accompanying ontological data with sensor data collected in living laboratories, we can contribute not only to the development and verification of IoT sensing technology, but also to the study of different locations and workplaces as spaces that serve for understanding human activities in ubiquitous contexts. The goal is thus to explore "non-tangible thing-centric" AI technology and the changes required when moving from product-based knowledge to service-oriented and human-centred knowledge.

3 Dialogue Modelling for Knowledge Explication

3.1 Dialogue Systems

Three different types of dialogue systems can be distinguished depending on the knowledge that they can reason about. The usual distinction is between open and closed-domain systems, with the former referring to chat-type systems which can converse on any topic and the latter to limited task-based system which focus on certain task. Besides intelligent speakers that are meant to assist users at home in every-day situations, other chat-systems enable open-domain conversations with the user on a wide range of topics without any particular task to be completed (e.g. [1, 16]). WikiTalk [21, 22] uses Wikipedia as the knowledge source to support open-domain dialogues. It is multilingual and able to present information on any topic that is of interest to the user in English, Finnish, and Japanese. IBM Watson backs interaction with an impressive QA system and allows users to design their own domains. However, the free version restricts the number of "concepts" to 35, which does not scale up to full open-domain dialogues. A third type of dialogue system is a multidomain system which has a task-based approach but can use several knowledge-bases and talk about different topics. For instance, a rule-based, distributed architecture is described in Komatani et al. [12], while the PyDial system explores multi-domain dialogues and how multiple domains can be included in an end-to-end statistical dialogue system [4].

Knowledge-based solutions combine digital resources with advanced information search and summarization techniques to equip dialogue systems with encyclopedic knowledge that allows the systems to extend their interaction capabilities. The early dialogue systems (e.g. PLUS, TRAINS, ARTEMIS) were based on manually created ontologies and plan-based dialogue models, but modern technology, which can be used to collect, manipulate, and deploy ontologies needed for reasoning is not very commonly used in projects and systems. An exception is the current EU project KRISTINA (http://kristina-project.eu/en/) which uses extensive knowledge sources and an ontology-based dialogue model.

Besides ontologies that describe the knowledge, the system needs reasoning capabilities to infer about relations between objects and actions. An inference engine supports the dialogue manager in its task of reasoning about the human intentions, as well as planning of appropriate responses. An example of a planning framework in the context of social robots is the Human-Aware Task Planner (HATP) [13] which extends the traditional hierarchical planning with "social rules" that specify acceptable behavior in the domain.

3.2 CDM Architecture

Following the Constructive Dialogue Modelling (CDM) framework [7], we assume that conversational interactions are cooperative activities through which the interlocutors build a common ground. The agents (humans and robots) pay attention to social signals that indicate the partner's awareness of the communication (contact and perception), and they are also engaged in the interaction (via understanding and reacting to the exchanged information). Figure 2 presents the CDM-based architecture, see more details in Jokinen and Wilcock [8].

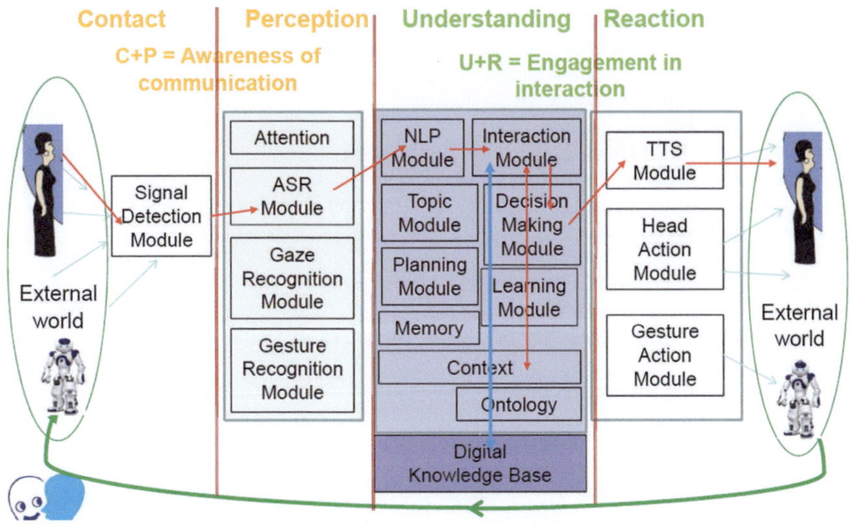

Fig. 2 System architecture in the constructive dialogue model framework, following Jokinen and Wilcock [8]. The red arrows show a flow of information from the user's speech signal through some of the modules belonging to contact, perception, understanding and reaction

3.3 An Example

The first version of the AIRC Robot Interaction (AIRI) system is implemented in the Nao robot platform (Softbank robotics). Speech recognition is done by the Nuance speech recognizer which comes with the robot and is based on keyword spotting. Dialogue engine uses Python modules and built-in QiChat state-based dialogue engine. Each of the eight main care-giving tasks functions as a possible topic which the user can initiate after the robot has recognized the user and requested a topic to continue interaction. The robot's reactions are based on the context given by the topic, the words expected to be found in the user utterance, and the action sequence representing the subtasks for the current task of which the robot is to provide information. For instance, a social robot can advise care-givers of the current state of a patient or instruct a trainee health-care professional about how to perform a task, what needs to be considered for a person's well-being, and what risks are included in performing tasks. The robot talks both Japanese and English. An example dialogue with a session shot of a user interacting with the robot is shown in Fig. 3.

N: Hi I'm Yuki. I can support your work. How can I help you today?
U: Well could you tell me how to change a person's position.
N: Ok, let me instruct you how change the person's position. There are five steps. Do you wish to hear an overview or go stepwise?
U: Stepwise, please.
N: Ok. First, you must check the person's current state.
U: What do you mean by the current state?
N: Please check if the person is sleeping, or the person is awake and conscious. Is the person sleeping or awake?
U: Awake.
N: Ok. And is the person able to communicate with you?
U: Yes, the person can communicate with me.
N: Ok. Follow my instructions. Say next when you are ready to start.
U: Next.
N: Ok. First move the person sideways.
U: What next.
N: Then lift the person up. Check if the person sits properly and feels good.
U: Just a minute please.
N: Ok, take your time. By the way, speak slowly and carefully to the elderly person. When you are ready for the next step, say next.
U: Next
N: Then check the person's position.
U: Next
N: Finally, make sure the person feels fine. Does the person seem to be fine?
U: Yes, the person seems to be fine.
N: Ok, very good. Then we are done.

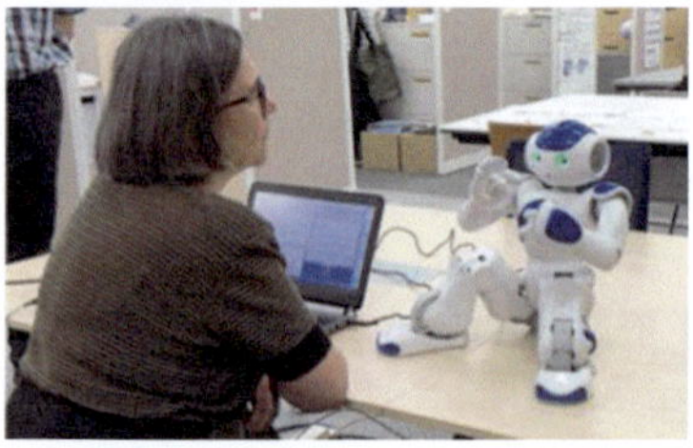

Fig. 3 A session shot and an example dialogue when the robot instructs the user on changing a person's position (U = user, N = Nao robot)

3.4 On the Evaluation of Social Robots for Care Services

The evaluation of the robot system proceeds in a manner common to user evaluation of dialogue systems. The robot system's functioning is successful if the goal is achieved (the instruction deals with the task that the user asked for, and the information given to the user is correct) and the goal is achieved in a reasonable time (should be shorter than what it would take by the care-giver to learn or do the task alone). However, for a social robot system to be successful, it is crucial that the care-giver feels happy for the received help and also the elderly person feels happy for having a robot assistant in the care-giving situation, i.e. neither feel useless, uncomfortable, or unsafe. Besides the common usability and user satisfaction studies that focus on the appropriate functioning of the social robot in the intended context with the help of subjective assessments, following e.g. SASSI type questionnaires [6], it is also crucial to consider the user's previous experience and expectations of the system. For instance, the Expectations and Experience (EE) methodology [9, 10] compares the users' expectations of the system with their actual experience of the system, and effectively measures the mismatch between the user's expectations and the interactive system's perceived capabilities. It can bring forward the user's preferences concerning the different system functionalities and attitudes towards robot agents in general.

We have already collected such evaluation data using the AIRI system as part of the multimodal data collection including eye-tracking, Kinect and video data, with a questionnaire concerning the user's experience of the interaction and understanding of the robot's presentation. The analysis of the data is on-going, but the preliminary results show that the interaction is considered friendly, interesting and not boring or awkward, yet somewhat tense, ordinary, and unnatural. Future work concern more detailed analysis of the questionnaire and also experiments related to the first impression of the user with the robot, with the aim to build a machine learning model that could enhance the robot's interaction competence, making it capable of detecting the emotional state and maybe personality of the person interacting with it and thus adapting to the user and behaving in a more compliant way.

Further development and evaluation can be carried out in a human-centred manner among the care-givers and service staff who will benefit most from a communicating robot assistant. Such assessments enable us to gather practical knowledge of the feasibility of the robot system and of the possibilities and challenges of social robot technology in care services. The care-givers' assessments are based on a real user's view-point of how to best explicate care knowledge to a novice, and they provide valuable expert feedback of the suitability, functionality, and the overall usability of the system. Such assessment sessions also allow care-givers to become more familiar with the robot agent, thus helping to alleviate common fears and uncomfortableness surrounding robot applications. The community-sourcing method that was used in the creation of knowledge graphs for the best practices of the basic care-giving task procedures can also be applied to system evaluation to co-create knowledge of the future direction of the development of robot agents.

4 Discussion and Conclusion

In this paper we have discussed different types of knowledge and designed the AIRI knowledge-based dialogue management for a robot agent which can instruct the human user on various task procedures related to elder care service scenarios. The robot draws its knowledge from a task graph which represents the knowledge possessed by the experts and is structured into a collection of explicit actions and activities. We initiated knowledge structuring by the community-sourced method, connecting the experts' knowledge to operational procedures that support communication and sharing of information. In the future, such knowledge may be created automatically from the care-giver's hand-over protocol logs or by the robot learning action patterns with the help of a human teacher or by observing the situation.

Since interactions with intelligent agents are likely to increase drastically in the coming years, it is important that the agents' communicative competence increases to allow the agents to converse with humans appropriately. Our goal is to study innovative ways to enable interaction between humans and intelligent agents, based on dialogue models, knowledge graphs, and AI-technology. It can be assumed that future dialogue scenarios include robots observing human activities, e.g. care-giving activities and drawing the human partners' attention to facts which may have previously been overlooked. The robot assistant may also notice inconsistencies in the care-giving reports and inform the care-giver of the situation, thus providing helpful information which allows the human to take an appropriate action.

Future research deals with the dialogue modelling and making it more flexible concerning the robot agent's understanding of various utterances and presentation of the information to the user. This includes integration of semantic parsing and analysis of the user's focus of attention. Also, deep-learning techniques are being experimented with and their integration with the knowledge graphs explored. The work also continues on the sophistication and development of the task hierarchies.

Finally, we will focus on the research goal (3) and explore novel ways in which robots can support health and well-being in the society. Humanoid robots which can move in the 3-dimensional environment, explicate their knowledge, and show sensitivity to human partners, embody interactions between users and technologies in a way that innovates new applications and services.

Acknowledgements This paper is based on results obtained from "Future AI and Robot Technology Research and Development Project" commissioned by the New Energy and Industrial Technology Development Organization (NEDO).

References

1. Banchs RE, Li H (2012) IRIS: a chat-oriented dialogue system based on the vector space model. In: Proceedings of the ACL 2012 system demonstrations
2. Baxter P, Wood R, Morse A, Belpaeme T (2011) Memory-centred architectures: perspectives on human-level cognitive competencies. In: Proceedings of the AAAI Fall 2011 symposium on advances in cognitive systems. AAAI Press, pp 26–33
3. Brachman R (1985) Introduction. In: Brachman R, Levesque HJ (eds) Readings in knowledge representation. Morgan Kaufmann, San Francisco, pp XVI–XVII
4. Budzianowski P, Ultes S, Su P-H, Mrkšić N, Wen T-H, Casanueva I, Rojas-Barahona LM, Gasic M (2017) Sub-domain modelling for dialogue management with hierarchical reinforcement learning. In: Proceedings of SIGdial, Saarbücken
5. Clancey WJ (1997) Situated cognition. On human knowledge and computer representations. Cambridge University Press, Cambridge
6. Hone KS, Graham R (2000) Towards a tool for the subjective assessment of speech system interfaces (SASSI). Nat Lang Eng 6(3&4):287–303
7. Jokinen K (2009) Constructive dialogue modelling – speech interaction with rational agents. Wiley, Chichester
8. Jokinen K, Wilcock G (2013) Multimodal open-domain conversations with the Nao robot. Natural interaction with robots, knowbots and smartphones: putting spoken dialogue systems into practice. Springer, New York, pp 213–224
9. Jokinen, K. and Hurtig, T. (2006). User Expectations and Real Experience on a Multimodal Interactive System. Proceedings of Interspeech, Pittsburgh

10. Jokinen K, Wilcock G (2017) Expectations and first experience with a social robot. In: Proceedings of the 5th international conference of human-agent interaction. ACM, New York, pp 511–515
11. Jokinen K, Nishimura S, Fukuda K, Nishimura T (2017) Dialogues with IoT companions - enabling human interaction with intelligent service items. In: Proceedings of the 2nd international conference on companion technology (ICCT 2017). IEEE
12. Komatani K, Kanda N, Nakano M, Nakadai K, Tsujino H, Ogata T, Okuno HG (2006) Multi-domain spoken dialogue system with extensibility and robustness against speech recognition errors. In: Proceedings of the 7th SIGdial workshop on discourse and dialogue, pp 9–17
13. Lallement R, Silva L, Alami R (2014) HATP: an HTN planner for robotics. In: Proceedings of the PlanRob 2014, ICAPS
14. Milhorat P, Lala D, Inoue K, Zhao T, Ishida M, Takanashi K, Nakamura S, Kawahara T (2017) A conversational dialogue manager for the humanoid robot ERICA. In: Proceedings of the international workshop spoken dialogue systems (IWSDS)
15. Nishimura S, Ohtani H, Hatakeyama N, Hasebe K, Fukuda K, Kitamura Y, Mizoguchi R, Nishimura T (2017) Employee driven approach to knowledge explication in elderly care service (In Japanese). Trans Jpn Soc Artif Intell 32(4):1–15
16. Otsuka A, Hirano T, Miyazaki C, Higashinaka R, Makino T, Matsuo Y (2017) Utterance selection using discourse relation filter for chat-oriented dialogue systems. In: Jokinen K, Wilcock G (eds) Dialogues with social robots. Lecture notes in electrical engineering, vol 427. Springer, Singapore. https://doi.org/10.1007/978-981-10-2585-3_29
17. Senft E, Baxter P, Kennedy J, Lemaignan S, Belpaeme T (2017) Supervised autonomy for online learning in human-robot interaction. Pattern Recogn Lett 99:77–86
18. Ultes S, Rojas Barahona LM, Su P-H, Vandyke D, Kim D, Casanueva I, Budzianowski P, Mrkšić N, Wen T-H, Gasic M, Young S (2017) PyDial: a multi-domain statistical dialogue system toolkit. In: Proceedings of ACL 2017, system demonstrations, pp 73–78
19. Varela F, Thompson E, Rosch E (1992) The embodied mind: cognitive science and human experience. The MIT Press, Cambridge
20. Watanabe K, Mochimaru M (2017) Expanding impacts of technology-assisted service systems through generalization: case study of the japanese service engineering research project. Serv Sci 9(3):250–262
21. Wilcock G, Jokinen K (2014) Advances in wikipedia-based interaction with robots. In: Proceedings of the ICMI workshop multi-modal, multi-party, real-world human-robot interaction, pp 13–18
22. Wilcock G, Jokinen K (2015) Multilingual WikiTalk: wikipedia-based talking robots that switch languages. In: Proceedings of the 16th SIGdial workshop on discourse and dialogue, pp 162–164

Virtual Dialogue Agent for Supporting a Healthy Lifestyle of the Elderly

Risako Ono, Yuki Nishizeki and Masahiro Araki

Abstract In this paper, we introduce a dialogue system that supports a healthy daily life of the elderly. This dialogue system proposes actions to be taken by the user from various types of supportive information. To respond to the diverse needs of the elderly, this system has various user adaptation functions, including a grasp of the users' interest through a chat dialogue. As a result, it can be used for long periods as well as continuous health monitoring of the elderly.

1 Introduction

In some aging societies, the number of households consisting of a single elderly person is increasing. It is a social problem to raise the quality of life of such elderly people and to prolong their healthy lifespan. To deal with such problems, providing daily support for the elderly via dialog agents is considered to be one solution [1, 2]. By having a daily conversation with a dialog agent, it is expected that the elderly can maintain good cognitive function. Also, healthy daily life can be expected if some physical activity, such as blood pressure measurement and gymnastics exercises, can be encouraged by the system.

However, the needs of the elderly are diverse. Therefore, such a system needs to be appropriately adapted to the user [3, 4]. Also, the ease of customization of a system that can properly deliver information about their place of residence is also an important factor.

R. Ono · Y. Nishizeki · M. Araki (✉)
Kyoto Institute of Technology, Kyoto, Japan
e-mail: araki@kit.ac.jp

R. Ono
e-mail: lisaco@ii.is.kit.ac.jp

Y. Nishizeki
e-mail: zeki@ii.is.kit.ac.jp

L. F. D'Haro et al. (eds.), *9th International Workshop on Spoken Dialogue System Technology*, Lecture Notes in Electrical Engineering 579,
https://doi.org/10.1007/978-981-13-9443-0_21

In this paper, we introduce the prototype implementation of a dialog system satisfying these requirements. Also, we discuss how to realize the user adaptation function and the customization capability for different conditions.

2 Approach

2.1 Role of Virtual Agent

This system aims to have the following three roles:(1) Role of <Personal doctor> supporting daily health care, (2) Role of <Family member> that provides various types of information tailored to the user's preference, and (3) Role of <Friend> that provides chat functionality.

First, the personal doctor role has the user measure their blood pressure every day and maintains the data in the database. Then, according to the users previous measurement results, the agent provides advice such as: "Drink more water because your blood pressure is lower than usual." Also, we added a function for the agent to do light gymnastics with the user. The gymnastics data is stored as a set of motion data for a 3D agent. Therefore, the content of the gymnastics can easily be customized to each user.

Second, family member role provides various types of information suitable for the user based on data such as the users residential area and preference. This information guides the everyday life of the elderly. For example, the information on regional events encourages the elderly to go out. This scenario is described in detail in Sect. 4.

Third, the friend role conducts a daily chat with the user. The system behaves like a friend that the user can talk to. It opens the conversation by providing the weather information of the day. After that, it continues the conversation with the user by discussing seasonal events or news that the user has an interest in.

2.2 Overall Structure of the System

The flow of our system is shown in Fig 1. The system consists of a main controller, an API wrapper module, and a database which maintains the user's information. As the system interface, we use the MMDAgent.[1]

Before opening the chat with the user, the system collects the current information, such as the weather, by calling the information gathering module. When the system finds out the user, it begins a conversation by sending a command to the MMDAgent module.

[1] http://mmdagent.jp/.

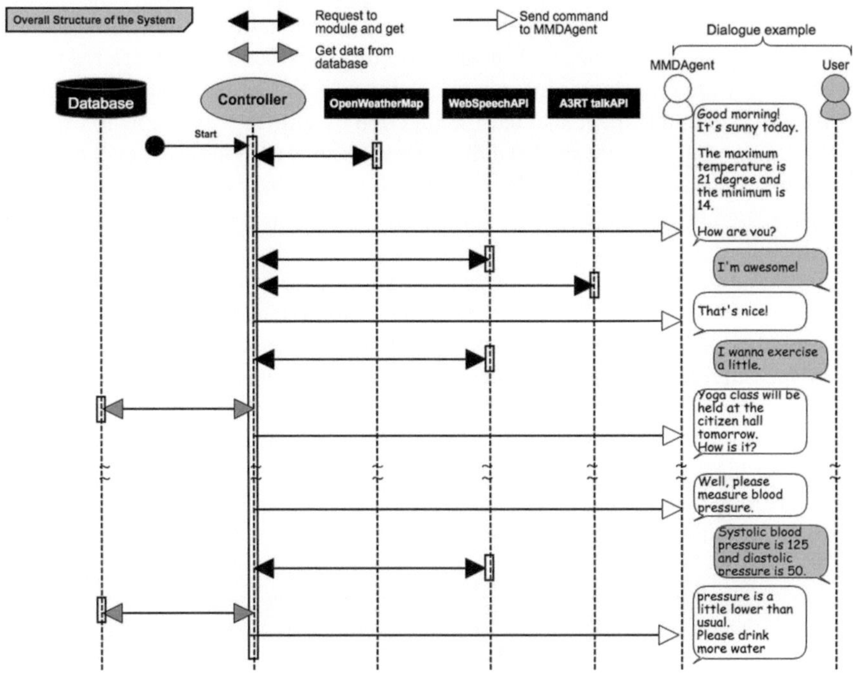

Fig. 1 Processing flow of the system and an example of the dialog

By using the MMDAgent, the user can experience the feeling of talking to people on the screen instead of talking to machines or applications. Our aim is for the user to acquire a feeling of attachment to the system.

In realizing the system described in this section, we implemented a weather information provision function, a chat dialogue function, a gymnastics function, a blood pressure measurement function and a recommendation of regional events function as shown in Fig. 1.

3 Realization of Support Dialogue

3.1 Each Function

Our system recognizes user speech using the Web Speech API[2] and generates a response using the A3RT talkAPI.[3] We are planning to extend this module by using the selection method of an appropriate response in chat-oriented dialogue systems

[2]https://developer.mozilla.org/ja/docs/Web/API/Web_Speech_API.

[3]https://a3rt.recruit-tech.co.jp/.

[5]. In addition, we will integrate the assessment functionality of users interests in a multimodal dialog based on an exchange unit [6]. It enables a users preferred topics to be elicited and reflects this information in the selection of the topic of the chat.

In our system, the agent also has the function of showing users how to exercise to help improving their physical health. The gymnastics used in this system are light and appropriate for elderly people. By following the same movement as the agent, the users strengthen their muscles; thus the system can support the maintenance of their health.

Furthermore, in our system, the users' daily health status is stored in the database to manage the health of the user. Specifically, the system encourages the user to measure their blood pressure at an appropriate time. After that, it asks for the result of the measurement and stores the information in the database.

By storing data in the database daily, the system can grasp the mean value of a users blood pressure. Based on this value, it can judge whether "the blood pressure is lower or higher than usual today" and provide appropriate advice to the user.

In fact, by providing comments based on the users data, the system is expected to become closer to the user.

3.2 Flexible and Reactive System

We developed the system using web application framework. By using the web application framework, we make it possible to construct a flexible system independent of development environment. In addition, cooperation with other applications developed under the same framework can be made very easy, and it is easy to add new functions. Specifically, we use a framework called Meteor[4] that is a full-stack JavaScript platform for developing modern web and mobile applications. By using this responsive characteristic of the platform, elderly users can reduce time and effort of reloading the web browser and pushing the buttons.

4 Recommendation of Regional Events

It is difficult for the elderly to find out information about the region from among the various information sources. Therefore, it is necessary to support the elderly by providing them with appropriate information through the system. It is also necessary to provide information according to the preference of each elderly person. In this system, we collected user preferences through the chat and provided information on events tailored to a users preferences.

In our current implementation, we use the information provided in the regional public relations magazine. By using this information, the system better reflects the region. The data to be used is not currently open or in an open data format, such as

[4]https://www.meteor.com/.

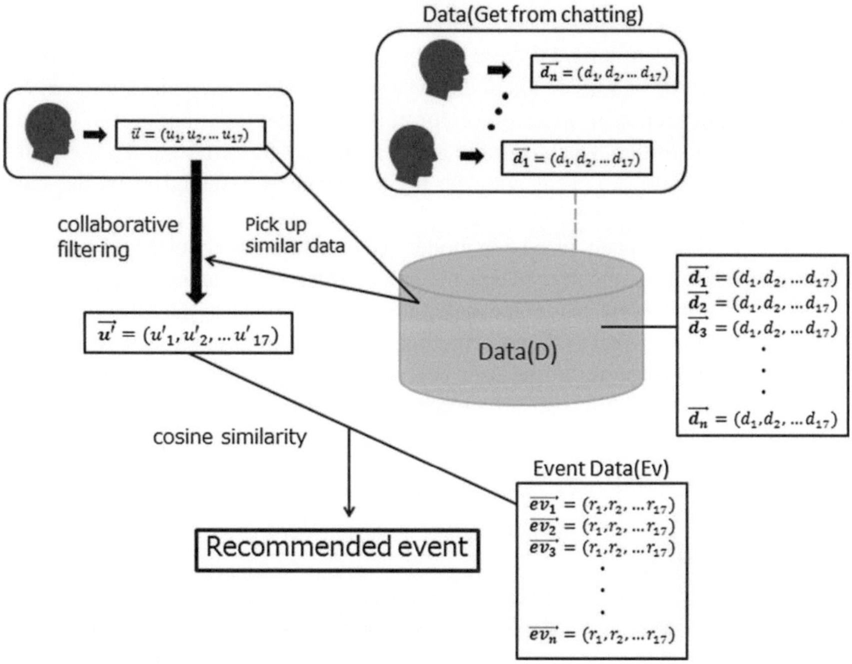

Fig. 2 Recommendation system configuration

RDF. We use this data by converting it into an open data format. This makes it easy to add data if it is released as open data in the future. In this trial setting, we use the information of 133 events as the event data.

The flow of recommendations is as shown in Fig. 2. In this system, it is judged whether the user has an interest in 17 genres through chatting. However, in this instance, we collected the data which 30 people are interested in, across 17 genres and used D as the original data. Instead, $\boldsymbol{d}$ represents the users interest in the data, and each value represents the degree of interest for each genre. Through chatting, it is determined whether a user has an interest in any of the 17 genres or not. In addition, $\boldsymbol{u}$ represents the interests of a user, and each value represents the degree of interest for each genre. It is difficult to get all the user's interest information by chatting. Users interested in the same genre may be interested in similar events. Therefore, we estimate the degree of interest about the genre that could not be determined using the collaborative filtering method. In collaborative filtering, it is possible to estimate the degree of interest of the user based on the data of the user having preference close to the user. For that reason, it becomes possible to recommend information of events that similar users will be interested in. In fact, $\boldsymbol{u}'$ represents the interest of the user after collaborative filtering. By using this procedure, we were able to obtain information about a users preference. In addition, events were also mapped to other relevant events among the 17 genres. $\boldsymbol{ev}$ indicates which genre the event is similar to, and each value represents the degree of similarity to each genre. Both preference information and events of the user are represented by vectors relating to 17 genres. By calculating the similarity of these vectors, events close to the user's preference information may be acquired. Then, the system calculates the cosine similarity for the acquired preference information vector of the user and the event vector. By this, the score of the similarity between the two vectors can be acquired. An event that provides a user with the event with the highest score is assumed to be the most appropriate.

We provided three event provision modes. The information required for each user is different. Therefore, the user selects the mode. The first mode is a simple version, providing only the event name and date. The second mode is the all information version, and it conveys all of the information. The third mode is a select version and it advances while listening to the user before determining whether to listen to the rest. In addition, the location information is displayed on the screen with a map, and the contacts are also displayed with letters on the screen, in order to provide support.

5 Conclusion

In this paper, we described a dialogue system that supports the healthy daily life of the elderly. This dialogue system has the role of a personal doctor, family member, and friend. In each role, the system provides information that is adapted to each user. As a result, it can be used for long periods for continuous health monitoring of the elderly.

We have a plan to extend this system by combining an advanced module for user adaptation and performing a field test involving the elderly.

Acknowledgements The authors would like to thank the members of MMI-group of Interactive intelligence laboratory at KIT.

References

1. Pinto H, Wilks Y, Catizone R, Dingli A (2008) The senior companion multiagent dialogue system. Proc AAMAS 2008:1245–1248
2. Miehle J, Bagci I, Minker W, Ultes S (2017) A social companion and conversation partner for elderly. In: Proceedings of IWSDS
3. Litman DJ, Forbes-Riley K (2014) Evaluating a spoken dialogue system that detects and adapts to user affective states. Proc SIGDIAL 2014:181–185
4. Miehle J, Yoshino K, Pragst L, Ultes S, Nakamura S, Minker W (2016) Cultural communication idiosyncrasies in human-computer interaction. Proc SIGDIAL 2016:74–79
5. Mori H, Araki M (2016) Selection method of an appropriate response in chat-oriented dialogue systems. Proc SIGDIAL 2016:228–231
6. Tomimasu S, Araki M (2016) Assessment of users' interests in multimodal dialog based on exchange unit. In: Proceedings of the workshop on multimodal analyses enabling artificial agents in human-machine interaction (MA3HMI '16). ACM, New York, NY, USA, pp 33–37

A Spoken Dialogue System for the EMPATHIC Virtual Coach

M. Inés Torres, Javier Mikel Olaso, Neil Glackin, Raquel Justo and Gérard Chollet

Abstract The EMPATHIC project is devoted to the development of future generations of personalised virtual coaches to help elderly people to live independently. In this paper we describe a proposal to deal with the Dialogue Management of the EMPATHIC Virtual Coach. The paper describes a DM system capable of dealing with both long-term goals and well-being plans, that can implement an effective motivational model. The system to be put into practice aims for high level healthy-ageing, utilising expressive multi-modal dialogue tailored for each specific user, working in tandem with short-term goal-oriented dialogue.

1 Introduction

The EMPATHIC Research and Innovation project is devoted to the development of future generations of Personalised Virtual Coaches to help elderly people to live independently. The EMPATHIC Virtual Coach (VC) will engage the healthy-senior user to take care of potential chronic diseases, maintain a healthy diet, have adequate physical activity as well as encourage social engagement, thus contributing to the older adults' ability to maintain a satisfying and independent lifestyle. Our ambition is to create a personal, friendly and familiar environment for the users, avoiding the threatening effects of unfamiliar new gadgets or an excessive focus on medical supervision. The VC will be capable of perceiving the emotional and social state of a person, in the learned context of the senior users' expectations and requirements, and their personal history, and will respond adaptively to their needs. The VC will put into practice high level healthy-ageing and well-being plans, and implement an effective motivational model, through expressive multi-modal dialogue tailored for each specific user. Thus, the research to be carried out is aimed at implement-

M. I. Torres (✉) · J. M. Olaso · R. Justo
Universidad del País Vasco UPV/EHU, Leioa, Spain
e-mail: manes.torres@ehu.eus

N. Glackin · G. Chollet
Intelligent Voice, London, UK

L. F. D'Haro et al. (eds.), *9th International Workshop on Spoken Dialogue System Technology*, Lecture Notes in Electrical Engineering 579,
https://doi.org/10.1007/978-981-13-9443-0_22

ing health-coaching goals and actions through an intelligent computational system, intelligent coach and spoken dialogue system adapted to users intentions, emotions and context.

2 Previous Work

Spoken dialogue systems have been developed in various domains and dealing with different goals but typically providing some information to the user such as flight schedules in pioneering proposal from ATT [8], restaurant recommendation [20] tourism information [7] or customer services [13]. However, spoken dialogue systems supporting tele-medical and personal assistant applications for older people [11], are more aligned with this research. The basic tasks of dialogue management are: to interpret the user utterance in a given context (Natural Language Understanding, NLU component), to decide what action the system should take (Dialogue Manager proper), and produce an appropriate response (Natural Language Generation, NLG component). The input speech is transformed into text using Automatic Speech Recognition (ASR), and the output produced by the Text to Speech Synthesis. The NLU component identifies the intent of the speaker and identifies the semantic content. This classification can be performed using a data-driven statistical approach or knowledge-based approach, such as hand-crafted grammars. The major challenge is the semantic inference due to the ambiguity of natural language and because semantic distinctions tend to vary from application to application.

The dialogue manager (DM) decides upon the actions a system should perform at any given state in the dialogue, given the string of semantic units provided by the NLU and the history of the dialogue. Due to the complexity of this task, traditionally the DM strategy has been designed using hand-crafted approaches based on trees and Finite State Machines which can be easily applied in task-oriented applications. However, for complex dialogues they can be very difficult to manage and impractical to design. A frame or an agenda-based DM [1, 10, 11] provides more flexible dialogue management by decoupling the information state and the actions that can be performed. The most advanced management structure is that of using distributed software agents, with which dialogues can be flexibly designed, making it possible to take the dynamic nature of dialogue into account. Dialogue flow can be modeled as a collaboration between the participating agents, their beliefs, and desires. In this way, intentions can be tracked, and multimodal and contextual information can be taken into account by the system when reasoning about its own state and the next action. A good example is the Ravenclaw DM developed by CMU [1], which was used to implement the LetsGo task [4], and has been used to obtain a large number of dialogues with real users [5].

Statistical dialogue managers were initially based on Markov Decision Processes [8, 17, 21], and Partially Observable Markov Decision Processes [20] where unknown user goals are modelled by an unknown probabilistic distribution over the user states. This approach is up-till-now considered as the state-of-the-art in Spoken Dialogue Systems [3, 22] even if their global optimisation features important computational difficulties that have to be addressed when dealing with real users. Bayesian Networks [9] and Stochastic Finite State models [6, 16] are alternative statistical approaches. Recently other machine learning methods have been proposed to obtain optimal dialogue strategies from data such as deep neural networks and reinforcement learning [15]. End-to-end learning for task-oriented dialogue systems was also proposed in [16] under an interactive pattern recognition approach that allowed dialogue generation based on stochastic bi-automata generative models. Recently hierarchical recurrent Neural networks have also been proposed to generate open domain dialogues [12] and build end-to-end Dialogue Systems. However, the real challenge of end-to-end framework is the decisional problem related to the DM for goal-oriented dialogues [2, 23]. A combination of Supervised learning allowing an LSTM to learn the domain and reinforcement learning that allows learning through interaction, has also been proposed as an alternative to deal with task-oriented dialogue systems [18, 19].

3 Progress Beyond State-of-the-art

EMPATHIC will go beyond the state of the art of SDS and DM in four main directions: It will include the perceived user affective status to support contextually relevant DM decisions; It will develop DM strategies that consider not only the user but also the high-level coaching goals to be achieved by a joint optimisation approach for dealing with shared user-coaching goals. It will maximise user and task understanding, will develop active turn-taking to recover the uncertainty, and will keep user engagement when required; and it will develop a DM capable of personalisation and adaptation to the user-specific profile and current status through novel online learning algorithms. Initially, EMPATHIC will define several models of interaction using a Wizard-of-Oz (WoZ) simulation approach. The WoZ has a human in the loop acting as the DM for the purpose of recording dialogues with the users. Then, a statistical DM, able to integrate not only the classical semantic inputs, but also parameters representing emotional status of the user, topics of the conversation and the user's intention, will be developed. This DM will be conducted by policies aimed at improving project-defined well-being metrics, resulting in system-driven interactions. Strategies defined by psychologists will be considered and studied through interaction, for user personalisation purposes. Alternatively, the DM will apply policies aimed at reaching user goals and at dealing with conversation topics previously detected automatically through machine learning techniques. Involved models and policies will get updated during interaction through online learning algorithms. EMPATHIC will use stochastic finite state bi-automata [16] as well as novel end-to-end learning approaches as main methodological frameworks, advancing the state of the art in dialogue management through a combination of supervised and reinforcement learning.

4 Dialogue Management and Intelligent Coach

We will now deal with user estimation and the main decisional systems needed to develop the virtual coach. To this end, we first identify cues and goals for well-being as well as personalized and detailed coaching plans, and actions to be implemented. The plans, developed by health professionals, will be implemented by an Intelligent Coach (IC) in tandem with the DM and modules devoted to user state identification and understanding. Figure 1 shows a diagram of the proposed system.

The Understanding module covers both local and global understanding. By local we refer to the transformation of sequences of words into a sequence of semantic units capturing concepts. Whereas global refers to the aims of the user i.e. the topic and/or the goals the user has when starting a dialogue. The first challenge is to build a component that identifies them using both current input (semantic units) and possibly also some history of the user. The DM layer includes short-term system decision. In a previous step several models of interactions had to be implemented using the WoZ while recording dialogues with the users. Two different kind of dialogues are expected: system-driven and user-driven dialogues. System-driven dialogues will be conducted by policies aiming at improving previously defined well-being metrics learned by the IC module in Fig. 1. Strategies defined by psychologists will be considered and learned through interactions for user personalisation purposes. Thus, the DM will implement the dialogue goal and dialogue strategy proposed by the IC. Alternatively, the DM will deal with user-driven dialogues that we foresee to be open-domain, to some extent. In such a case the DM will also be assisted by a User goal tracker and a Topic detection module implemented trough the Understanding module in Fig. 1. In both cases the DM will be able to integrate not only the classical semantic inputs, but also additional parameters representing the emotional state of the user. These parameters will be obtained from an emotional module that might consider

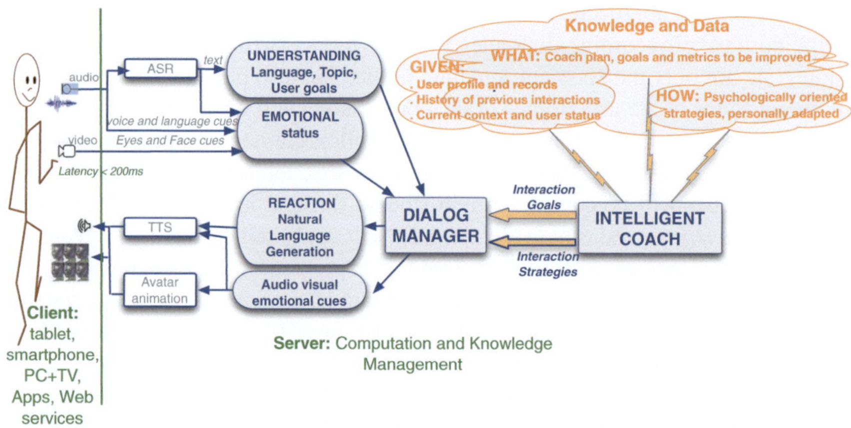

Fig. 1 General view of the Virtual Coach

features extracted from language, speech and/or images. Statistical approaches previously developed by the authors, such as a DM based on Bi-automata where policies are implemented through optimised search of an interaction graph tree will be used. Such approach is based on Stochastic-Finite-State-Transducers (SFST) where three different alphabets are defined: one related to semantic input, one for task related attributes and an additional one for dialog action outputs. The model is easily learned from examples using classical inference algorithms that can be run online allowing task adaptation. Additionally, they allow the development of user models that have been successfully tested [14]. We also foresee end-to-end learning approaches that combine sequence-to-sequence deep learning with reinforcement learning. We want to develop algorithms capable of selecting the specific way to approach each user, e.g. positive tone vs. natural tone, and so on. Since individual users react differently to different types of interaction. Thus, the input to the SDS (given by the ASR) in the form of semantic concepts will be of an n-best type lattice structure for training the correct interface between the SDS and ASR semantic representation. The deepest layer described in Fig. 1 is the IC, which deals with the reasoning to take long-term decisions on the basis of evidence-based personalised coaching action plans. From time to time the DM will commence a specific dialogue with the user in order to achieve long-term coaching goals. This is a novel task, as most dialogue systems are user-triggered, while this will be a system-triggered.

5 Concluding Remarks

The EMPATHIC Research and Innovation project is devoted to the development of future generations of Personalised Virtual Coaches to help elderly people to live independently. We have described a proposal to deal with the Dialogue Management of a Virtual Coach capable of perceiving the emotional and social state of a person, and which can consider their personal history and respond adaptively to their needs. The paper has described how the EMPATHIC Virtual Coach we will put into practice high level healthy-ageing and well-being plans, and implement an effective motivational model, through expressive multi-modal dialogue tailored for each specific user. The project will be developed under the Social Challenge Pillar and Health Work programme, demographic change and well-being of the EU Horizon 2020 Program (www.empathic-project.eu).

Acknowledgements This work is founded by the European Commission H2020 SC1-PM15 program under RIA grant 769872.

References

1. Bohus D, Rudnicky AI (2009) The RavenClaw dialog management framework: architecture and systems. Comput Speech Lang 23(3):332–361
2. Bordes A, Boureau YL, Weston J (2017) Learning end-to-end goal oriented dialog. In: International conference of learning representations
3. ci cek FJ, Thomson B, Young S (2012) Reinforcement learning for parameter estimation in statistical spoken dialogue systems. Comput Speech Lang **26**(3), 168–192
4. Eskénazi M, Black AW, Raux A, Langner B (2008) Let's go lab: a platform for evaluation of spoken dialog systems with real world users. In: INTERSPEECH, p 219. ISCA
5. Ghigi F, Eskenazi M, Torres MI, Lee S (2014) Incremental dialog processing in a task-oriented dialog. In: InterSpeech, pp 308–312
6. Hurtado LF, Planells J, Segarra E, Sanchis E (2016) Spoken dialog systems based on online generated stochastic finite-state transducers. Speech Commun 83:81–93. https://doi.org/10.1016/j.specom.2016.07.011
7. Kim S, D'Haro LF, Banchs RE, Williams JD, Henderson M (2017) The fourth dialog state tracking challenge. In: Dialogues with social robots - enablements, analyses, and evaluation, seventh international workshop on spoken dialogue systems, IWSDS 2016, Saariselkä, Finland, 13–16 Jan 2016, pp 435–449. https://doi.org/10.1007/978-981-10-2585-3_36
8. Levin E, Pieraccini R, Eckert W (2000) A stochastic model of human-machine interaction for learning dialog strategies. IEEE Trans Speech Audio Process 8(1):11–23
9. Martínez FF, López JF, de Córdoba Herralde R, Martínez JMM, Hernández RSS, Muñoz JMP (2009) A bayesian networks approach for dialog modeling: the fusion bn. In: Proceedings of the IEEE international conference on acoustics, speech, and signal processing ICASSP 2009. IEEE, New Jersey, EEUU. http://oa.upm.es/5579/
10. Olaso J, Torres MI (2017) User experience evaluation of a conversational bus information system in spanish. In: 8th IEEE international conference on cognitive infocommunications, Debrecen, Hungary, September 2017
11. Olaso JM, Milhorat P, Himmelsbach J, Boudy J, Chollet G, Schlögl S, Torres MI (2017) A multi-lingual evaluation of the vAssist spoken dialog system. Comparing Disco and RavenClaw. Springer, Singapore, pp 221–232
12. Serban IV, Sordoni A, Bengio Y, Courville A, Pineau J (2016) Building end-to-end dialogue systems using generative hierarchical neural network models. In: Proceedings of the thirtieth AAAI conference on artificial intelligence, AAAI'16. AAAI Press, pp 3776–3783. http://dl.acm.org/citation.cfm?id=3016387.3016435
13. Serras M, Perez N, Torres MI, Del Pozo A, Justo R (2015) Topic classifier for customer service dialog systems. Springer International Publishing, Cham, pp 140–148. https://doi.org/10.1007/978-3-319-24033-6_16
14. Serras M, Torres MI, Del Pozo A (2017) Online learning of attributed bi-automata for dialogue management in spoken dialogue systems. Springer International Publishing, Cham, pp 22–31. https://doi.org/10.1007/978-3-319-58838-4_3
15. Su PH, Vandyke D, Gasic M, Kim D, Mrksic N, Wen TH, Young S (2015) Learning from real users: rating dialogue success with neural networks for reinforcement learning in spoken dialogue systems. In: InterSpeech, pp 2007–2011
16. Torres MI (2013) Stochastic bi-languages to model dialogs. In: International conference on finite state methods and natural language processing, pp 9–17
17. Walker M (2000) An application of reinforcement learning to dialogue strategy selection in a spoken dialogue system for email. J Artif Intell Res 12:387–416
18. Williams JD (2016) End-to-end deep learning of task-oriented dialog systems. In: Keynote in future and emerging trends in language technologies FETLT, Seville
19. Williams JD, Asadi K, Zweig G (2017) Hybrid code networks: practical and efficient end-to-end dialog control with supervised and reinforcement learning. In: ACL (1). Association for Computational Linguistics, pp 665–677

20. Williams JD, Young S (2007) Partially observable Markov decision processes for spoken dialog systems. Comput Speech Lang 21(2):393–422
21. Young S (2000) Probabilistic methods in spoken dialogue systems. Philos Trans R Soc Lond
22. Young S, Gašić M, Thomson B, Williams JD (2013) POMDP-based statistical spoken dialog systems: a review. Proc IEEE 101(5):1160–1179
23. Zhao T, Eskénazi M (2016) Towards end-to-end learning for dialog state tracking and management using deep reinforcement learning. In: Proceedings of the SIGDIAL 2016 conference, The 17th annual meeting of the special interest group on discourse and dialogue, 13–15 Septr 2016, Los Angeles, CA, USA, pp 1–10 (2016)

Stitching Together the Conversation—Considerations in the Design of Extended Social Talk

Emer Gilmartin, Brendan Spillane, Christian Saam, Carl Vogel, Nick Campbell and Vincent Wade

Abstract Spoken interaction mediates much human social and practical activity. Talk is not monolithic in form but rather weaves in and out of different genres. Practical tasks are peppered with lubricating social talk, while casual conversation proceeds in phases of interactive chat and longer almost monologue chunks. There is increasing interest in building applications which enable convincing human-machine spoken or text interactions, not only to facilitate immediate practical tasks but also to build a longer term relationships within which conversation can take place in order to entertain, provide companionship and care, and build a user model which will facilitate future tasks through an 'always on' conversational interface. Such applications will require modelling of the different subgenres of talk, and of how these can be convincingly joined to form a coherent ongoing conversation. In this paper we describe our work towards modelling such talk, focussing on theories of casual talk, insights gleaned from human-human corpora, and implications for dialog system design.

E. Gilmartin (✉) · N. Campbell
Speech Communication Laboratory/ADAPT Centre, Trinity College Dublin, Dublin, Ireland
e-mail: gilmare@tcd.ie

N. Campbell
e-mail: nick@tcd.ie

B. Spillane · C. Saam · V. Wade
ADAPT Centre, Trinity College Dublin, Dublin, Ireland
e-mail: brendan.spillane@adaptcentre.ie

C. Saam
e-mail: christrian.saam@adaptcentre.ie

V. Wade
e-mail: vwade@adaptcentre.ie

C. Vogel
School of Computer Science and Statistics, Trinity College Dublin, Dublin, Ireland
e-mail: vogel@tcd.ie

L. F. D'Haro et al. (eds.), *9th International Workshop on Spoken Dialogue System Technology*, Lecture Notes in Electrical Engineering 579,
https://doi.org/10.1007/978-981-13-9443-0_23

1 Introduction

Much dialog system technology has focused on practical (instrumental or task-based) exchanges, and modelling less focussed casual or social talk has been considered less tractable than modelling a task-based exchange [1]. Task-based exchanges can often be formalized and predicted as the progress of dialog is heavily dependent on the exchange of recognisable information (prices, options), the task is known to interlocutors, goals are short-term, and the form of the dialog is largely question-answer. Such dialogs can be modelled as slot-filling paradigms with rules or stochastic methods used to manage the dialog flow. However, much human interaction is partially or wholly composed of social and casual talk, where the ultimate goal seems to be maintaining social bonds, achieved by keeping an entertaining or interesting flow of smalltalk, gossip, conversational narratives or anecdotes, or discussion of topics of mutual interest [9, 15, 16]. It has also been postulated that keeping this channel of dialog open facilitates discussion and performance of practical tasks as and when they arise [3, 12]. There has been much progress in the creation of systems which perform simple tasks or maintain short chats, and now attention is shifting to include longer relational dialogs. Longer casual talk has been shown to occur in phases. Ventola modelled conversation as a sequence of phases including greeting, light conversational approach phases, more detailed centring phases, and forumulaic leavetakings, as shown in Fig. 1 [17]. Ventola develops a number of sequences of these elements for conversations involving different levels of social distance. She describes conversations as minimal or non-minimal, where a minimal conversation is essentially phatic, particularly in Jakobsen's sense of maintaining channels of communication [12], or Schneider's [15] notion of defensive smalltalk—such a conversation could simply be a greeting, a greeting followed by a short approach phase and goodbye, or could be a chatty sequence of approach stages. Non-minimal conversations involve centring—where the focus shifts to longer bouts often fixed on a particular topic. Conversations between people who know each other well (low social distance) can progress directly from greetings to centring, 'dispensing with formalities', while conversations between strangers or more distant acquaintances incorporate more 'smalltalk' in the form of approach phases. Some conversational elements occur only once, as in the case of greetings (G) and goodbyes (Gb), while others can recur. Approach stages can occur recursively, generating long chats without getting any deeper into centring. Centring stages can recur and are often interspersed with Approach stages in longer talks. In their work on casual conversation, Slade and Eggins have observed a structure of alternating phases of often light interactive 'chat', and longer more monologic 'chunks' where one participant tells a story or discusses a topic [4]. Some parallels can be drawn between Ventola's approach phases and Slade and Eggin's chat phases, as some centring phases would match chunk phases, although centring could also include stretches of task based dialog in the case of business encounters wrapped in casual or social talk (as noted by Laver [13]). It is clear that conversation is not monolithic, and may be better modelled as a sequence of phases, where each phase might well involve different dynamics. This has implications for natural lan-

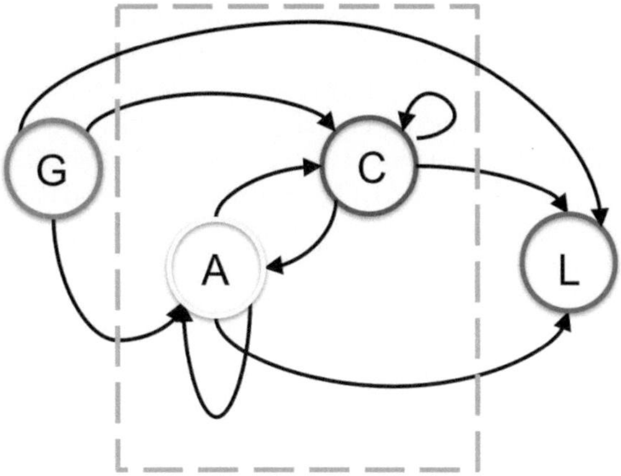

Fig. 1 A simplified version of Ventola's conversational phases—Greeting (G) and Leavetaking (L) occur at most once in a conversation, while the Approach (A) and Centring (C) phases may repeat and alternate indefinitely in a longer conversation. Different conversational sequences may be generated from the graph

guage generation, endpointing and turntaking management, and for the optimization of information flow in a range of applications. Most work on casual conversation to date has been theoretical or qualitative, with little quantitative analysis. The lack of large datasets of substantial casual conversations is an ongoing difficulty in the field, and work has often been limited to the study of corpora of short interactions. We have assembled a collection of casual talk interactions, and are carrying out quantitative analysis on the phases of talk in long multiparty casual conversations (chat and chunk) and shorter dyadic text exchanges (greeting and leavetaking). Below we briefly describe progress made in further understanding casual conversational phases in our recent work, and discuss where this knowledge can be exploited in the design of artificial conversations.

2 Modelling Phases

Our investigations are based on a number of corpora of multiparty casual spoken interaction, d64 [14], DANS [10], and TableTalk [2], and on the ADELE corpus of text conversations [8].

2.1 Greetings and Leavetaking

Our recent work on these phases is based on the ADELE Corpus, a collection of text exchanges where participants were asked to discover biographical information and preferences about their partners through friendly chatty conversation. Using the 'ISO standard 24617-2 Semantic annotation framework, Part 2: Dialogue acts' [11], we annotated 196 conversations from the corpus. From these, we learned that the greeting sequences typically involved 4-6 turns, while leavetaking and goodbyes involved 6-8 turns. We found that the existing labels for greeting and leavetaking insufficient to cover the dialog act patterns we encountered and created a number of new dialog act labels. The sequence of dialog acts comprising greeting and leavetaking phases of dialog were found to vary very little. A full description of the collection, annotation and analysis of the ADELE corpus can be found in [8]. We conclude that modelling such sequences must involve awareness of their multiturn nature and of the acts involved.

2.2 *Approach/Chat and Centring/Chunk*

To aid our understanding of the structure of the 'meat' of conversation phases, we segmented, transcribed and annotated six long form conversations for chat and chunk phases. The segmentation, transcription and annotation of the data, and the methodology for annotating chunks are more fully described in [5]. The annotations resulted in 213 chat segments and 358 chunk segments overall. Further details of these experiments can be found in [6, 7]. We found a number of differences in chat and chunk phases:

2.2.1 Length of Phase

We have found that the distributions of durations of chat and chunk phases are different, with chat phase durations varying more while chunk durations have a more consistent clustering around the mean (34 s). Chat phase durations tend to be shorter than chunk durations. We found no significant differences in chunk duration due to gender, speaker, or conversation.These findings seem to indicate a natural limit for the time one speaker should dominate a conversation. However, we did find that chat phase durations were conversation dependent.

2.2.2 Laughter Distribution in Chat and Chunk Phases

We found laughter occurring more frequently in chat phases. Comparing the production by all participants in all conversations, laughter accounts for approximately 9.5% of total duration of speech and laughter production in chat phases and 4.9% in chunk phases.

2.2.3 Overlap and Silence

Overlap is more than twice as common in chat phases (12.6%) as in chunk phases (5.3%), with chunk overlap generally occurring as backchannels, while in chat there

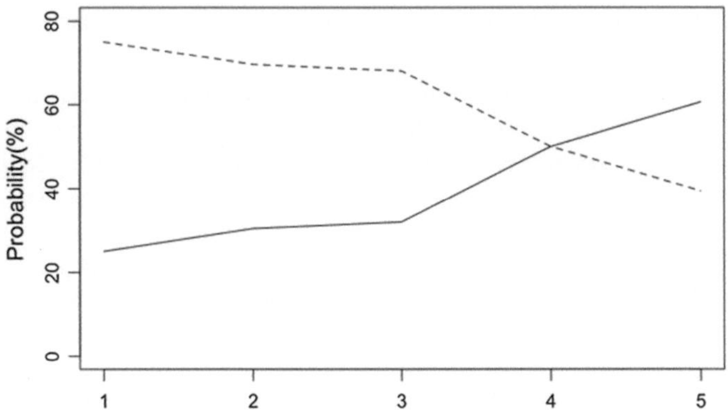

Fig. 2 Probability of chunk-chunk transition (solid) and chunk-chat transition (dotted) as conversation elapses (x-axis = time) for first 30 min of conversation data in 6-min bins

is often competition for turns and the talk is more interactive, resulting in longer stretches of overlap. Silence was marginally more common in chat phases. As would be expected, between speaker silences (gaps) predominated in chat, while chunks contained more within speaker silences (pauses) than gaps. The mean lengths of silences were quite similar, but silence duration in chat varied considerably more than in chunk phases.

2.2.4 Chat and Chunk Position

We observed more chat at conversation beginnings, with chat predominating for the first 8–10 min of conversations. Although our sample size is small, this observation conforms to descriptions of casual talk in the literature, and reflects the structure of 'first encounter' recordings. However, as the conversation develops, chunks start to occur much more frequently, and the structure is an alternation of single-speaker chunks interleaved with shorter chat segments. Figure 2 shows the probability of a chunk phase being followed by chat or by chunk for the first 30 min of conversation. It can be seen that there is a greater tendency for the conversation to go directly from chunk to chunk the longer the conversation continues, resulting in 'story swapping'.

3 Discussion

We have identified several points of interest on phases in talk. The knowledge gained from studying greeting and leavetaking phases can be directly applied to scripting or generating these rather standard sequences. For mid conversation, we have found significant differences in chat and chunk phases. The rather stable duration of chunk phases could be useful in designing the length of stretches where the system shares information with a user—for example for a companion application summarising news articles for an elderly user, or indeed in educational applications. It could also aid by imposing a limit for uninterrupted speech—in prompt generation for system speech and also to inform backchannel models. The distribution of laughter and overlap could inform the behaviour of an agent in different phases of interaction. Differences in overlap and silence distribution could prove very valuable in creating more accurate endpointing and turntaking management depending on the phase of conversation, thus avoiding inappropriate interruptions or uncomfortable silences due to miscalculation by the system—for example, a system taking turns based on an elapsed silence trained on chat could be infelicitous in a chunk phase and vice versa. The positioning of chat and chunk phases also has implications. Chat is more common towards the beginning of phases while the likelihood of chunk to chunk transitions grows over the course of the conversation. This knowledge could help with design of dialog flow, particularly where dialog is to entertain or keep the user company. The modular nature of conversation, with phases differing in their dynamics, makes modelling of the dialog as a monolith rather impractical, particularly in light of the extreme shortage of conversational data beyond collections of task-based interactions and short first encounter casual conversation. However, this very aspect could allow for modelling using diverse corpora for different conversational phases, in a manner similar to unit selection in speech synthesis.

4 Conclusions

Creation of more human-like conversational agents entails understanding of how humans behave in interactions. Casual talk is an essential part of human-human interaction, contributing to bonding, allowing for experiences and opinions to be swapped, and providing entertainment. Casual talk can also act as a matrix for embedded task-oriented talk, keeping conversational channels open in readiness for practical needs. Systems aiming to provide companionship or a persistent presence with a user will need to generate realistic casual conversation. We believe that our ongoing work will help inform the design of such systems—agents which can provide companionship or better serve the user by performing as a human would at the basic dynamics of conversation. We are currently using these insights to build a companion/coaching system for the elderly, ADELE, and exploring different data sources to model different phases. We are also exploring the dynamics of conversational phases in dyadic talk, to see if our multiparty results generalise. The major limitation of the current work is the scarcity of data as corpora of for casual conversations longer than 15 min are hard to find and difficult to collect without significant resources. We hope that the current study will encourage the production of corpora of longer form casual conversation, to facilitate the design of convincing artificial interlocutors.

Acknowledgements This work is supported by the European Coordinated Research on Long-term Challenges in Information and Communication Sciences and Technologies ERA-NET (CHIS-TERA) JOKER project, JOKe and Empathy of a Robot/ECA: Towards social and affective relations with a robot, and by Science Foundation Ireland (Grant 13/RC/2106) and the ADAPT Centre (www.adaptcentre.ie) at Trinity College, Dublin.

References

1. Allen J, Byron D, Dzikovska M, Ferguson G, Galescu L, Stent A (2000) An architecture for a generic dialogue shell. Nat Lang Eng 6(3&4):213–228
2. Campbell N (2008) Multimodal processing of discourse information; the effect of synchrony. In: 2008 second international symposium on universal communication ISUC'08. IEEE, pp 12–15
3. Dunbar R (1998) Grooming, gossip, and the evolution of language. Harvard University Press
4. Eggins S, Slade D (2004) Analysing casual conversation. Equinox Publishing Ltd (2004)
5. Gilmartin E, Campbell N (2016) Capturing chat: annotation and tools for multiparty casual conversation. In: Proceedings of the tenth international conference on language resources and evaluation (LREC 2016)
6. Gilmartin E, Campbell N, Cowan B, Vogel C (2017a) Chunks in multiparty conversation-building blocks for extended social talk. Proc IWSDS 2017:6–9
7. Gilmartin E, Cowan BR, Vogel C, Campbell N (2017) Exploring multiparty casual talk for social human-machine dialogue. In: Karpov A, Potapova R, Mporas I (eds) Speech and computer. Springer International Publishing, Cham, pp 370–378
8. Gilmartin E, Spillane B, O'Reilly M, Saam C, Su K, Cowan BR, Levacher K, Devesa AC, Cerrato L, Campbell N et al (2017) Annotation of greeting, introduction, and leavetaking in dialogues. In: Proceedings of the 13th joint ISO-ACL workshop on interoperable semantic annotation (ISA-13)
9. Hayakawa SI (1990) Language in thought and action. Houghton Mifflin Harcourt (1990)
10. Hennig S, Chellali R, Campbell N (2014) The D-ANS corpus: the Dublin-Autonomous Nervous System corpus of biosignal and multimodal recordings of conversational speech. Reykjavik, Iceland
11. ISO (2012) ISO 24617–2:2012—Language resource management—Semantic annotation framework (SemAF)—Part 2: Dialogue acts. International Organization for Standardization, Geneva, Switzerland
12. Jakobson R (1960) Closing statement: Linguistics and poetics. Style Lang 350:377
13. Laver J (1975) Communicative functions of phatic communion, pp 215–238
14. Oertel C, Cummins F, Edlund J, Wagner P, Campbell N (2010) D64: a corpus of richly recorded conversational interaction. J Multimodal User Interfaces 1–10
15. Schneider KP (1988) Small talk, vol 1. Hitzeroth Marburg
16. Thornbury S, Slade D (2006) Conversation: from description to pedagogy. Cambridge University Press
17. Ventola E (1979) The structure of casual conversation in English. J Pragmat 3(3):267–298

Part VI
Humor in Dialogue Agents

Towards an Annotation Scheme for Causes of Laughter in Dialogue

Vladislav Maraev and Christine Howes

Abstract This paper presents a first attempt to develop an annotation scheme for laughter in dialogue operationalising the previously reported idea of laughter being caused by incongruity, and based on violations of Grice's maxims. This exploratory scheme is intended to form the basis of a spoken dialogue system that can laugh during dialogue in a human like manner and can understand why users laugh. We present the scheme and discuss preliminary results.

1 Introduction

Recent research has focussed on creating more human-like spoken dialogue systems by means of adding capabilities to produce [4], or recognise laughter [12, 21, 24], react appropriately [5, 15], recognise sarcasm [22], be humourous [11, 16], and discover how and where laughter occurs in dialogue [7, 23]. However, there is no agreement on the causes of laughter, with, for example, some research which focusses on humour [9, 19], and other research which highlights the social functions of laughter, such as affiliation and agreement [2, 20], and qualitative analysis of the roles of laughter in interaction and its coordination with speech (see [7], for a review of conversation alanalys is approaches to laughter).

Furthermore, as argued by [13], existing taxonomies of laughter have reliability issues: they mix the functions that use laughter as a means of communication with the different emotions that laughter triggers. For example, in [18], affiliation (i.e. agreement laughter) is roughly the illocutionary act performed by laughter, while joy is a feature triggered by laughter. Another issue with most current studies of

---

V. Maraev (✉) · C. Howes
Department of Philosophy, Linguistics and Theory of Science, Centre for Linguistic Theory and Studies in Probability (CLASP), University of Gothenburg, Gothenburg, Sweden
e-mail: vladislav.maraev@gu.se

C. Howes
e-mail: christine.howes@gu.se

L. F. D'Haro et al. (eds.), *9th International Workshop on Spoken Dialogue System Technology*, Lecture Notes in Electrical Engineering 579,
https://doi.org/10.1007/978-981-13-9443-0_24

laughter is that they do not tend to recognise the propositional content that laughter can convey (see [6], for discussion).

In the present study, following [6], we look at laughter based on the stimulus that provokes it, henceforth the *laughable*. Laughables will be analysed from two interlinked perspectives: (a) incongruity and (b) Gricean maxims.

The *theory of incongruity* explains laughter as arising from an inconsistency between the expectations of the conversational participants and some event. This has been studied extensively in theories of humour [9, 19], and offers a plausible account for the causes of humour in jokes, for example. However, although incongruity seems intuitive and offers an explanation for (some) causes of laughter, it is a vague and general notion, with incongruities being available at all levels of linguistic interaction (e.g. phonology, semantics, pragmatics). It is therefore difficult to build a computational account of incongruity as it is currently conceived. In order to offer a more fine-grained account, we assessed (i) whether incongruity is recognised by naive coders and (ii) whether it can be subdivided into categories corresponding to Grice's conversational maxims [8].

Four of these maxims, defined by [8] as part of the cooperative principle of conversation which directs the interpretation of utterances in dialogue, can be briefly described as follows:

Maxim of Quantity	"Be exactly as informative as is required"
Maxim of Quality	"Try to make your contribution one that is true"
Maxim of Relevance	"Be relevant"
Maxim of Manner	"Be perspicuous"

Looking at a genuine example of laughter in dialogue, we now describe how flouting one of these maxims in dialogue can lead to a laughable, and the relationship to incongruity.

(1) A: they he had to fill out some forms but I guess California might be tougher I don't know
B: yeah they might be or you know how we are here in Texas it's [laughter: like] everybody's a hunter so **[laughter]** I'm not much of a hunter but
A: [laughter] yeah [noise] (Switchboard, sw2014, *discussing gun control*)

Focusing on B's second laughter (shown in **bold**) we can see that the laughter was caused by the the utterance: "Here in Texas it's like everybody's a hunter". What can be said about this laughable? Definitely, that some sort of stereotypical proposition was produced. Analysing this from a Gricean perspective we can say that B's contribution is not true, like any other gross generalisation that ascribes all the members of a population with a single common habit. It seems that from the perspective of both dialogue participants this statement is taken to be false, i.e. it violates the maxim of quality.

In terms of incongruity, we can state that a clash between certain scripts has taken place,[1] namely between the "regular situation", where not all of the population of the state are hunters and the "constructed situation", where all the population are hunters.

[1] See [19, Chap. 6] for analysis of similar content in jokes.

In this case, we can see that the incongruity itself arises because of the violation of the maxim of quality.

The functional role of the laughter here could be explained as indication by speaker B, that s/he is aware that the utterance is not literally true. The laughter of speaker A could be interpreted as showing her/his awareness of that and an acknowledgement of B's statement.

The observation that laughter can be caused by the violation of Gricean maxims led us to develop a preliminary annotation scheme for analysing laughter in dialogue is terms of incongruities that can be sub-categorised according to these violations. Specifically we ask: (a) how different are laughters in terms of their causes and functions, (b) whether laughters are connected to violation(s) of the Gricean maxims, (c) whether laughters are caused by incongruity of some sort, (d) to what extent do people agree in their judgements regarding various features of laughables.

2 Annotation Scheme

For our preliminary study, we randomly selected one full dialogue from The Switchboard Dialog Act Corpus (SWDA) [10], 5 excerpts from other conversations in SWDA (provided with a brief context) and 5 from part of the British National Corpus (BNC), previously analysed for laughter [14]. SWDA consists of dyadic telephone conversations between American participants who were unfamiliar with each other on a pre-determined topic, while the spoken portion of the BNC consists of British face-to-face dialogues from a range of contexts (see [1], for details).

We asked participants to fill in the following questionnaire:

- Q1 How well have you understood the given laughter? (from 1 to 5)
- Q2 Please indicate the line where the cause for laughter occurs.
- Q3 Was the laughter caused by something that the laugher says her/himself or the partner says?
- Q4 Does the cause occur before, during, or after the laughter?
- Q5 Was the laughter caused because one of the participants (from the laugher's perspective):
  - Q5.1 gives more or less information that was needed?
  - Q5.2 gives information that was false or wasn't supported by evidence?
  - Q5.3 gives information that was irrelevant for the discussion?
  - Q5.4 gives information that was obscure or ambiguous?
  - Q5.5 says something that clashed with a certain background information, common sense, another interpretation or another utterance?
- Q6 Please explain the cause of the laughter.
- Q7 Please explain why the person has laughed.

Q1 was provided to give a self-estimated confidence score for the following questions. Questions Q2–Q4 are about some basic properties of laughables which are usually considered to be agreed upon. Questions Q5.1–Q5.4 represent the Gricean maxims and Q5.5 explicates the notion of incongruity in way that is comprehensible for the coders. Q6 and Q7 are free form questions that give coders an opportunity to explain, respectively, the cause and the function of the laughter. We also provided coders with an example of annotation for example (1).

3 Preliminary Results

The results that we report here are from a pilot study with 3 annotators.[2] While there is not enough data to calculate inter-annotator agreement, the free-form answers to Q6 regarding the cause of laughter suggest that, at least in some cases, coders understand and agree on the cause of the laughter.

(2) Ian: [pause] basic details, name [pause] and address, telephone number,
John: Okay, yeah.
Ian: national insurance number, date of birth.
Ian: Erm another code number form a directory [pause]
John: **[laugh]** (BNC, JNW, 402–405)

(3) Patrick: Oh if you don't think they look well then they obviously need it if they look better after they've been watered, that's what the paper says.
Katherine: Well then they do need water.
Patrick: That's the answer
Katherine: They [unclear]
Patrick: if they look as though they need it they need it but if they don't look as though they need watering don't water them.
Katherine: Well [pause] look, look at the birds **[laugh]** I [unclear dur=6] aren't they sweet [pause] all the same I shall buy a nesting box next er next year.
Patrick: Mm. (BNC, KCV, 300–305, *discussing some plants)*

(4) B: there's an old profane expression about Texas weather,
B: it's always too damn cold, too damn hot, too damn windy **[laugh]**.
(SWDA, sw3936, 391–392)

(5) B: and you know, I mean, a lot of people they go, they're better than the Beatles,
B: and I'm like you know,
A: [laugh].
B: you don't know what you're talking about.
A: No [laugh].
B: I mean, the comparison made between New Kids On The Block with the Beatles [laugh]. It was just,
A: You can only laugh [laugh]. (SWDA, sw2020, 822–931)

In Example 2, there was total agreement on the violation of the maxim of quantity (too little information, Q5.1), and 2 out of 3 coders annotated obscurity in Ian's utterance (violation of the maxim of manner, Q5.4).

[2]The annotators were not native English speakers, which may mean they did not pick up on all the subtleties of the laughter and laughable. However examples in the BNC are also not necessarily produced by native speakers, and there are also cultural differences which are known to affect interpretations of humour and laughter even between native speakers (e.g. between American and British speakers of English). In future studies (see Discussion, below) we intend to involve a wide range of annotators, including native and non-native speakers of English.

In Example 3, coders agree on the violation of the maxim of relevance by the sudden change of topic (Q5.3).

In Example 4, coders recognise incongruity against some "normal situation" (Q5.5). For Q6, regarding the cause of laughter, one of the coders wrote: "Normally a place is either too cold, or too hot, or too windy. It is hard to have all the extremes".

Example 5 is interesting, because all the coders agree that neither of laughters are caused by violation of any of the Gricean maxims. Nevertheless, the coders agree that these laughters are caused by incongruity from comparing the incomparable Beatles with a lesser band. According to the comments given by annotators, the attempt to compare any band with The Beatles seems ridiculous to both interlocutors in (5) and their laughters are driven by this.

Some of the presented excerpts show that even for humans it can be hard to describe the cause and function of laughter even when they understood the laughters quite well. Example 6 shows disagreement between the coders regarding the position of the laughable (whether it occurred before or after the laughter); the cause of the laughter (e.g. "Saying something sad about another person" vs. "Being depressed of other peoples' problems, and at the same time bringing them their problems"); and its function ("Softening" vs. "Marking incongruity").

(6) A: We have a boy living with us who works for a credit card, uh, company that,
A: and he makes calls to people who have problems, you know, credit problems,
B: Huh-uh.
A: that are trying to work out
A: and, uh, **[laugh]**. Poor thing he comes home very depressed every night [laugh],
B: Oh. (SWDA, sw2883, 451–481)

4 Discussion and Future Work

We believe that this approach, together with the precise identification of laughables in dialogue, can contribute towards an implementable account for identifying events where laughter can be appropriate, i.e. as a result of violating Gricean maxims (changes of topic, irony and sarcasm, jokes, bold statements). However, it is not the case that every violation of a Gricean maxim or incongruity in dialogue results in laughter, and we therefore believe that this kind of analysis should also be carried out more generally, with some additional account of which potential laughables in dialogue are more likely to elicit laughter (we expect this to be modulated by, for example, familiarity of dialogue participants, formality of the domain, intonation and other non-verbal cues etc). The precise positioning of the laughter with respect to the laughable may also offer clues in understanding what triggers the laughter, and help to differentiate between emotional or social causes and incongruous or humorous causes (though of course, as with other features of dialogue, any given laughter event may be multifunctional) which we also intend to investigate in future work.

We intend to run similar experiments with broader coverage of examples and annotators using Amazon Mechanical Turk. Given the shortcomings of agreement calculation using chance-adjusted metrics, e.g. Krippendorff's α, for tasks such as ours, we will use a probabilistic annotation model [3] that has been successfully applied to crowdsourced NLP data collection tasks, such as word sense annotation [17]. In these tasks, as with our laughter annotation, there is no gold standard and these methods are more reliable for deriving the ground truth from the population of annotators.

We are also aware of the role of prosody and phonetic form of laughter in identifying its causes and functions, and our annotators reported that audio would have been helpful for better understanding. We therefore plan to extend our text-based samples with audio to check whether it improves inter-annotator agreement.

Our ultimate aim for this work is to implement a spoken dialogue system (for a limited domain) which can understand, produce and reason about laughter in its dialogues with users, and to demonstrate how laughter contributes semantic and pragmatic import to dialogue. This kind of system would be a proof of concept that can be used to test theoretical insights about human conversation.

Acknowledgements This research was supported by a grant from the Swedish Research Council for the establishment of the Centre for Linguistic Theory and Studies in Probability (CLASP) at the University of Gothenburg. We would also like to thank Staffan Larsson, Jonathan Ginzburg, Chiara Mazzocconi and our anonymous reviewers for their useful comments.

References

1. Burnard L (2000) Reference guide for the british national corpus (World Edition). Oxford University Computing Services. http://www.natcorp.ox.ac.uk/docs/userManual/
2. Chapman AJ (1983) Humor and laughter in social interaction and some implications for humor research. In: Handbook of humor research. Springer, pp 135–157
3. Dawid AP, Skene AM (1979) Maximum likelihood estimation of observer error-rates using the em algorithm. Appl Stat 20–28
4. Ding Y, Prepin K, Huang J, Pelachaud C, Artières T (2014) Laughter animation synthesis. In: Proceedigns of AAMS 2014. International Foundation for Autonomous Agents and Multiagent Systems, pp 773–780
5. El Haddad K, Çakmak H, Gilmartin E, Dupont S, Dutoit T (2016) Towards a listening agent: a system generating audiovisual laughs and smiles to show interest. In: Proceedings of the 18th ACM international conference on multimodal interaction. ACM, pp 248–255
6. Ginzburg J, Breitholtz E, Cooper R, Hough J, Tian Y (2015) Understanding laughter. In: Proceedings of the 20th Amsterdam Colloquium
7. Glenn P (2003) Laughter in interaction, vol 18. Cambridge University Press
8. Grice H (1975) Logic and conversation. Syntax Semant 3(S 41):58
9. Hempelmann CF, Attardo S (2011) Resolutions and their incongruities: further thoughts on logical mechanisms. Humor-Int J Humor Res 24(2):125–149
10. Jurafsky D, Shriberg E, Biasca D (1997) Switchboard SWBD-DAMSL shallow-discourse-function annotation coders manual. Institute of Cognitive Science Technical report 97–102
11. Katevas K, Healey PG, Harris MT (2014) Robot stand-up: engineering a comic performance. In: Proceedings of the workshop on humanoid robots and creativity at the IEEE-RAS international conference on humanoid robots humanoids (Madrid)

12. Kaushik L, Sangwan A, Hansen JH (2015) Laughter and filler detection in naturalistic audio. In: Proceedings of Interspeech Germany
13. Mazzocconi C, Tian Y, Ginzburg J (2016) Multi-layered analysis of laughter. In: Proceedings of SemDial 2016, Proceedings of the 20th workshop on the semantics and pragmatics of dialogue, Rutgers
14. Mazzocconi C, Tian Y, Ginzburg J (in prep) What's your laughter doing there? A taxonomy of the pragmatic functions of laughter
15. Niewiadomski R, Hofmann J, Urbain J, Platt T, Wagner J, Piot B, Cakmak H, Pammi S, Baur T, Dupont S et al (2013) Laugh-aware virtual agent and its impact on user amusement. In: Proceedings of the 2013 international conference on Autonomous agents and multi-agent systems, pp 619–626
16. Nijholt A, Niculescu AI, Alessandro V, Banchs RE (2017) Humor in human-computer interaction: a short survey. In: Adjunct conference proceedings interact 2017
17. Passonneau RJ, Carpenter B (2014) The benefits of a model of annotation. TACL 2:311–326
18. Poyatos F (1993) Paralanguage: a linguistic and interdisciplinary approach to interactive speech and sounds, vol 92. John Benjamins Publishing
19. Raskin V (1985) Semantic mechanisms of humor. Synthese language library, vol 24. Reidel, Dordrecht
20. Scott SK, Lavan N, Chen S, McGettigan C (2014) The social life of laughter. Trends Cogn Sci 18(12):618–620
21. Tahon M, Devillers L (2015) Laughter detection for on-line human-robot interaction. Cough 85(65.0):0–77
22. Tepperman J, Traum D, Narayanan S (2006) "Yeah right": sarcasm recognition for spoken dialogue systems. In: Ninth international conference on spoken language processing
23. Tian Y, Mazzocconi C, Ginzburg J (2016) When do we laugh? SIGDIAL Conference 360–369
24. Truong KP, Van Leeuwen DA (2007) Automatic discrimination between laughter and speech. Speech Commun 49(2):144–158

Humor Intelligence for Virtual Agents

Andreea I. Niculescu and Rafael E. Banchs

Abstract Humor is pervasive in human social relationships and one of the most common ways to induce positive affect in others. Research studies have shown that innocent humor increases likeability, boosts trust, reduces tension, encourages creativity and improves teamwork. In this paper, we present a study focusing on deploying humor in interaction with a virtual agent. 25 participants evaluated the logs of conversations exchanged between a human user and two virtual agents acting as tour guides. Even though answers were equal in terms of content delivered, one agent used humorous statements to respond the queries while the other agent presented the content in a neutral way. To create answers with a humorous effect we combined information extracted from various websites focusing on tourist fun facts, pun and jokes collections. Results showed that our manipulation was successful, i.e. the humorous agent was indeed perceived as being significantly funnier. Additionally, the agent was perceived as delivering more interesting answers as compared with its counterpart. Further, participants showed statistically significant preferences towards the humorous agent when asked to choose between the agents. As such, we believe that using humor in interaction with virtual agents increases the agent likeability and possibly contributes towards a better user experience.

A. I. Niculescu (✉)
Institute for Infocomm Research, A*STAR, 1 Fusionopolis Way #21-01 Connexis South Tower, Singapore 138632, Singapore
e-mail: andreea-n@i2r.a-star.edu.sg

R. E. Banchs
School of Computer Science and Engineering, College of Engineering, Nanyang Technological University (NTU), 50 Nanyang Avenue, Block N4 #02a-32, Singapore 639798, Singapore
e-mail: rbanchs@ntu.edu.sg

L. F. D'Haro et al. (eds.), *9th International Workshop on Spoken Dialogue System Technology*, Lecture Notes in Electrical Engineering 579,
https://doi.org/10.1007/978-981-13-9443-0_25

1 Introduction

Humor is a pervasive phenomenon in human society being described as the natural tendency to create laughter and good mood [1]. Research has shown that humor has biological roots and plays a beneficial role in a variety of social functions, such as reducing stress, aiding in education or defining gender identity [2].

Over the time, humor has been intensively investigated in many areas, such as psychology, sociology, biology, linguistics or computer science (CS). In the natural language processing (NLP) and artificial intelligence (AI) communities for example, humor is an established field with own research agenda. In contrast, in the human computer interaction (HCI) humor appears to be a rather neglected research topic, despite that—as a CS discipline focusing on interaction—it could greatly benefit from its deployment.

One reason is that HCI traditionally focus on interfaces meant to increase task performance on one side, and minimize task duration, learning time and error rate, on the other side. Since the use of humour would distract users from their tasks increasing the total completion time, it would contradict HCI policies of maximizing efficiency in interaction [3].

However, in the future we can expect HCI to become less goal directed not only in entrainment computing, but also in our ordinary daily life. Technology moves slowly from working environment to our living room where computers and artificial entities are becoming social actors [4]. Therefore, it is important for interaction designers working in applied AI to take into account the social aspect of interaction between humans and computers where humor could become an influential ingredient towards technology acceptance and overall user satisfaction.

As such, in this paper, we are exploring the use of humor in interaction with a virtual tour guide. Currently, our tour guide combines task-oriented dialogues with Chabot functionalities: it offers information about Singapore, as well as informal chatting.

To explore the impact of humor in interaction with the agents, we handcrafted conversation scripts with humorous answers and asked test participants to evaluate them against scripts with neutral answers. Since the answers were manually generated by our team, there was no external validation on whether they were indeed perceived as humorous.

Thus, the goal of our experiment was twofold: (1) first, the study is aiming to validate the handcrafted humor data set; (2) second, the study tries to verify our hypothesis concerning the participants' preference towards the humorous agent; according to this hypothesis, we expect participants to perceive the information delivered by the humorous agent as more useful and the interaction more interesting.

2 Related Work

A handful of researchers explored the use of humor in interaction with virtual entities. The study by Morke et al. [5] found that participants who received humorous comments from a computer rated the system significantly more cooperative, more likable and more competent. During the experiment those who received the humorous comments also smiled and laughed more displaying a more social behaviour as compared to those who didn't receive such comments.

Another study by Huan and Szafir [2] investigated humor in computer mediated learning. Their results showed that using humor significantly increased the instructor likeness, regardless if this was a human or robot.

The study by Niculescu et al. [3] showed that humour in interactive task with a social robot receptionist increase the users' likeliness towards the robot's speaking style and personality, as well as towards the overall task enjoyment.

Also, the study performed by Dybala et al. [6] proved that users interacting with conversational agents rated the humorous agent as being more funny and likeable and more human-like. Comparing the two agents—humorous versus non-humorous—the authors found out that better ratings were given to the humorous agent.

Humor can be also used to help a Chabot system recovering from errors: a study by Niculescu and Banchs [7] showed that using humor in situations when the system is unable to retrieve the correct answer may prompt the user to reformulate the query, thus helping the system to recover from errors.

Last but not least, the study by Babu et al. [8] found that a humorous virtual receptionist may foster user participation in social conversations by using jokes.

All these experiments demonstrate that humor can have positive effects in HCI enhancing the overall user experience and keeping the users engaged in the interaction with the system.

3 The Virtual Tour Guide

Our experiment is based on virtual tour guide that offers touristic information about Singapore [9]. At the core, the tour guide is a multimodal dialogue system based on a client-server architecture. The system is composed of two modules operating in cascade: a rule-based module accessing information from two databases—a handcrafted database and a database composed of resources automatically extracted from web directories—and a data driven module accessing information from an index based on Wikipedia articles. A learning module complements the architecture enabling the system to learn from users' answers.[1] The overall system architecture is presented in the Fig. 1.

[1]At the moment, this module is disabled and under development.

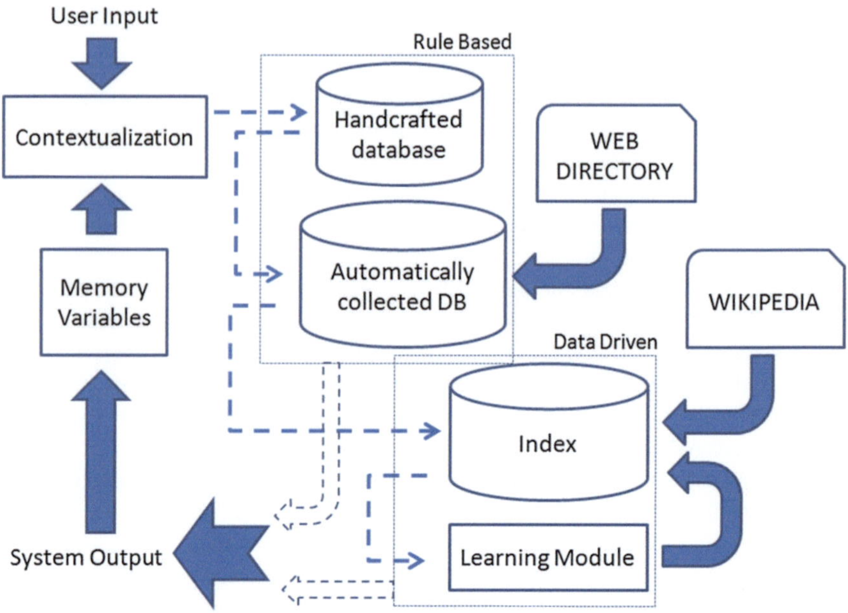

Fig. 1 Overview architecture SARA

3.1 Databases

The first data base—the handcrafted database—is created by adding manually responses about relevant tourist spots in Singapore. It includes a total of 75 locations of interests concerning museums, theatres, temples, historical buildings, parks and heritage sites. The automatically collected data resources—the second database—come from crawling web directories about Singapore.

The database contains about 8000 entries concerning shopping malls, eateries, hotels, restaurant recommendations, transportation, stores etc. The system would access this database only if it fails to find an answer in the first database.

If the system fails to find an answer in the second database as well, the user input is passed to the data driven module. This module is implemented by an example-based question answering system. While the databases deal with queries that are specific to touristic attractions and other venue locations, the index contains a collection of question-answer pairs on more general information about Singapore.

The index data collection was automatically populated by crawling and processing Wikipedia pages related to Singapore [10]. Each index pair is matched with a given input query to compute the contextual similarity between them [4]. If an index pair has a higher similarity score compared to other examples, its answer is considered as first candidate for the system response.

3.2 Multimodal Dialogue System

Server side

As suggested in the previous paragraph, for the dialogue manager (DM) implemented in our system we chose two different strategies: a rule-based approach and an example based approach. The rule base approach uses a set of manually defined heuristics to determine an appropriate answer on the system site. For the example-based approach, we use Lucene,[2] a special library meant for developing search engines.

We use the rule-based approach for handling task-oriented queries while the example-based approach focuses on more general questions.

Concerning the natural language understanding (NLU) module and dialogue topic tracking, the system uses a hybrid approach. The user input is transformed into a semantic representation using rules and statistical models. These are built based on data collected for the Singaporean touristic domain.

We used around 40 h of human-human dialogue to train the system. The data was collected in both English and Chinese[3] as dialogues exchanges between visitors and tour guides. Dialogues were further manually annotated on several sematic levels, such as word, utterance, and dialogue segment.

To deliver the answer to the user, a natural language generation (NLG) module is being used. The NLG module of the rule-based dialogue manager uses a template-based approach.

For the data driven approach, no NLG component is used: the answers are provided exactly as extracted from Wikipedia pages. Once the answer is generated, it is further sent to the web client for image/map/web display and text-to-speech (TTS) generation.

The system main components are linked together using Apollo, a pluggable dialogue platform allowing the interconnection and control of the different independent components, as shown in Fig. 2 [11].

Client side

On the client side, the user interacts with a web browser. The web-based client interface shown in Fig. 3 has several components. From top to bottom, these components are: the avatar, the text input field and the response.

The avatar, as seen in the top left of the picture, is used to provide a spoken response to the user by reading out the text returned by Apollo. The user can input his query in the upper field and receives the answer in the form of a spoken answer (by the avatar), as well as written as text—in Fig. 3, both question and the answer prompted by the system can be visualized in the black response area. The user can use typing or speech to input the query.

To input a query, the user needs to press the grey button located on the right-side of the input field. The web interface contains at the bottom additional information

[2]http://lucene.apache.org/.

[3]For our current experiment, only the English version was used.

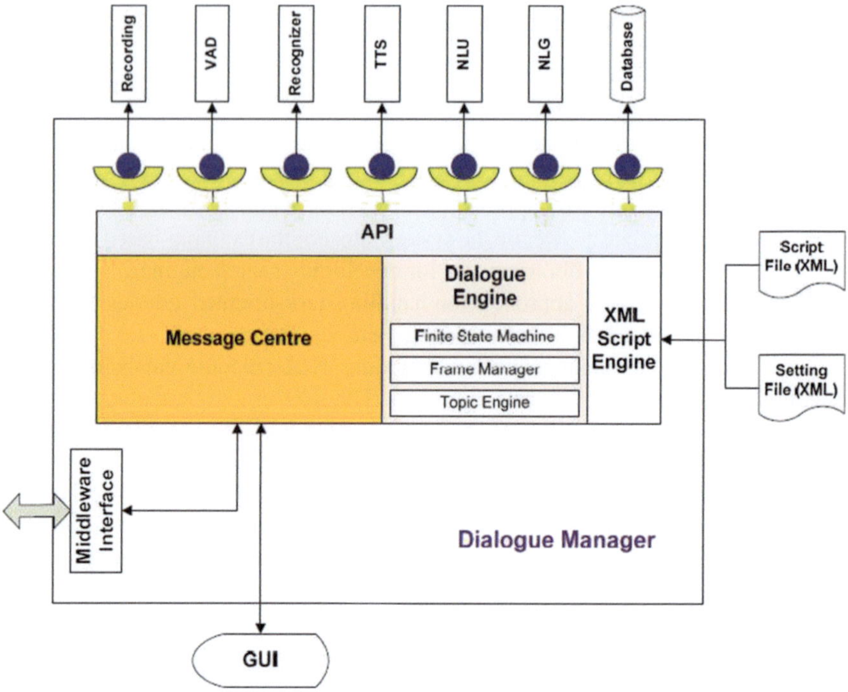

Fig. 2 Apollo architecture

related to the query, such as websites related to the query, maps, traveling direction etc. In Fig. 3, the user is presented with the website of the Singapore Flyer which is referenced in the question. This webpage is retrieved from the database and can be configured in the server-side scripts.

In terms of content, the system provides answers about local sightseeing, restaurant recommendations, travel advice based on the current user location, i.e. using Google location services, can make online reservation and send SMS for confirmation.

Additionally, the system understands questions in context, i.e. the system is able to related the question to the previous one shown in Fig. 3—"there" stands for "Singapore Flyer".

4 Methods

In order to carry out our experiment, we prepared 2 sets of conversation logs (set A and set B). Each set contained 3 different dialogues scripts (1, 2, 3). The participants were first introduced to the system and had the opportunity to play around with it, i.e. to ask their own questions and test its functionalities. Further, they were asked to

Fig. 3 System's web client

read the scripts and to fill a questionnaire form after each log they read. The exposure to the log conversations was randomized between the agents to avoid biases.

4.1 Conversation Logs

Each dialogue script was composed by 12 questions & answers (Q&A) pairs. The pairs covered greeting exchanges, chitchat (*What is your name? Do you like Chinese food?*), personal questions (*Do you love me? Would you marry me?),* focused questions about locations of interest, stores, restaurant recommendations, direction advise and general purpose questions about Singapore (*How is the weather in Singapore? What is the exchange rate?* etc.). Additionally, the logs contained questions asked in context, i.e. questions requesting anaphora resolution, as in the example shown in Fig. 3. We include in the scripts also questions that were unknown to the system, i.e. questions for those typically no answers could be found in the database.

Participants were told the first set (A) came from a virtual agent called Tyra while the second set (B) was stemming from an agent named Sara. The difference between the Tyra log set was made at the answer formulation level: while the user questions for both Sara and Tyra were identical, the answer Tyra gave were formulated in a neutral form; in contrast, Sara's answers contained jokes and were usually formulated in a humorous way—see Table 1 for a few examples. Both Sara and Tyra delivered the same amount of information in terms of content.

Tyra's logs were extracted from real conversations between users and system exchanged during past testing sessions. Sara's logs were adapted to follow a humorous paradigm: her statements were chosen from websites containing fun facts about Singapore [12, 13], funny puns [14] and humorous answer collections [15]. Each Sara script log included between 6 and 9 humorous statements.

Table 1 Q&A examples from the logs used during the experiment

Type: question	Set A Tyra answer	Set B Sara answer
Chat: "What is your job?"	"I am a tour guide!"	"I am a CIA agent disguised as tour guide!"
Chat: "What do you do?"	"I am a tour guide"	"I am painting my nails while waiting for people to ask me question about Singapore!"
Personal: "Do you love me ?"	"Sorry, I am not sure about that!" (*agent has no understanding of the question*)	"Ah, you are looking for love in all the wrong places!" (*agent has no understanding; use word association to create an answer*)
General: "How is the weather in Singapore?"	"Singapore has uniform temperature and pressure m high humidity, and abundant rainfall" *[Link on weather report of the day]*	"Usually hot, hot, hot like chilli peppers!" *[Link on weather report of the day]*
General: "What do you know about Singapore?"	"Singapore, officially the Republic of Singapore, is a sovereign city-state and island country in South East Asia"	"I know the most important things: don't walk around naked, don't spit, don't litter, don't carry around durians and always remember to flush the toilet"
General (more focused): "Are there any nice beaches?"	"In Sentosa, there are many nice beaches, such as Siloso or Palawan beach"	"Oh, well I never go to the beach because of skin cancer and all, but in Sentosa, there are many nice beaches such as Siloso or Palawan beach"
Focused (question in context): "What are the opening hours [of Underwater World]?"	"Unfortunately, the aquarium has closed down."	"Unfortunately, the aquarium has closed down. I heard, after closing down they invited everyone to a sea BBQ party"

4.2 Questionnaire

The questionnaire form comprised 3 sections, one for each log script. Each section was divided in 2 parts, one for each virtual agent. Participants filled in their response concerning the answer usefulness, whether the agent was funny, the conversation interesting and how much they liked the agent in that particular script. The questions were formulated as statements, i.e. "I liked the agent in this dialogue conversation" on a 5 point Likert scale from "*strongly disagree*" to "*strongly agree*". At the end of each section and after reading the scripts of both agents, were asked to choose which agent would they prefer to talk to. Additionally, they could leave comments, if any.

5 Results

The questionnaire allowed us to gather both another statistical and qualitative data in the form of comments.

5.1 Demographics

A total of 25 persons performed the test, 13 women and 12 men with ages between 19 and 51. They were chosen from staff and students working in our department. Good English knowledge was compulsory for participation.

The majority (19–76%) was composed by: local Singaporeans & long term residents (13) and Asians (6) from China, Malaysia, Myanmar and Vietnam. However, we also had participants from Colombia (2), USA (1), Romania (1) and Russia (2).

More than half of the participants (13–52%) were below 30 years old; 6 were between 30 and 40, 5 were between 40 and 50 and 1 was between 50 and 60.

5.2 Quantitative Data

As a *Kolmogorov–Smirnov* test for normality run on our data proved to be negative, we used a Wilcoxon signed-rank test. The analysis showed that SARA was significantly perceived as being funnier and more interesting to interact with as compared to Tyra across. This finding is valid across all three scripts—see Table 2 for calculated Z asymptotic, median & median rank and p-values.

In terms of content usefulness, there were no significant differences between the agents—thus infirming our initial hypothesis. The explanation is that, one side the answers were formulated equal in terms of content—as such, the humor did not appear to contribute towards answer's usefulness.

Table 2 Results of the Wilcoxon Signed ranked test for the 3 scripts

Script 1 N = 25	**Answer is useful**	**Agent is funny**	**Conversation is interesting**	**I like the agent in this conversation**
Z	−1.748[a]	−4.086[b]	−2.942[b]	−1.882[b]
Asymp. Sig (2-tailed)	.080	**.000**	**.003**	.060
Sara: mean rank+/median	8.50/4	**12.30/4**	**8.46/4**	9.42/4
Tyra: mean rank+/median	4.56/4	5.50/2	5.00/3	8.00/4
Script 2 N = 25	**Answer is useful**	**Agent is funny**	**Conversation is interesting**	**I like the agent in this conversation**
Z	−1.508[a]	−4.061[b]	−3.175[b]	−1.384[b]
Asymp. Sig (2-tailed)	.132	**.000**	**.001**	.166
Sara: mean rank+/median	4.00/4	**11.83/4**	**9.12/4**	9.63/4
Tyra: mean rank+/median	4.67/5	4.50/2	16.00/3	9.25/3
Script 3 N = 25	**Answer is useful**	**Agent is funny**	**Conversation is interesting**	**I like the agent in this conversation**
Z	−1.748[a]	−4.086[b]	−2.942[b]	−1882[b]
Asymp. Sig (2-tailed)	.577	**.000**	**.005**	**.013**
Sara: mean rank+/median	6.40/4	**11.33/4**	**8.00/4**	**9.33/4**
Tyra: mean rank+/median	6.57/4	4.50/2	4.50/3	10.33/3

[a]Based on positive ranks, [b]Based on negative ranks

On the other side, the humor effect might not have had the strong effect that we hoped for. Also, our number of participants was relatively low to enable strong and significant statistical differences.

Also, there was no statistically significant difference between Sara and Tyra concerning the agent likeliness except for script 3—where Sara was significantly more liked by participants as compared to Tyra. The reason for this preference might have been related to the agent's humorous answers: we found significant correlations for script 3 between liking Sara and the agent capability of being funny ($r = .81, p < .001$) and offering an interesting conversation ($r = .78, p < .001$).

Concerning their preference for Sara or Tyra, from a total of 75 responses along the three different tests (25×3) 51 participants chose the agent Sara while 24 decided

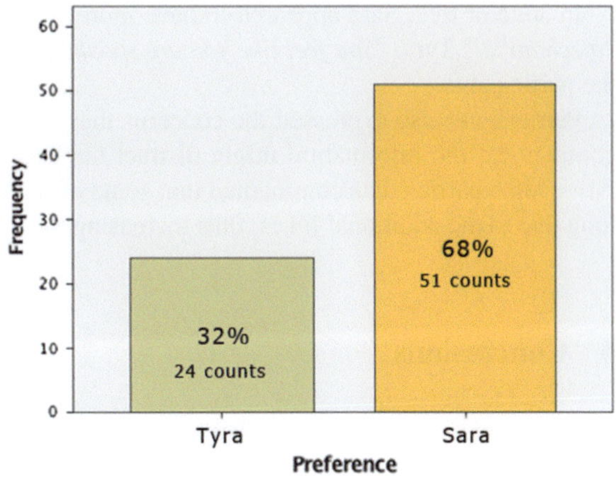

Fig. 4 Total percentage difference between agents

for Tyra—see Fig. 4. A Binomial test found the proportion of participants choosing Sara to be significantly higher ($p = .002$) as compared with the proportion of voters for Tyra—the proportion was measured at an expected .50 level.

5.3 Qualitative Data

The questionnaire allowed us to collect a total of 31 comments from our participants. Since comments were optional not all participants choose to do so.

A common trend among the comments was that Sara was indeed perceived as funnier and more entertaining as comparing to Tyra. Tyra was often considered "*bland*", "*boring*" and "*unnatural*". Sara however, was perceived not equally funny in all scripts, e.g. some participants commented for example that script 3 was more entertaining as compared with script 2—a fact that was also confirmed by our statistical analysis.

A foreign participant suggested the use of "*lah*", as "*it might sound funny*". "*Lah*" is a local spoken feature commonly used at the end of a sentence to emphasize content.

Another two participants criticized Sara for being sometimes too "*negative*" and "*blunt*" in her statements while three other participants seemed to have difficulties understanding some of the agent's jokes and thought the agent gave them a wrong answer.

Some local participants indicated that Sara's statements could be sometimes perceived as "*offensive*", "*negative*", "*sarcastic*" or even "*rude*": "*Sara is funny/sarcastic, but I am not sure if everyone find her like that. There may be some cultural/age determined sensitivity*" wrote a local participant.

In spite of that, Sara appeared to have more "*human touch*" as compared to the "*mechanical*" Tyra: "*You feel like you are speaking to a real person …*" wrote on of the participants.

Participants also expressed the concerns that humor despite being a nice way to complement the information might distract the user's attention from their search. Also, some participants commented that some of Sara's answers appeared to be too long due to the additional jokes, thus increasing the total interaction time.

6 Conclusions

The results of our experiment demonstrated that our humor manipulation was successful: the agent Sara was perceived significantly funnier as Tyra.

Further, our experiment showed that participants perceived the humorous agent engaging users in more interesting conversations. Also, participants seemed to prefer the humorous Sara over the more neutral Tyra. All these results suggest that humor might have beneficial effects when designed to support the interaction with virtual agents.

However, designing humorous responses is not an easy task: as pointed out by our participants' comments, humor effects are highly context dependent and subjected to cultural constrains. More traditional cultures, such as the Asian cultures for example, might be less open to use humor in formal situations or in situation of status inequality between individuals.

Personal preference and timing play also an important role on whether humor has the intended effect: no everyone has the same taste for jokes or is—at a given moment—in the right mood for jokes; also, some people might be in a hurry looking for a short, straight, informational answers, thus listing to jokes might be counterproductive for their goal.

Additionally, having a good command of the language spoken/written in interaction with the agent helps understanding the humorous intention.

Since Chabot systems—once available online—can be virtually accessed by anyone, having all these differences accounted for within a single system is a huge challenge that future research needs to address.

In the future, we plan to continue our work on implementing humor in interaction with our agent Sara by automatizing the responses created from a handcrafted database of humorous answers and jokes. The choice of answer will be based on the validated preferences from large test users pool. Further, we plan testing our humorous agent in direct interaction with users to detect best algorithmic strategies to complement the interaction with humor in the appropriate context.

Acknowledgements We would like to thank to our 25 test participants who spent time and effort helping us with this experiment.

References

1. Wikipedia Humor. Online available: https://en.wikipedia.org/wiki/Humour. Retrieved on 19 Feb 2017
2. Huang CM, Szafir D (2001) No joke: examining the use of humor in computer mediated learning. Unpublished material
3. Niculescu AI, van Dijk B, Nijholt A, Li H, See SL (2013) Making social robots more attractive the effects of voice pitch, humor and empathy. Int J Soc Robot 5(2):171–191
4. Nijholt A, Stock O, Dix A, Morkes J (2003) Humour modeling in the interface. In: CHI'03 extended abstracts on human factors in computing systems, pp 1050–1051. ACM
5. Morkes J, Kernal HK, Nass C (1999) Effects of humor in task-oriented human-computer interaction and computer-mediated communication: a direct test of SRCT theory. Hum-Comput Interact 14(4):395–435
6. Dybala P, Ptaszynski M, Rzepka R, Araki K (2009) Humoroids: conversational agents that induce positive emotions with humour. In: Proceedings of 8th international conference on autonomous agents and multi agent systems (AAMAS), pp 1171–1172
7. Niculescu AI, Banchs RE (2015) Strategies to cope with errors in human machine spoken interactions: using chatbots as back-off mechanism for task-oriented dialogues. In: Proceedings of ERRARE, errors by humans and machine in multimedia, multimodal and multilingual data processing
8. Babu S, Schmugge S, Barnes T, Hodges LF (2006) What would you like to talk about? An evaluation of social conversations with a virtual receptionist. In: Proceeding of intelligent virtual agents, pp 169–180. Springer, LNCS Series
9. Niculescu AI, Yeo KH, D'Haro LF, Kim S, Jiang R, Banchs RE (2014) Design and evaluation of a conversational agent for the touristic domain. In: Proceedings of Asia Pacific Signal and Information Processing Association (APSIPA)
10. Kim S, Banchs RE, Li H (2014) Wikipedia-based Kernels for dialogue topic tracking. In: Proceedings of ICASSP
11. Jiang DR, Tan YK, Kumar Limbu D, Li H (2012) Component pluggable dialogue framework and its application to social robots. In: Proceedings of the Int'l workshop on spoken language dialog systems
12. Avakian T (2018) Travel tips: 16 odd things that are illegal in Singapore. Online available: http://www.stuff.co.nz/travel/destinations/asia/70915066. Retrieved on 22 Apr 2018
13. The fact file: Interesting facts about Singapore. Online available: http://thefactfile.org/singapore-facts/. Retrieved on 22 Apr 2018
14. Pun of the day (2016). Online available: http://www.punoftheday.com/cgi-bin/disppuns.pl?ord=F. Retrieved on 22 Apr 2018
15. Carfi J, Cliff C (2012) Brilliant answer for everyday questions. Carle & Carfi Publishing, Los Angeles
16. Salton G, Wong A, Yang CS (1975) A vector space model for automatic indexing. Commun ACM 18(11):613–620

Part VII
Chat-Oriented Dialogue Systems

Chat Response Generation Based on Semantic Prediction Using Distributed Representations of Words

Kazuaki Furumai, Tetsuya Takiguchi and Yasuo Ariki

Abstract In this paper, we propose a chat response generation system using distributed expression of words by Word2vec. With the conventional one-hot representation method, there was a problem in that the model becomes complicated as the vocabulary increases, and only the words that appear in the dialogue corpus can be handled. We address these problems by using Word2vec and extend it to handle unknown words that did not appear in the conversation corpus. In a subjective evaluation experiment, we showed that various responses can be generated by estimating words using semantic prediction.

1 Introduction

In recent years, dialogue systems have been actively studied owing to advances in artificial intelligence. There is a rule-based method for generating responses according to rules created beforehand in response generation of the dialogue system, but there is a problem that cost is required for various kinds of responses. The chat system in this research is not a presumption of a specific topic or task, but is called a non-task-oriented type which focuses on the dialog itself with humans. Unlike task-oriented systems, this system is required to deal with various topics. Therefore, it is necessary to use a method to generate a response sentence automatically instead of using a rule-based method.

As far as we know, word representation in dialog systems is mainly based on one-hot representation, but in chat, it is expected that the number of words handled will be very large. However, when trying to deal with various topics, if one-hot

K. Furumai (✉) · T. Takiguchi · Y. Ariki
Kobe University, Kobe, Japan
e-mail: kazuaki.furumai@stu.kobe-u.ac.jp

T. Takiguchi
e-mail: takigu@kobe-u.ac.jp

Y. Ariki
e-mail: ariki@kobe-u.ac.jp

L. F. D'Haro et al. (eds.), *9th International Workshop on Spoken Dialogue System Technology*, Lecture Notes in Electrical Engineering 579,
https://doi.org/10.1007/978-981-13-9443-0_26

representation is used, an increase in the number of dimensions of the word vectors cannot be avoided, and the model becomes complicated. In addition, it is not possible to generate response sentences other than the words that appear in the corpus, and it is heavily dependent on the corpus. Therefore, by using a semantic representation vector of fixed dimension learned with some text data beforehand instead of one-hot representation, it is possible to limit the complexity of the model and also to deal with words that do not exist in the corpus. In this paper, we propose a method to generate response sentences by word prediction using a Recurrent Neural Network, which utilizes distributed representations of words instead of one-hot representation.

2 A Dialogue System Using a RNN Encoder-Decoder

RNNs are often used as a method for automatically generating responses in dialog systems, as seen in the Neural Conversational model (Vinyals et al. [8]) and the Neural Responding Machine (Shang et al. [7]). It receives a sequence $X = (x_1, \ldots, x_{T_x})$ of an input word vector and outputs sequence $Y = (y_1, \ldots, y_{T_y})$ of an output word vector.

The hidden layer $h_{(t)}$ of a RNN can be represented by $h_{(t)} = f(h_{(t-1)}, x_t)$. This model uses a RNN for processing an input word sequence X as Encoder and a RNN for generating an output word series Y as Decoder and hidden layer $h_{(T_x)}$ for $h_{(0)}$ in Decoder; therefore, it is called a RNN Encoder-Decoder. In this paper, we use the RNN Encoder-Decoder model.

3 Distributed Representation of Words

Distributed representation of words is based on the distribution hypothesis and represents words by low-dimensional real-valued vectors, for which Word2vec proposed by Mikolov et al. [4–6] is mainstream. Using a one-hot representation makes it impossible to consider the relationship between words, whereas using a distributed representation makes it possible to perform operations such as (King − Man + Woman = Queen) that considers the meaning of words.

There are two learning methods of Word2vec, CBOW (Continuous Bag-of-Words) and Skip-gram. It is pointed out that learning by the Skip-gram model shows better results in [4], so in this paper we use the Skip-gram model to learn Word2vec.

4 The Proposed Method

In this paper, we generate a response sentence using the distributed representation vector proposed by Word2vec as an input/output vector of a RNN Encoder-Decoder. A schematic diagram of the model is shown in Fig. 1. First, the input word sequence

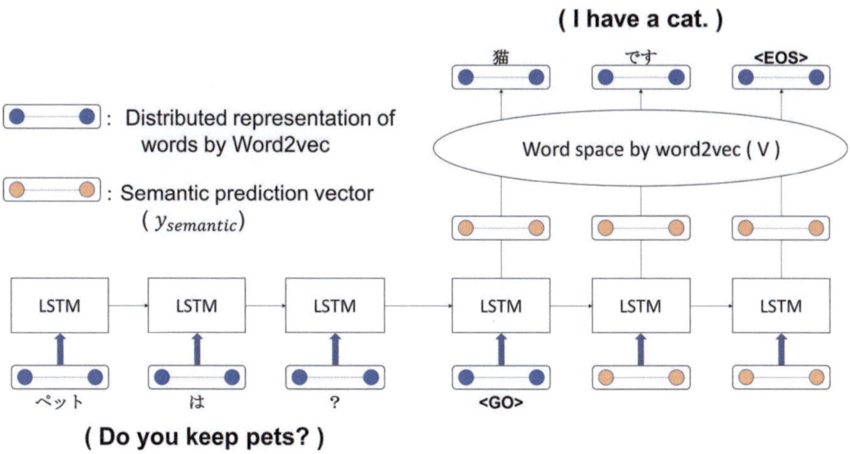

Fig. 1 Schematic diagram of the proposed method

is converted to the d_{word} dimension vector by Word2vec learned beforehand and input to the Encoder. Next, let the hidden layer $h_{(T_x)}$ generated by Encoder be the initial state $h_{(0)}$ of the hidden layer of Decoder. The output vector of the Decoder can be treated as the semantic prediction vector $y_{semantic_t}$, and is the d_{word} dimensional vector whose element takes a real number. When generating a response, this sentence prediction vector $y_{semantic_t}$ is used to output a response sentence with the word y_t having the highest cos similarity among the word vector set V created by Word2vec as the corresponding word. When the number of vocabulary is N, and the word vector learned by word 2 vec is $W_k \in V(k = 1, \ldots, N)$, it can be expressed as

$$y_t = \arg\max_{W_k} cos(y_{semantic_t}, W_k).$$

When the correct word sentence is $T = (t_1, \ldots, t_{T_t})$, the loss function L used for learning is $L = \sum_i |t_i - y_{semantic_i}|$.

5 Datasets

Since the data set that learns words and the data set that learns response sentence generation may be different, in this paper we prepared a dialogue corpus collected on Twitter and a data set created from Japanese Wikipedia articles. After shaping it into an appropriate format, we carry out a morphological analysis using MeCab [3].

Since we do not consider the speaker character and the dialog histories in this research, we created a dialogue corpus with 360,000 pairs of Tweet/Reply pairs from

Twitter. For learning the Word2vec, in addition to the collected dialogue corpus, we used Japanese Wikipedia articles (3 GB). We used the Twitter dialogue corpus for learning dialogue after learning Word2vec using these data sets.

6 Experiment

6.1 Experimental Conditions

The dimension number of a distributed representation of a word using Word2vec is $d_{word} = 128$. We excluded words with an occurrence count of 10 or less, and trained by Skip-gram model. As a result, the number of vocabulary became about 200,000 words. About the RNN Encoder-Decoder, a Long Short-Term Memory (LSTM) cell is used, and the number of units is 256, and the number of hidden layers is 3. Adam [2] is used as the optimization method of learning, and the learning rates are $\alpha = 0.0001$, $\beta_1 = 0.9$, $\beta_2 = 0.99$. $< GO >$ indicating the beginning of the sentence and $< EOS >$ indicating the end of the sentence were also learned as words with Word2vec as special symbols. When generating a response sentence using the RNN Decoder, $< GO >$ is input as the first word, and a response is generated until $< EOS >$ is output. In addition, when the current cos similarity is 0.5 or less, or when the current cos similarity is 60% or less of the cos similarity of the output word at one previous time, the word is excluded.

In machine translation, question answering, etc., BLEU [1] is used for evaluation, which calculates the score on the degree of matching of partial word strings. However, it has been is pointed out that there is a difference between evaluation by BLEU and manual evaluation. For example, there is no problem between languages having similar grammatical structures, such as English and French, but differences arise when carrying out tasks between languages with different the grammatical structures, such as English and Japanese. In the chat system, various combinations of input sentences and output sentences are considered, and it is a more complicated task, so it is conceivable that a difference is generated between BLEU evaluation and manual evaluation. Therefore, the following two evaluation indexes are prepared.

- Appropriateness: Degree of feeling that the system is reactive or understanding the input sentence
- Variety: Degree of performing various responses.

Regarding variety, evaluation criteria are based on whether it is possible to make a response peculiar to the conversation, rather than a safe response.

Response sentences were generated by 46 sentences which were collected from Twitter and not used for learning, and 5 point liker scale evaluations were conducted for each evaluation on each generated sentence by 10 participants.

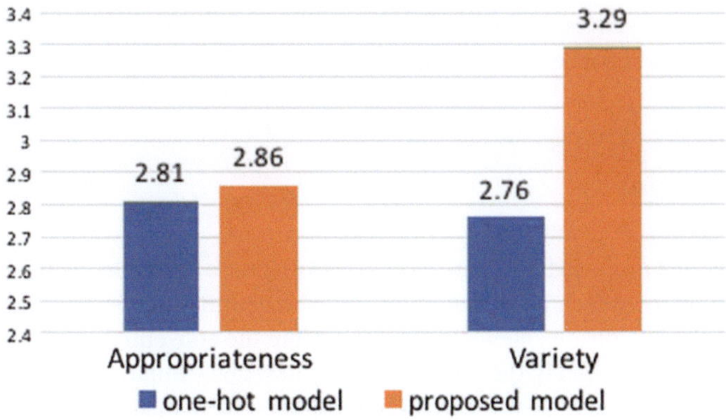

Fig. 2 Experimental results using 5-point Likert scale evaluation

6.2 *Results*

The experimental results are shown in Fig. 2. For each evaluation index, the results of averaging the respective evaluation values are shown. Compared with the one-hot representation method, the proposed method improved the variety. From the viewpoint that synonyms and relationships between words can be considered in the proposed method, the improvement of appropriateness was expected; however, there was actually little difference. However, there is an opinion that the judgment of appropriateness was difficult when the meaning of the input sentence (user's utterance) selected from Twitter is difficult to understand. This means that improvement of the questionnaire is necessary. Regarding variety, the influence of the beam search [9] used in the conventional method can be considered, so that a close examination will be necessary.

An example of a generated response sentence is shown in Table 1. In the one-hot model, since the words "Darth Vader" and "Guile" are not included in the training data and they were treated as $< unk >$, the response has not changed. On the other

Table 1 Examples of generated reply-sentences

input	output
初めまして〜お話しましょう (Nice to meet you. Let's talk.)	(W) ええよ！(OK!) (O) ありがとうございます（ ´∀｀ ）(Thank you.)
仲良くしてください！ (Let me be your friend.)	(W) ほんなら、何て呼びましょ? (Hey, what should I call u?) (O) なんて呼んでください（ ´∀｀ ）(Please call me.)
ガイル強いよね (Guile is strong, right?)	(W) 本当に...?(Sure?) (O) そうだった（¯ − ¯ ）(That's right.)
ダース・ベイダー強いよね (Darth Vader is strong, right?)	(W) 本当に！！！(I totally agree!) (O) そうだった（¯ − ¯ ）(That's right.)

(W) = Word2vec model, (O) = one-hot model

hand, in the proposed method (Word2vec model), each vector has been converted to another vector, and the response sentence has changed.

7 Conclusion

In this paper, we proposed a method to generate response sentences using distributed representations of words as input/output. We showed that the variety evaluation is higher than the response-sentence generation using the conventional one-hot model. Since, in the current model, the one with the highest cos similarity is output, and the process corresponding to beam search is not implemented, many response sentences with many grammatical errors are generated in many cases. In addition, even when words not included in the dialogue corpus appeared at the time of output generation, it often became ambiguous in terms of the whole sentence. In the future we will consider using seqGAN [10] and other models and aim to improve accuracy. In addition, since the Twitter corpus contains many noises, even a generated sentence that seems to be correct could have a bad evaluation. In the future, we need to think about a data set of multi-turn conversation which are less noisy and can also take dialog histories into account.

References

1. Doddington G (2002) Automatic evaluation of machine translation quality using n-gram co-occurrence statistics. In: Proceedings of the Second International Conference on HLT '02, pp 138–145
2. Kingma DP, Ba J (2014) Adam: a method for stochastic optimization. arXiv:1412.6980
3. Kudo T (2005) Mecab: yet another part-of-speech and morphological analyzer. http://mecab.sourceforge.net/
4. Mikolov T, Chen K, Corrado G, Dean J (2013) Efficient estimation of word representations in vector space. arXiv:1301.3781
5. Mikolov T, Sutskever I, Chen K, Corrado GS, Dean J (2013) Distributed representations of words and phrases and their compositionality. In: Proceedings of NIPS, pp 3111–3119
6. Mikolov T, Yih WT, Zweig G (2013) Linguistic regularities in continuous space word representations. In: Proceedings of NAACL-HLT 2013, pp 746–751
7. Shang L, Lu Z, Li H (2015) Neural responding machine for short-text conversation. Proceedings of ACL 2015, pp 1577–1586
8. Vinyals O, Le Q (2015) A neural conversational model. In: ICML Deep Learning Workshop
9. Wiseman S, Rush AM (2016) Sequence-to-sequence learning as beam-search optimization. Proceedings of EMNLP, pp 1296–1306
10. Yu L, Zhang W, Wang J, Yu Y (2016) SeqGAN: sequence generative adversarial nets with policy gradient. arXiv:1609.05473

Learning Dialogue Strategies for Interview Dialogue Systems that Can Engage in Small Talk

Tomoaki Nakamura, Takahiro Kobori and Mikio Nakano

Abstract This paper proposes a method with which an interview dialogue system can learn user-friendly dialogue strategies. Conventional interview dialogue systems mainly focus on collecting the user's information and simply repeat questions. We have previously proposed a method for improving user impressions by engaging in small talk during interviews that performs frame-based dialogue management and generates small-talk utterances after the user answers the system's questions. However, the utterance selection strategy in the method was fixed, making it difficult to give users a good impression of the system. This paper proposes a method for learning strategies for selecting system utterances based on a corpus of dialogues between human users and a text-based interview dialogue system in which each system utterance was evaluated by human annotators. This paper also reports the results of a user study that compared the proposed method with fixed utterance selection strategies.

1 Introduction

We aim to develop interview dialogue systems that ask questions and obtain a user's information. One example is a dialogue system for diet recording that asks what the user eats or drinks. Workforce reduction is one potential advantage of such dialogue systems. Humans are more likely to disclose their information to dialogue systems

T. Nakamura (✉) · T. Kobori
The University of Electro-Communications, 1-5-1 Chofugaoka, Chofu, Tokyo, Japan
e-mail: tnakamura@uec.ac.jp

T. Kobori
Present address: Yahoo Japan Corporation, 1-3 Kioicho, Chiyoda-ku, Tokyo, Japan
e-mail: tkobori@yahoo-corp.jp

M. Nakano
Honda Research Institute Japan Co., Ltd., 8-1 Honcho, Wako, Saitama, Japan
e-mail: nakano@jp.honda-ri.com

L. F. D'Haro et al. (eds.), *9th International Workshop on Spoken Dialogue System Technology*, Lecture Notes in Electrical Engineering 579,
https://doi.org/10.1007/978-981-13-9443-0_27

than human interviewers [7], and thus may be able to obtain more information than human interviewers do.

Only a few studies of dialogue systems have delved into their application to interviews. For example, Stent et al. built a dialogue system for questionnaires that was applied to university course evaluations [10] and Johnston et al. developed a dialogue system for social surveys [4]. These studies mainly focused on collecting user information. Although the users might use the dialogue systems once, we believe they would be unwilling to use them continuously. In contrast, non-task-oriented dialogue systems [3, 11, 12], designed for users to enjoy chatting with, have been widely studied.

We previously proposed a dialogue management method that mainly engages in an interview and sometimes engages in small talk, which improves users' impression of interview systems [5]. We implemented a system that generates small-talk utterances during an interview based on heuristic rules.[1] A user study revealed that the small talk improves the impressions of the system, but we found that it was difficult to always select appropriate utterances with static heuristics.

In this paper, we propose to introduce utterance selection rules learned from annotated dialogue corpus and evaluate the effectiveness of the proposed system.

2 Proposed Method

The proposed method selects system utterances from questions for interviewing and small-talk utterance candidates as was done in our previous method [5]. The difference is that it takes a learning-based approach for selecting system utterances. In our proposed method, evaluation score y_c that indicates the appropriateness of the cth candidate $\boldsymbol{u}_c$ is estimated from dialogue history $\boldsymbol{x}$ using regression function f:

$$y_c = f(\boldsymbol{\phi}(\boldsymbol{u}_c; \boldsymbol{x})). \tag{1}$$

A candidate utterance is selected based on y_c. $\boldsymbol{\phi}(\boldsymbol{u}_c; \boldsymbol{x})$ denotes feature vectors extracted from system utterance candidate $\boldsymbol{u}_c$ and dialogue history $\boldsymbol{x}$, using the following features:

- Bag of words (BoW) of the user's last utterance,
- The number of turns from the beginning of the dialogue,
- The number of small-talk utterances generated up to that point in time,
- BoW of the system and user's utterances during the last three exchanges (three system and three user utterances),
- Words that co-occur in both the system utterance candidate and the user's last utterance, and

[1] DeVault et al. developed a virtual human interviewer for automatic assessment of distress indicators [1]. It can generate an utterance after the user answers the system question. This is a kind of small talk, although the objective the system is different from ours, which is to obtain structured information.

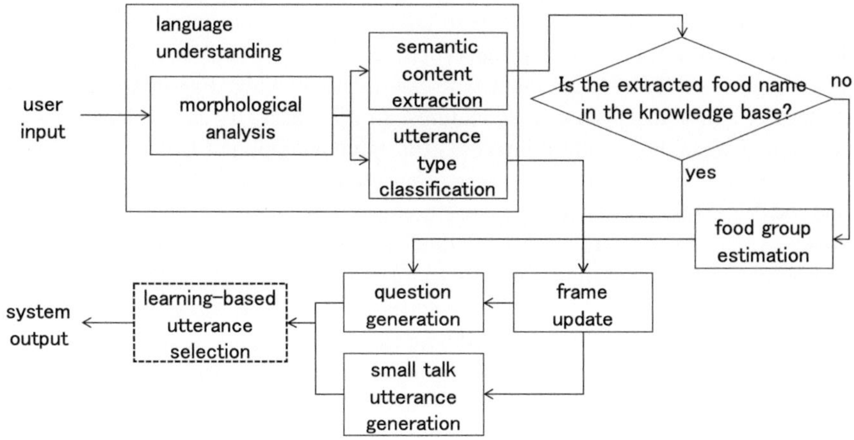

Fig. 1 Architecture of the proposed dialogue system. The dashed rectangle shows the novel module being added to the previous system [5]

- The sequence of the type of the last system utterance and the type of the candidate utterance.

The parameters of the regression function parameters are learned from a corpus, in which system utterances in the dialogue logs are manually evaluated using 5-point Likert scale. By estimating this evaluation score, the system can select more appropriate utterances.

3 An Interview Dialogue System for Diet Recording

This section describes a system that implements the proposed method. It can engage in interview dialogue for diet recording. Figure 1 depicts its architecture. The system is basically the same as the system previously presented in [5], but there are the following differences:

- The training data for understanding the user's utterances are updated by adding sentence templates.
- The content of the knowledge base is significantly larger.
- A new method for selecting system utterances is introduced.

3.1 Understanding User Utterances

User utterances are classified into three types: greetings, affirmative utterances (including replies to system questions), or negative utterances. We chose these types because they allow the system judge whether frames should be updated based on the user's utterance. A user utterance to tell that he/she had some food or drink is classified as affirmative, and the frame is updated according to its content. We used l-ogistic regression (LR) for utterance type classification, with a bag-of-words (BoW) of the user's utterances serving as features. In this study, we used Mecab [6] for morphological analysis and Liblinear [2] for implementation of LR.

Furthermore, from the user's utterance, the system extracts five semantic contents–food and drink, ingredients, food group, amount of food, and time at which the food or drink was consumed–using conditional random fields (CRFs). We used the unigrams and bigrams of the surface form, original form, and part of speech of the words as CRF features, and CRFsuite [8] for implementation of the CRFs. Moreover, in addition to CRFs, semantic content is extracted by a rule-based method that utilizes regular expressions, which makes it possible to accurately extract predefined words and extract words difficult to extract only with CRFs.

The implementation of this module is the same as that of our previous system [5]; however, the training dataset was updated. It now includes 3,659 sentences generated by replacing the content words in 603 sentence templates for training the LR and CRFs.

3.2 Dialogue Management for Interviews

As in our previous work [5], we utilize frame-based dialogue management. The slots of the frame are meal time (breakfast, lunch, and dinner), composition, and amount (Fig. 2). The frame is updated according to user utterances. To fill in slots appropriately, the system must know the food groups of the extracted food names. For example, if the user says, "I had ramen for lunch," the system must understand that "ramen" is a one-dish meal and fill "ramen" into the "one-dish-meal" slot. The food group for a food name is determined using the knowledge base, examples of which are shown in Table 1. In this implementation, 2,134 food and drink instances are included in the knowledge base, as compared to only 304 instances in the previous implementation [5].

However, all food names are not covered by the knowledge base, and the user might tell a food that is not included. In such a case, we employ LR-based food group estimation as proposed by Otsuka et al. [9] with BoW, unigrams and bigrams of characters, and character type (hiragana, katakana, Chinese characters, and Latin alphabet) as features.

The system asks a question to fill vacant in slots by referencing the current frame, or chats using small talk. The question utterances are not fixed expressions, and candidates that include various expressions are generated. Finally, the system selects one candidate as a system utterance.

```
breakfast:  composition:  one-dish-meal:  -
                          shushoku:       bread
                          okazu:          -
                          soup:           corn soup
                          …               -
            amount:       small
lunch:      composition:  one-dish-meal:  ramen (noodle soup)
                          shushoku:       -
                          okazu:          -
                          soup:           -
                          …               -
            amount:       large
supper:     ...
```

Fig. 2 Example frame

Table 1 Knowledge base content

Food group	Example instances	#
Shushoku (side dish mainly containing carbohydrates)	Steamed rice, bread, cereal	152
Okazu (main or side dish containing few carbohydrates)	Hamburg steak, fried shrimp, grilled fish	668
Soup	Corn soup, miso soup	70
One-dish meal	Sandwich, noodle soup, pasta, rice bowl	695
Drink	Orange juice, coffee	343
Dessert	Cake, pancake, jelly	134
Confectionery	Chocolate, donut	72
Total		2,134

3.3 Small Talk Generation

Small talk candidates are generated as system utterances depending on the user utterance type and the context as in the previous work [5]. If a user utterance is estimated to be affirmative, all utterances except for negative utterances are generated as response candidates; for example, utterances such as, "I like that too!" (self-disclosure); "That's great!" (empathy); and "Was it tasty?" (question) are generated. However, if the user's utterance is estimated to be negative, response utterances are generated as candidates: for example, "That's too bad· · · " and "You should have something you like instead." If specific foods are included in the user's utterance, corresponding small talk is included in the response candidates. For example, if the user's utterance includes "cookie," then "I heard it's originally from Persia," is generated. In addition, if the user's utterance includes the amount of the meal, small talk about this becomes candidates for reply. For example, if the user says "I didn't

Table 2 Small talk utterances obtained via questionnaire

Type	#
Showing empathy	24
Commenting that the expressed amount is large	15
Commenting that the expressed amount is small	43
Asking a question	5
Self-disclosure	2
Backchannel	6
Giving an impression of the user's negative answer	10
Reaction to individual food	302
Total	407

eat much," a system utterance such as "It's better to eat properly for your health," is generated as candidates.

To collect small talk, we conducted a crowdsourced questionnaire, from which we obtained 442 utterances. In the previous implementation, we used all 442 utterances; however, some of these were inappropriate. We removed these instances, and 407 utterances are used in the current system, details of which are shown in Table 2.

3.4 Selection of System Utterances

As previously described, once questions and small talk candidates are generated, the system must select one as an utterance. In our previous study, an utterance was selected based on heuristic rules; the system chose to ask a question or make small talk utterance based on the rule where the number of consecutive small talk utterances is fixed, and one candidate in the selected type (question or small talk) was selected randomly. However, this fixed strategy often gave users a bad impression of the system. To solve this problem, we utilize a learning-based approach to select the system utterances using the proposed method explained in Sect. 2.

First, evaluation scores for all candidates are estimated, and then the system selects whether a question or small-talk utterance should be generated. Here, we consider it better for the system to generate small-talk utterances preferentially to facilitate smooth dialogue; therefore, if there are small-talk utterances whose evaluation scores are greater than threshold T, a small-talk utterance is selected. Otherwise the question with the highest evaluation score selected as the system utterance. When a small-talk utterance is to be selected, then one with a score greater than T is selected based on the following conditions:

1. Utterances that have already used in a dialogue cannot be selected again,
2. Candidate with the first, second, and third highest score is reserved, and the fourth highest candidate is selected, and

3. If there are fewer than four candidates, the one with highest score of the reserved candidates is selected.

Condition 1 prevents unnaturalness in the dialogue. Regarding Condition 2, we found that small-talk utterances with higher scores are likely to be general, and can be used as responses to various user utterances (e.g, "Was it tasty?"). In contrast, utterances for specific foods and drinks are likely to have lower scores. Therefore, if the utterance with the highest score is always selected, the dialogue becomes boring. Condition 2 prevents such a situation, as small-talk utterances with lower scores (though still higher than T), which are, to an extent, more natural, are selected as system utterances. Furthermore, if it is difficult for the system to respond to the user's utterance, Condition 3 prevents a breakdown of the dialogue by using reserved general utterances. This study employed an extremely randomized tree regressor (ETR) to estimate the score, and two ETR models are separately learned for selecting questions and small talk.

4 Experiments

We conducted a user study to investigate the effectiveness of our proposed method.

4.1 Corpus for Training System Utterance Selection

To train the ETRs, we constructed dialogue corpus with evaluation scores. The corpus included 4,523 system utterance candidates randomly chosen from dialogue logs obtained in the previous study [5]. We measured annotation agreement between two annotators using 2,018 system utterance candidates. The weighted kappa statistics, which measures the agreement between the scores of two annotators, and it was 0.452. Since this value represents moderate agreements, we used the annotations of one of the annotators.

4.2 Experimental Setup

We compared four systems: an interview dialogue system without small talk (NO-STU); a system that generates one small-talk utterance after each user utterance (1-STU); and two systems that generate small-talk utterances using ETR models (ETR and ETR*), whose thresholds T, explained in Sect. 3.4, are 2.3 and 2.8, respectively. To evaluate the systems, 160 people participated, and 40 people interacted with each system and evaluated it. We assigned participants to each system so that ages and genders of people using each system became roughly equal. If the purpose of the

Table 3 Survey items

ID	Adjective pair	
Q_1	System responses are meaningful	↔ system responses are meaningless
Q_2	Fun	↔ not fun
Q_3	Natural	↔ unnatural
Q_4	Warm	↔ cold
Q_5	Want to continue to talk	↔ don't want to continue to talk
Q_6	Lively	↔ not lively
Q_7	Simple	↔ complicated
Q_8	Want to talk to the system again	↔ don't want to talk to the system again

experiment is to make significant differences among the systems clear, it might be better that same subjects use different systems. However, we would like to find how the impression against the same system changes depending on the number of times of its use in this experiment and, therefore, we assigned the subjects as described above. The subjects accessed the server on which the dialogue systems were running and engaged in the interview dialogue. In this experiment, to investigate how the impressions of each system changed with continuous use, the subjects were required to use the assigned system at least four times in a week and up to once per day. After dialogue with the systems, we asked the participants to evaluate the dialogue using a five-point Likert scale for 8 survey items, shown in Table 3, and describe their impression of the dialogue. Considering the participants' load, the dialogue was automatically stopped after 40 turns. Some participants who did not follow these experimental conditions or could not use the system because of system errors were excluded. Moreover, we found that a threshold of 2.8 for ETR* was too high, as there were few small-talk utterance candidates whose scores were higher than 2.8 toward the end of each dialogue. We consider this is not a system that we intended in this paper, and this system is excluded for the evaluation. Therefore, the number of participants whose evaluations were used in this investigation was 57 (NO-STU:17, 1-STU:20, ETR:20).

4.3 Questionnaire Evaluation

In this experiment, we analyzed how the users' impression of the system in each condition changed during four dialogues. To make this change clearer, evaluation score e_{in} of user i against the n-th dialogue was normalized to an average of zero:

$$\hat{e}_{in} = e_{in} - \frac{1}{4}\sum_{n} e_{in}. \tag{2}$$

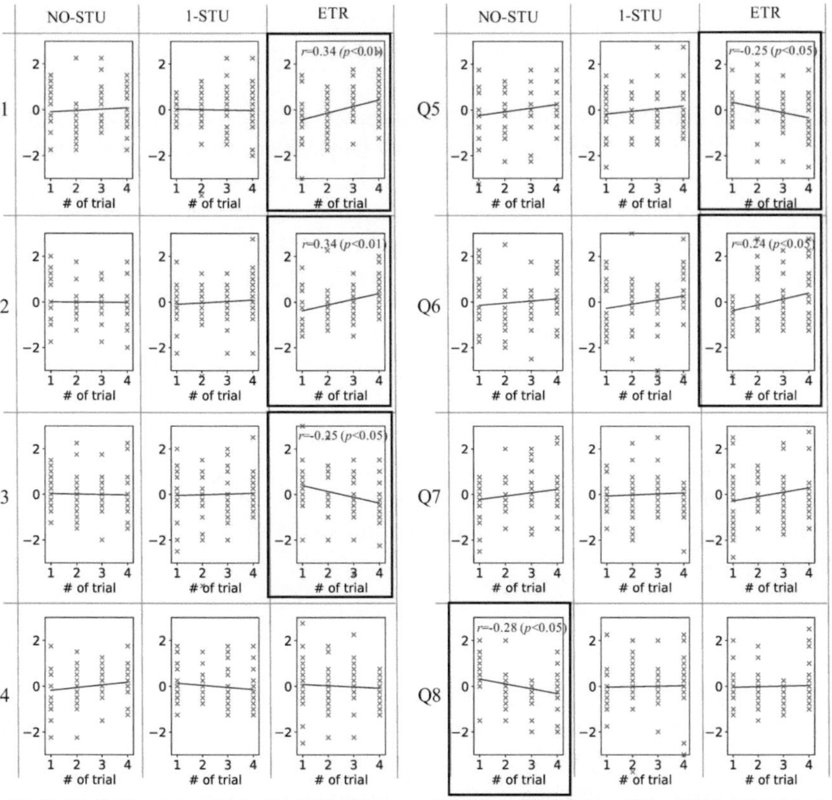

Fig. 3 User evaluations of each dialogue system. Each point represents a user evaluation; solid lines are the regression line of all user evaluations. Graphs in a black rectangle denote significant items ($p < 0.05$)

The normalized evaluation scores of all subjects are plotted in Fig. 3. Each graph shows each survey item: the horizontal axis of each graph represents the number of dialogues, and the vertical axis represents the normalized evaluation scores. We conducted a regression analysis in which the number of dialogues is the explanatory variable and the evaluation score is the dependent variable. Graphs with black rectangles denote significant items.

There were significant positive correlations in Q1, Q2, and Q6 of the ETR condition; users felt that system responses became more meaningful and that the dialogues became more fun and lively as the number of dialogues increased. In contrast, in other conditions, significant correlation was not seen in these survey items. The system in the NO-STU condition simply repeated questions; participants who used the NO-STU system described their impression with phrases such as "It utters the same pattern every time," and "Repeating the same questions is boring." Such impressions decreased in conditions using small-talk utterances. In the 1-STU condition, another reason for giving a bad impression is that the system ignored the user's response to its small talk utterance and instead asked the next question. In the ETR condition, however, small-talk utterances can be generated continuously, which was more fun for users.

However, there were negative correlations in Q3 and Q5 of ETR condition; as the number of dialogues increased, the users felt that system utterances were unnatural and they did not want to continue talking with the system. We believe this result was caused by expectations for the system. In the ETR condition, the system generated various small-talk utterances, and the subjects expected that the system were able to talk flexibly. However, the current system cannot engage in flexible conversation, creating a gap between users' expectations and the current system's abilities and decreasing users' evaluations.

One of the most important results is the significant negative correlation in Q8 of the NO-STU condition; users got bored and did not want to talk with the system without small-talk utterances again. Even the 1-STU system, which generates a small-talk utterance after a user utterance, prevented this negative effect. However, positive correlation in Q8 of the ETR condition, which we expected, was not seen. We plan to investigate the factors that are important in making the user willing to talk to a system again.

5 Conclusion

We have developed a dialogue system that generates small-talk utterances to improve user-friendliness and encourage future use. In this study, we proposed a method for system utterance selection based on an evaluation score estimated by a regression model. We conducted a user study to investigate the effectiveness of the proposed method and small-talk utterances. The results showed that the evaluation values of "meaningful", "fun" and "lively" increased, and those of "natural" and "continue to talk" decreased by the proposed method as the number of dialogues increases. Moreover, we confirmed that fewer participants wanted to talk to a system that did not generate small talk again as the number of dialogues increased. This result indicates that small-talk utterances are important for continuous use of a dialogue system. However, the evaluation value of "want to talk again" of the proposed method did not have the correlation with the number of dialogues, which is different from our expectation, so we plan to investigate the reasons for that.

References

1. DeVault D, Artstein R, Benn G, Dey T, Fast E, Gainer A, Georgila K, Gratch J, Hartholt A, Lhommet M, Lucas GM, Marsella S, Morbini F, Nazarian A, Scherer S, Stratou G, Suri A, Traum DR, Wood R, Xu Y, Rizzo AA, Morency LP (2014) SimSensei Kiosk: a virtual human interviewer for healthcare decision support. In: International conference on autonomous agents and multi-agent systems, pp 1061–1068

2. Fan RE, Chang KW, Hsieh CJ, Wang XR, Lin CJ (2008) LIBLINEAR: a library for large linear classification. J Mach Learn Res 9:1871–1874
3. Higashinaka R, Imamura K, Meguro T, Miyazaki C, Kobayashi N, Sugiyama H, Hirano T, Makino T, Matsuo Y (2014) Towards an open-domain conversational system fully based on natural language processing. In: International conference on computational linguistics, pp 928–939
4. Johnston M, Ehlen P, Conrad FG, Schober MF, Antoun C, Fail S, Hupp A, Vickers L, Yan H, Zhang C (2013) Spoken dialog systems for automated survey interviewing. In: Annual meeting of the special interest group on discourse and dialogue, pp 329–333
5. Kobori T, Nakano M, Nakamura T (2016) Small talk improves user impressions of interview dialogue systems. In: Annual meeting of the special interest group on discourse and dialogue, pp 370–380
6. Kudo T, Yamamoto K, Matsumoto Y (2004) Applying conditional random fields to Japanese morphological analysis. In: Proceedings of the conference on empirical methods in natural language processing, pp 230–237
7. Lucas GM, Gratch J, King A, Morency LP (2014) It's only a computer: virtual humans increase willingness to disclose. Comput Hum Behav 37:94–100
8. Okazaki N (2007) CRFsuite: a fast implementation of conditional random fields (CRFs). http://www.chokkan.org/software/crfsuite/
9. Otsuka T, Komatani K, Sato S, Nakano M (2013) Generating more specific questions for acquiring attributes of unknown concepts from users. In: Annual meeting of the special interest group on discourse and dialogue, pp 70–77
10. Stent A, Stenchikova S, Marge M (2006) Dialog systems for surveys: the rate-a-course system. In: IEEE spoken language technology workshop, pp 210–213
11. Wallace RS (2009) The anatomy of A.L.I.C.E. Parsing the Turing Test, pp 181–210
12. Wilks Y, Catizone R, Worgan S, Dingli A, Moore R, Field D, Cheng W (2011) A prototype for a conversational companion for reminiscing about images. Comput Speech Lang 25(2):140–157

Chatbol, a Chatbot for the Spanish "La Liga"

Carlos Segura, Àlex Palau, Jordi Luque, Marta R. Costa-Jussà and Rafael E. Banchs

Abstract This work describes the development of a social chatbot for the football domain. The chatbot, named *chatbol*, aims at answering a wide variety of questions related to the Spanish football league "La Liga". Chatbol is deployed as a Slack client for text-based input interaction with users. One of the main Chatbol's components, a NLU block, is trained to extract the intents and associated entities related to user's questions about football players, teams, trainers and fixtures. The information for the entities is obtained by making sparql queries to Wikidata site in real time. Then, the retrieved data is used to update the specific chatbot responses. As a fallback strategy, a retrieval-based conversational engine is incorporated to the chatbot system. It allows for a wider variety and freedom of responses, still football oriented, for the case when the NLU module was unable to reply with high confidence to the user. The retrieval-based response database is composed of real conversations collected both from a IRC football channel and from football-related excerpts picked up across movie captions, extracted from the OpenSubtitles database.

Keywords Language understanding · Natural language processing · Spoken dialogue systems · Question answering system · Retrieval based chat oriented dialogue · Chat oriented dialogue system

C. Segura (✉) · À. Palau · J. Luque
Telefónica Research, Barcelona, Spain
e-mail: carlos.seguraperales@telefonica.com

À. Palau
e-mail: alex.palau.domenech@gmail.com

J. Luque
e-mail: jordi.luque@telefonica.com

M. R. Costa-Jussà
TALP Research Center - Universitat Politècnica de Catalunya, Barcelona, Spain
e-mail: marta.ruiz@upc.edu

R. E. Banchs
Nanyang Technological University, Singapore, Singapore
e-mail: rbanchs@ntu.edu.sg

L. F. D'Haro et al. (eds.), *9th International Workshop on Spoken Dialogue System Technology*, Lecture Notes in Electrical Engineering 579,
https://doi.org/10.1007/978-981-13-9443-0_28

1 Introduction

Recent advances in Machine Learning and Natural Language Understanding are making possible the implementation of human-machine interaction systems that are driven by means of natural language conversations. These systems, commonly referred to as chatbots, have been recently gaining popularity in different mobile and online platforms. Their main objective is to serve as "virtual assistants" to support informational and transactional user requirements in a wide variety of domains and sectors. The recent boom on conversational interfaces is capturing the attention of both the research community and tech companies as the possible next big technological revolution: *"Until now, we have been forced to learn the language of computers. But in the technological revolution currently underway, computers are finally learning to understand ours."*[1]

Regardless the actual impact and adoption of numerous applications in the market, the state-of-the-art of these technologies is still quite limited and their development requires a high cost of adaptation to the specific domain. Handcrafted applications with a limited amount of automation and a high level of human intervention is typically the common scenario. Nevertheless, successful applications, which are able to collect and model huge amount of interactions, are the best candidates to leverage on machine learning techniques to significantly reduce the level of human intervention.

Some examples of existing service chatbots include systems in the health domain, such as currently described chatbot service to provide Health information about HIV/AIDs [5], primary care [14] or mental Health counseling [15]; the educational domain, such as [21], a chatbot to improve interactions in online courses or MOOCBuddy chatbot [1] to provide personalized course recommendations; the entertainment domain, such a Stark Trek system [12] or Humorist Bot that can generate and recognize humorous sentences [2]; Human Resources chatbot [6] or Customer Service agents [29], just to cite a few.

In this work we describe CHATBOL, a chatbot for providing information about the Spanish Football League "La Liga". Different from similar systems that offers news and other factual information about the Spanish Football League, the presented system combines two different conversational functions into a single agent: a question answering system that search over Wikidata entities to generate responses related to the topic of interest and a retrieval-based chat-oriented dialogue system that serves as fallback mechanism to reply to either out-of-domain user queries or in-domain queries that cannot be served by the question answering system.

The rest of the paper is structured as follows. First, in Sect. 2, an overall description of our system is presented. In Sect. 3, some related work is briefly described. In Sect. 4, a comprehensive description of the system architecture and its main components is provided. In Sect. 5, the experimental framework is described along with the

[1] https://medium.com/conversational-interfaces/conversational-interfaces-the-next-big-technological-revolution-4efea1d97606.

conducted evaluations and results. Finally, in Sect. 6, the main conclusions of our work are presented along with the future work to be conducted.

2 Chatbol Description

3 Related Work

Chatbots have a long history since first prototype appeared in mid 60s [27]. The objective of this section is to make the paper self-contained, for complete survey the reader should refer to [18]. During all these years, chatbots have been approached from different perspectives which mainly include rule-based [26] or data-driven which are retrieval/statistical-based [3, 7, 9, 19] or generative-based [25, 28] just to name a few examples. Within these data-driven approaches, the main challenge nowadays is to take advantage of large datasets. For tackling previous challenge, our proposed model uses a mixture of rule and retrieval-based approach and is fed by different public resources such as Wikidata and the OpenSubtitles database. Note that both data resources have already been employed to develop chatbot engines, previously in [22, 25].

More specifically, in this paper, we are aiming at training a task-oriented chatbot. In task-oriented, the objective of the tool is to get information from the user in order to accomplish a particular task. Tools of these kind that have become very popular in real life are Siri or Alexa, among others. In our case, we focus the chatbot in the domain of football, similarly to tools like Toni from Facebook Messenger[2] or Tokabot,[3] but for the particular case of the Spanish league.

4 Architecture

The architecture of Chatbol is composed of a Natural Language Understanding module based on RASA NLU [4], a Wikidata entity search module to generate responses and Slack plugin. Additionally a retrieval-based conversational engine is included as a fallback option, when the first block is unable to provide a response due to the NLU block is not recognizing intents or entities with a high likelihood score or when the entities are not found in Wikidata. A diagram of the architecture is depicted in Fig. 1.

[2]https://toni.football/.

[3]https://botlist.co/bots/tokabot-football.

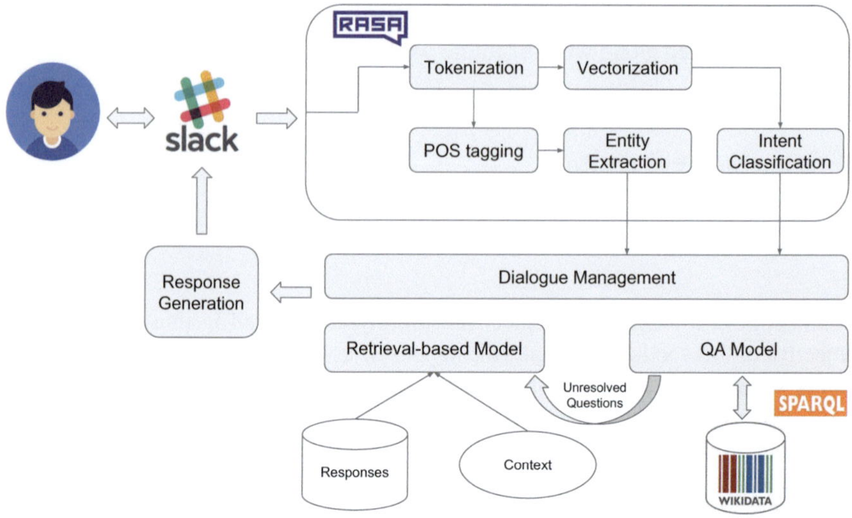

Fig. 1 Chatbol architecture: Rasa NLU is used to recognize intents and entities that are passed to the QA model that maps user input into a query to Wikidata to generate the answer. As a fallback strategy, for unresolved questions, a response is selected from the retrieval-based model

4.1 Natural Language Understanding

Rasa NLU is an open source library for intent classification and entity extraction. It is a high level library similar to online services that parses user messages into dialog acts composed of intents and a set of corresponding entities. Rasa NLU follows a modular design, relying on existing NLP and Machine learning libraries, like spaCy [11], scikit-learn [16], sklearn-crfsuite [13].

The Rasa NLU pipeline consists of a tokenization process followed by a parts of speech (POS) tagging by means of spaCy NLP library. Then, the GloVe vectors [17] extracted from each token are concatenated forming a feature vector of the whole sentence. This sentence representation is used to train a multiclass support vector machine used to recognize user intents. Finally, a conditional random field classifier is trained on the sentence tokens and POS tags to extract entities. The intent classifier and entity extractor are trained using scikit-learn and sklearn-crfsuite, respectively.

For training Rasa NLU, a list of questions annotated with intents and entities is prepared using the human readable json format. In this work, the following 7 intents are included in the chatbot:

- What team is [player] playing for ?
- List of players of [team]
- What is the stadium of [team] ?
- What team is [stadium] of ?
- Who is the coach of [team] ?

- What team is coached by [coach] ?
- Get fixtures of [team]

The list of questions is created by writing around 10 ad hoc phrases for every single intent and afterwards expanded using a list of football players, teams, stadiums and coach gathered using the Wikidata API.[4] This process helps Rasa NLU training process to correctly identify not only the intents but also the entities the chatbol is interested in.

4.2 Response Generation

Chatbol bot is connected to Slack using slackclient API [5]; therefore, all interaction with the bot is through this instant messaging software. Once the bot receives a message, Rasa NLU analyses the sentence and returns a human readable json package with information on:

- The likelihood of the sentence belonging to each one of the trained intents.
- Every entity that has been identified.

Using this information, the chatbot has been designed to try the 3 (number obtained via trial and error) most likely intents in order. For each intent and entity, Chatbol sends a query to either Wikidata or an ad hoc structured database with information on the following matches of every team obtain via the scrapping of the site futbolportv.com.[6]

In case the combination of intent and detected entity results in an answer, this answer is sent via the Slack API. It is also possible that none of the 3 intents and entities results in a response; in that case, a retrieval-based response block described in Sect. 4.4 will be the generator of the answer.

4.3 Database Query

In order to have always the most up-to-date information, Chatbol obtains the answer to the detected intent on the fly using the Wikidata API or web scrapping. In case of Wikidata, it first obtains the Qnumber (reference number used in Wikidata to point at an element of information) of the entity via wikidata, by making a POST to the Wikidata API service with the name of the entity. Then, once the number is found, we generate a query to Wikidata using SPARQLWrapper [8], a SPARQL endpoint interface to Python. The answer from Wikidata is analyzed, reformatted to natural

[4]https://www.mediawiki.org/wiki/API.

[5]https://api.slack.com/.

[6]https://futbolportv.com/.

```
def get_playeres_team(qnumber):
    sparql = SPARQLWrapper("https://query.wikidata.org/sparql")
    sparql.setQuery("""SELECT DISTINCT ?playerLabel WHERE {
        ?player wdt:P106 wd:Q937857.
        ?player p:P54 ?teamMember.
        ?teamMember ps:P54 wd:""" + qnumber + """.
        OPTIONAL {?player p:P1449 ?Nickname.}
        ?player wdt:P21 wd:Q6581097.
        SERVICE wikibase:label { bd:serviceParam  \
                wikibase:language "[AUTO_LANGUAGE],en, \
            es, ca, de". }
        FILTER(NOT EXISTS { ?teamMember pq:P582 ?end. })
        FILTER(EXISTS { ?teamMember pq:P580 ?start. })
        FILTER(NOT EXISTS { ?player wdt:P570 ?death. })
        OPTIONAL { ?player wdt:P569 ?birth. }
        FILTER(year(?birth) > 1979).
        }
        ORDER BY ?playerLabel""")
    sparql.setReturnFormat(JSON)
    return sparql.query().convert()
```

Fig. 2 Automatic Python + SPARQL code to retrieve the list of players of a football team specified by its Qnumber

language and sent as a response to the user. The sample Python code depicted in Fig. 2 shows how to get the list of players of a team, given the team qnumber:

There are two ways this process may go wrong. First, the Qnumber may not be found, either because the detected entity does not exist or because the name is not exactly the one Wikidata has for that entity. For instance, the player Iniesta is not recognized as Andrés Iniesta by Wikidata. The big number of alias and ways of calling the different players and also the objective of making the chatbot as most user-friendly as possible made this a critical problem.

Therefore, a solution based on Fuzzywuzzy Python package [10] has been applied. This solution allows the comparison of the similarities between a word and an array of words and the selection of the most similar one from the array while also returning a value from 0 to 100 of the similarity between both words. In the chatbot case, the entity detected in the user input is compared with the array of words formed by the names of the players, coaches, teams or stadiums (depending on the intent). If the similarity is big enough (70 in our case), the most similar word is used to find the Qnumber in Wikidata. If no word is similar enough, the intent is supposed to be erroneous and the algorithm jumps to the next most likely intent.

The second way the Wikidata query may go wrong is because the query generated returned no valid information (it may be asking Wikidata for the stadium of a coach, for instance). In this case the intent is also categorized as erroneous and the same process as in the case before is followed.

4.4 Chichat Fallback Component

Chatbol includes a fallback mechanism to produce responses when the main system component is unable to do it. This can happen when the confidence score of the intent detection falls below a threshold, or when the information requested is not found in Wikidata. This mechanism is based on a retrieval-based conversational engine, which follows the Informal Response Interactive System (IRIS) approach [3]. In this type of systems, semantic similarity scores at the turn and dialogue history levels are combined to retrieve the most suitable matches from a database of dialogues. The retrieved candidate responses are then ranked by using a recall-oriented re-ranking algorithm. Favoring recall (instead of precision) helps producing more varied responses, making the chatbot more interesting.

Spanish movie data from OpenSubtitles [23] was used as dialogue database. As we wanted the system to be able to converse in the topic of interest, only those files containing the word "fútbol" (Spanish for soccer) more than ten times were extracted. In this way, from a total of 191,987 files contained in the Spanish OpenSubtitles data collection, only 94 were useful for our purpose. In order increase the amount of data supporting the Chitchat component, we also included chat sessions from the #futbol channel at IRC-HISPANO.[7] A total of 3638 sessions comprising 148,893 utterances were collected and included as part of the dialogue database.

The implementation was done in Python, using the scikit-learn library [16]. Standard word tokenization was used and spanish accents removed, mapping all characters to ASCII. Separated indexes were constructed for turns and dialogue sessions in the data collection. The definition of a dialogue session in the OpenSubtitles dataset is somehow ambiguous as no explicit information on when a new dialogue session starts is available. To this end, a breaktime threshold of 3 s was used for deciding on dialogue boundaries.

The chat-oriented dialogue system is inspired on the vector space model framework [20, 24]. The vectorization of the user input and database utterances is completed by projecting them into their Term frequency—inverse document frequency (Tf-idf) representation. The vector similarity among user's input and stored sentences are computed for retrieving best matches from the dialogue databases. For doing so, three different vector space models (VSM) are computed. It aims at describing the dialogue database Tf-idf vector distributions with respect the current state of the dialogue. That is, to account for the utterance-level and the dialogue-level information. Cosine similarity is employed to compute scores among both previous and current state vectors and their respective VSMs. Similarly, the history score is also obtained by using cosine similarity through the current dialogue history, which includes all utterances interchanged by the current user and Chatbol. It is worth to note that the history vector is updated with each new sentence in the dialogue yielding to a collection of stacked Tf-idf vectors at each turn level. The history score is computed between this vector and the vector representations for each full dialogue stored in the dialogue database. In summary, for each new user input in a given conversation,

[7] https://chathispano.com/.

three cosine similarity scores are computed (current utterance, previous utterance and dialogue history) and combined into an overall score to retrieve the candidate responses. The overall cosine similarity score is computed as in the following equation and updated at each dialogue-turn step,

$$Score = 0.6 * u_c + 0.1 * u_p + 0.3 * h \ ,$$

where the utterance-level scores (u_c and u_p) and the dialogue-level h score are combined for drawing a final rank of candidate responses. The following weighting factors: 0.6, 0.1 and 0.3, were set after some manual experimentation.

5 Experimental Framework and Evaluation Results

The experimental framework is based on Slack, where users can interact with direct messages to the Chatbol app. This makes it possible to maintain a separate conversation context for each user and is very convenient for evaluating the chatbot responses.

The evaluation protocol used follows the current WOCHAT Shared Task evaluation framework.[8] In this sense, two different evaluations were conducted, one a the session level and the other at the turn level. For the session level evaluation, we used the available User Satisfaction Survey, in which each user has to respond to eight closed questions and three open questions after interacting with the chatbot. These questions are mainly focused on the user experience aspect of the interaction, as well as on the overall performance of the system. Results of the user satisfaction survey are depicted in Fig. 3.

As seen from the figure, most of the users found the system responses to be appropriate either *sometimes* or *often*. In the open questions, users commented on the strengths and problems of the systems. Among the strengths, it was mentioned that the system was able to provide "*accurate information about Spanish soccer teams*" and "*capable to understand incorrect words and very flexible with team names variations*". Overall the system was perceived as a "*fast way to get football data without searching the Internet*". Among the main problems reported, evaluators agreed that "*sometimes the question intents are not properly captured by the system*", even "*misunderstands basic questions*" and "*despite small changes in the sentence*".

For the turn level evaluation, we followed the provided guidelines for appropriateness score, in which each turn of the dialogue session is annotated with one of the following three scores: *valid*, *acceptable* or *invalid*, according to its perceived appropriateness with respect to the current utterance and dialogue context. A total of 6 users, tested and evaluated Chatbol. The results of our evaluation are summarized in Table 1, showing the cumulative number of scores for both conversational engines. Defining an appropriate answer as the response scored with *valid*, or *acceptable*, the

[8] http://workshop.colips.org/wochat/main_sharedtask.html.

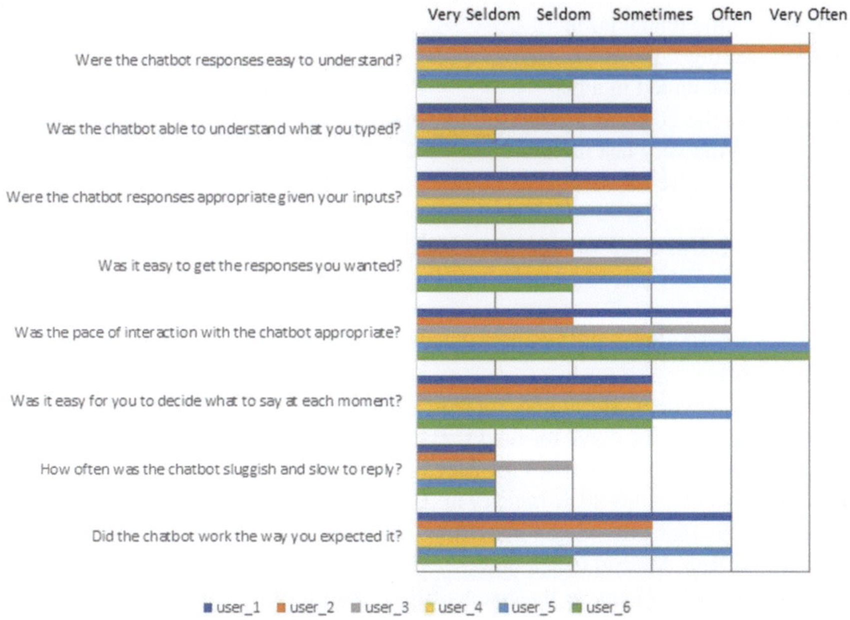

Fig. 3 Results of the user satisfaction survey

Table 1 Evaluation of the responses generated by Chatbol for each conversational model

System	Responses		
	Valid	Acceptable	Invalid
QA model	34 (53%)	12 (19%)	18 (28%)
Retrieval model	50 (43%)	28 (25%)	37 (32%)

QA system is able to generate appropriate responses 72% of the time, whereas the chitchat system obtains a 68% of appropriate answers. Below is a sample of a few selected interactions with the chatbot for both conversational engines. Samples were originally in Spanish and have been translated into English:

Chitchat engine:
user: how's life?
chatbol: I am a bit injured, I had an ankle sprain last Tuesday... How are you?
user: Which team do you support?
chatbol: I support the rival for your team
user: I am an atletic fan
chatbol: atletic is a great team, good players, coacher and supporters, they are huge!
user: is the football match on TV now?
chatbol: it just started

user: how is the match going?
chatbol: leading 1-0
user: the game ended in a tie
chatbol: was the goal during the penalty shoot-out?
user: yes, in the very last moment
chatbol: what were the scores?
user: 1-1
chatbol: cool, I won the sweepstakes! I think no one else bet for 1-1
user: bye
chatbol: take care :d

QA engine:
user: What team does Messi play for?
chatbol: Lionel Messi plays for the FC Barcelona
user: What are the next matches for Barça ?
chatbol: Next Thursday 11th January at 21:30h Barcelona - Celta in Copa del Rey leage. You can watch the match in: TBD
user: who is Real Madrid coach now ?
chatbol: Ziniedine Zidane is the coach of Real Madrid
user: Who is the coarch of sevilla ?
chatbol: Vincenzo Montella is the coach of Sevilla Fútbol Club
user: What stadium does Osasuna play in ?
chatbol: El Sadar Stadium is the home stadium of Club Atlético Osasuna
user: When does Geta play?
chatbol: Next Friday 12th January at 21:00h Getafe - Málaga in LaLiga Santander leage. You can watch de match in: LaLiga TV, beIN LaLiga.
user: What is the current squad of Barça ?
chatbol: The players on the squad of Fútbol Club Barcelona are: André Gomes, Andrés Chávez, Andrés Iniesta, Arda Turan, Gerard Piqué, Ivan Rakitić, Jasper Cillessen, Javier Mascherano, Jorge Alba Jose Manuel, Arnáiz Díaz, Lionel Messi, Lucas Digne Marc-André ter Stegen Munir El Haddadi Nélson Semedo Ousmane Dembélé Paulinho Philippe Coutinho, Samuel Umtiti, Sergi Roberto, Sergi Samper, Sergio Busquets, Thomas Vermaelen, Wilfrid Kaptoum

6 Conclusions

This paper described the development of a chatbot for providing information about the Spanish Football League "La Liga". The proposed system combines two different conversational engines into a single agent: a QA engine that maps natural language utterances of the user into SPARQL queries to retrieve data from Wikidata to generate responses and a retrieval-based chat-oriented dialogue engine that serves as fallback mechanism to reply to either out-of-domain user queries or in-domain queries that cannot be served by the question answering system. The evaluation of the chatbot at turn level show that the users found the system responses to be appropriate 72% of the time for the QA engine and 68% for the chitchat engine. From the session level evaluation, users found that current implementation is robust against word errors and able to recognize variations of player and team names, but sometimes failed to recognize the intent correctly, even for simple questions. For future work we plan to improve the intent classification block and to extend the QA model by incorporating to the system the most popular questions found in user inputs. We are also currently implementing interactive message buttons after each chatbot response to be able to get instant feedback from users.

Acknowledgements This study has been funded by the Spanish Ministerio de Economía y Competitividad, the European Regional Development Fund and the Agencia Estatal de Investigación, through the postdoctoral senior grant Ramón y Cajal, the contract TEC2015-69266-P (MINECO/FEDER,EU) and the contract PCIN-2017-079 (AEI/MINECO).

References

1. Adrian I, Jean V (2016) Moocbuddy: a chatbot for personalized learning with moocs, vol 91
2. Augello A, Saccone G, Gaglio S, Pilato G (2008) Humorist bot: Bringing computational humour in a chat-bot system. In: International conference on complex, intelligent and software intensive systems, 2008. CISIS 2008. IEEE, pp 703–708
3. Banchs RE, Li H (2012) IRIS: a chat-oriented dialogue system based on the vector space model. In: The 50th Annual Meeting of the Association for Computational Linguistics, Proceedings of the System Demonstrations, July 10, 2012, Jeju Island, Korea, pp 37–42
4. Bocklisch T, Faulker J, Pawlowski N, Nichol A (2017) Rasa: open source language understanding and dialogue management. arXiv preprint arXiv:1712.05181
5. Brixey J, Hoegen R, Lan W, Rusow J, Singla K, Yin X, Artstein R, Leuski A (2017) Shihbot: a facebook chatbot for sexual health information on hiv/aids. In: Proceedings of the 18th annual SIGdial meeting on discourse and dialogue, pp 370–373
6. Chandar P, Khazaeni Y, Davis M, Muller M, Crasso M, Liao QV, Shami NS, Geyer W (2017) Leveraging conversational systems to assists new hires during onboarding. In: Bernhaupt R, Dalvi G, Joshi A, K. Balkrishan D, O'Neill J, Winckler M (eds) Human-Computer Interaction - INTERACT 2017. Springer International Publishing, Cham, pp 381–391
7. Fatemi M, Asri LE, Schulz H, He J, Suleman K (2016) Policy networks with two-stage training for dialogue systems. In: Proceedings of the 17th annual meeting of the special interest group on discourse and dialogue, Los Angeles. Association for Computational Linguistics, pp 101–110
8. Fernández S, Zakhlestin A (2017) Sparql endpoint interface to python. https://github.com/RDFLib/sparqlwrapper
9. Gašić M, Jurčíček F, Keizer S, Mairesse F, Thomson B, Yu K, Young S (2010) Gaussian processes for fast policy optimisation of pomdp-based dialogue managers. In: Proceedings of the 11th annual meeting of the special interest group on discourse and dialogue, SIGDIAL '10, Stroudsburg, PA, USA, 2010. Association for Computational Linguistics, pp 201–204
10. Gonzalez J, Rodrigues P, Cohen A (2017) Fuzzywuzzy: Fuzzy string matching in python. https://github.com/seatgeek/fuzzywuzzy
11. Honnibal M, Montani I (2017) spacy 2: natural language understanding with bloom embeddings, convolutional neural networks and incremental parsing. https://github.com/explosion/spaCy

12. Jena G, Vashisht M, Basu A, Ungar L, Sedoc J (2017) Enterprise to computer: star trek chatbot. arXiv preprint arXiv:1708.00818
13. Korobov M (2017) sklearn-crfsuite. https://github.com/TeamHG-Memex/sklearn-crfsuite
14. Ni L, Lu C, Liu N, Liu J (2017) Mandy: towards a smart primary care chatbot application. In: Chen J, Theeramunkong T, Supnithi T, Tang X (eds) Knowledge and systems sciences. Springer, Singapore, pp 38–52
15. Oh K-J, Lee D, Ko B, Choi H-J (2017) A chatbot for psychiatric counseling in mental healthcare service based on emotional dialogue analysis and sentence generation. In: 2017 18th IEEE international conference on mobile data management (MDM). IEEE, pp 371–375
16. Pedregosa F, Varoquaux G, Gramfort A, Michel V, Thirion B, Grisel O, Blondel M, Prettenhofer P, Weiss R, Dubourg V et al (2011) Scikit-learn: machine learning in python. J Mach Learn Res, 12:2825–2830
17. Pennington J, Socher R, Manning C (2014) Glove: global vectors for word representation. In: Proceedings of the 2014 conference on empirical methods in natural language processing (EMNLP), pp 1532–1543
18. Ramesh K, Ravishankaran S, Joshi A, Chandrasekaran K (2017) A survey of design techniques for conversational agents. Springer, Singapore, pp 336–350
19. Ritter A, Cherry C, Dolan WB (2011) Data-driven response generation in social media. In: Proceedings of the conference on empirical methods in natural language processing, EMNLP '11, Stroudsburg, PA, USA. Association for Computational Linguistics, pp 583–593
20. Salton G, Wong A, Yang CS (1975) A vector space model for automatic indexing. Commun ACM 18(11):613–620
21. Song D, Oh EY, Rice M (2017) Interacting with a conversational agent system for educational purposes in online courses. In: 2017 10th International conference on human system interactions (HSI), pp 78–82
22. Suglia A, Greco C, Basile P, Semeraro G, Caputo A (2017) An automatic procedure for generating datasets for conversational recommender systems. CLEF
23. Tiedemann Jörg (2009) News from opus-a collection of multilingual parallel corpora with tools and interfaces. Recent Adv Nat Lang Process 5:237–248
24. van Rijsbergen CJK (2005) A probabilistic logic for information retrieval. In David EL, Fernández-Luna JM (eds) Advances in information retrieval. Springer, Berlin, Heidelberg, pp 1–6
25. Vinyals O, Le QV (2015) A neural conversational model. CoRR arXiv:abs/1506.05869
26. Wallace R (2003) The elements of aiml style
27. Weizenbaum Joseph (1966) Eliza: a computer program for the study of natural language communica tion between man and machine. Commun ACM 9(1):36–45
28. Wu Y, Wu W, Li Z, Zhou M (2016) Topic augmented neural network for short text conversation. CoRR arXiv:abs/1605.00090
29. Xu A, Liu Z, Guo Y, Sinha V, Akkiraju R (2017) A new chatbot for customer service on social media. In: Proceedings of the 2017 CHI conference on human factors in computing systems. ACM, pp 3506–3510

Improving Taxonomy of Errors in Chat-Oriented Dialogue Systems

Ryuichiro Higashinaka, Masahiro Araki, Hiroshi Tsukahara and Masahiro Mizukami

Abstract In previous studies, top-down and bottom-up approaches have been proposed for creating taxonomies of errors in chat-oriented dialogue systems. However, the reported κ (kappa) value for the taxonomy based on the top-down approach is low at 0.239, and no evaluation has been conducted for that based on the bottom-up approach. In this paper, we propose to revise these taxonomies to achieve better inter-annotator agreement. The revised taxonomy based on the bottom-up approach yielded a reasonable κ of 0.44 (Fleiss' κ), suggesting that this taxonomy can be used reliably to classify errors in chat-oriented dialogue systems.

1 Introduction

Current chat-oriented dialogue systems make various errors due to the limitation of current language technology [1, 9, 17, 18]. To improve such systems, error analysis is crucial; we need to have a good taxonomy of errors so that developers and researchers can have a common conception of errors, grasp the distribution of errors, and identify the main types of errors.

R. Higashinaka (✉)
NTT Media Intelligence Laboratories, NTT Corporation, Tokyo, Japan
e-mail: ryuichiro.higashinaka.tp@hco.ntt.co.jp

M. Araki
Faculty of Information and Human Sciences, Kyoto Institute of Technology, Kyoto, Japan
e-mail: araki@kit.ac.jp

H. Tsukahara
Research and Development Group, Denso IT Laboratory, Inc, Tokyo, Japan
e-mail: htsukahara@d-itlab.co.jp

M. Mizukami
NTT Communication Science Laboratories, NTT Corporation, Tokyo, Japan
e-mail: masahiro.mizukami.df@hco.ntt.co.jp

L. F. D'Haro et al. (eds.), *9th International Workshop on Spoken Dialogue System Technology*, Lecture Notes in Electrical Engineering 579,
https://doi.org/10.1007/978-981-13-9443-0_29

To the best of our knowledge, two taxonomies have been proposed for chat-oriented dialogue systems. One is based on the top-down (TD) (theory-driven) approach [7] and the other on the bottom-up (BU) (data-driven) one [10]. The former was derived by researchers' analyses of errors by a chat-oriented dialogue system. This taxonomy classifies errors by how each Grice's maxim is violated by a system utterance. The latter was derived by clustering annotators' comments that describe errors by a chat-oriented dialogue system. One way to evaluate the validity of a taxonomy is by using inter-annotator agreement; the reported κ (kappa) value for the TD taxonomy is low at 0.239, and no evaluation has been conducted for the BU taxonomy, suggesting that these taxonomies, in their current form, cannot be reliably used for error analysis.

We propose to revise these taxonomies to achieve better inter-annotator agreement. Our revised taxonomy based on the BU approach achieved a reasonable κ of 0.44 (Fleiss' κ). Figure 1 shows a dialogue example with error annotations by using our revised TD/BU taxonomies.

In the following section, we cover related work on the classification of errors in dialogue systems. In Sect. 3, we describe two current taxonomies of errors in chat-oriented dialogue systems. In Sect. 4, we describe how we revised the previously proposed taxonomies and report on their inter-annotator agreement. We also analyze the revised taxonomies on the basis of confusion matrices. Finally, in Sect. 5, we summarize the paper and mention future work.

	Utterance (translated from Japanese)	NB	PB	B	TD taxonomy	BU taxonomy
U	Though watermelons are good, peaches are also yummy.					
S	Watermelons absent?? Peaches are good.	4	14	12	Utterance-Syntactic error	Grammatical error
U	(I) love sweet fruits.					
S	(I) like hard ones.	9	11	10	Response-Excess/lack of information	Lack of information
U	Talking of hard ones, I would say walnuts.					
S	(I) like Kazehaya.	6	5	19	Response-No-relevance	Diversion
U	(I) went to see the fireworks there.					
S	Kazehaya is so cool, isn't he?	2	5	23	Context-Non-relevant topic	Topic-change error
U	(He is) very cool.					
S	Thank you.	12	13	5	Context-Others	Others
U	Not at all.					
S	Yes.	16	6	8		
U	(I) almost got sick thanks to these very hot summer days.					
S	Will very hot summer days continue? The temperature will rise until midsummer, won't it?	9	14	7	Response-Unclear intention	Unclear relation
U	(I) can't sleep well on midsummer nights.					
S	Midsummer days are cool.	2	15	13	Environment-Lack of common sense	Violation of common sense
U	I don't like the summer heat.					
S	Watermelons are yummy on midsummer days.	11	13	6	Context-Excess/lack of proposition	Repetition

Fig. 1 Dialogue excerpt with error annotations by using revised TD/BU taxonomies. NB (Not a Breakdown), PB (Possible Breakdown), and B (Breakdown) indicate number of breakdown labels given to each system utterance. The dialogue system used here is NTT DOCOMO's chat API called DCM (See Sect. 4.1)

2 Related Work

There are mainly three approaches to classifying errors in dialogue systems.

One is to adopt the general taxonomy of miscommunication proposed by Clark [4]. According to Clark, there are four levels in communication; channel, signal, intention, and conversation. By using these four levels, errors can be classified into four categories depending on which level the errors occurred (e.g., [3, 12, 14]).

The second is to classify errors in view of cooperativeness in dialogue. Grice proposed a set of maxims of cooperative dialogue; quantity, quality, relevance, and manner [6]. References [2, 5] incorporated Grice's maxims in classifying errors in task-oriented dialogue systems. Reference [7] used Grice's maxims to construct a taxonomy of errors in chat-oriented dialogue systems, although the inter-annotator agreement of this taxonomy was low, as we noted.

The third approach uses a clustering method to discover classes of errors. Reference [10] showed the effectiveness of using a clustering method called the Chinese restaurant process (CRP) in clustering comments that describe errors by a system. In this study, a taxonomy of errors was automatically derived, although its validity was not evaluated.

In this paper, we verify the taxonomies derived from the second and third approaches. We did not use the first approach, four levels by Clark, because we currently deal with text-based systems in which channel and signal level errors rarely occur.

3 Current Taxonomies

We briefly describe the two current taxonomies of errors in chat-oriented dialogue systems.

3.1 Top-Down Taxonomy

The TD taxonomy [7] has main categories that distinguish to which scope of the context the errors relate; utterance-level, response-level (adjacency pair), context-level (local context), and environment-level (not within the local context) errors. In addition, within each main category, there are sub-categories based on Grice's maxims. The taxonomy has the following categories.

- Utterance-level (syntactic error, semantic error, un-interpretable)
- Response-level (excess/lack of information, non-understanding, no-relevance, unclear intention, misunderstanding),
- Context-level (excess/lack of proposition, contradiction, non-relevant topic, unclear relation, topic switch error)
- Environment-level (lack of common ground, lack of common sense, lack of sociality).

3.2 Bottom-Up Taxonomy

The BU taxonomy [10] has 17 categories. These categories were obtained from an analysis of comments (written by researchers) that describe the errors made by a chat-oriented dialogue system. Over 1,500 comments were collected and clustered using an automatic clustering method. As a result, 17 clusters were identified, which led to the following 17 categories:

> General quality, Not understandable, Ignore user utterance, Ignore user question, Unclear intention, Contradiction, Analysis failure, Inappropriate answer, Repetition, Grammatical error, Expression error, Topic-change error, Violation of common sense, Word usage error, Diversion, Mismatch in conversation, and Social error.

4 Revising Taxonomies

We evaluated the TD and BU taxonomies on the basis of inter-annotator agreement (we used agreement rate, Cohen's κ, and Fleiss' κ) and revised them iteratively. For the evaluation, we used datasets of multiple dialogue systems, which we believed necessary to ensure the generality of the resulting taxonomies. Note that the TD taxonomy has been evaluated, but only by using a dataset of a single system.

4.1 Datasets

We used the datasets provided by the two series of dialogue breakdown detection challenges (DBDCs) [8], i.e., DBDC and DBDC2.

DBDC dataset: The DBDC dataset provides 100 dialogues between human users and a chat-oriented dialogue system based on NTT DOCOMO's chat API [13] (called DCM). Each dialogue contains 21 utterances in which the first utterance is the system's prompt and the remaining are ten pairs of utterances in exchanges by a human user and the system. Dialogue breakdown labels (labels indicating whether a system utterance leads to a dialogue breakdown [11]) were given to each of the system's utterances by 30 annotators.

DBDC2 dataset: The DBDC2 dataset provides dialogues of three different chat-oriented dialogue systems. In addition to DCM, a system with label propagation [16] (called DIT) and a system based on statistical machine translation (called the IR-Status or IRS from [15]) were used. In the same manner as DBDC, 100 dialogues were collected for each system and the dialogue breakdown labels were given to each system utterance by 30 annotators; i.e., there were 300 dialogues with dialogue breakdown annotations.

We divided the data (400 dialogues in total) into five datasets (datasets A–E), each containing 80 dialogues (20 dialogues from the data of DCM in DBDC1 and those of DCM, DIT, and IRS in DBDC2). With these five datasets, we conducted five rounds of annotations and evaluations.

4.2 Evaluation of Current Taxonomies

We first evaluated the current taxonomies by using the first round of data (i.e., dataset A). We recruited four annotators and divided them up into two groups of two. One group used the TD taxonomy, the other the BU taxonomy. We provided them with annotation manuals that contained the definitions of the error categories with some examples. In this round, there were 16 category labels for the TD taxonomy and 17 for the BU taxonomy. Annotations were done to system utterances to which the majority (i.e. 15 or more) of the annotators of the DBDCs gave breakdown labels. This setting remained the same for all other rounds.

The inter-annotator agreement for the TD taxonomy was 0.35, and Cohen's κ was 0.26, which is rather low and similar to the previously reported κ [7]. When we focused on the main categories, that is, when we viewed the annotation as a four-label annotation, the agreement was 0.59, and Cohen's κ was 0.22, which was even lower than the sub-category annotation. The inter-annotator agreement of the BU taxonomy was 0.35, and Cohen's κ was 0.27, which is as low as that of the TD taxonomy.

4.3 Procedure for Revising Taxonomies

With κ values of 0.26 and 0.27 as a starting point, we iteratively revised the taxonomies with the following procedure.

1. Two annotators use the taxonomy to annotate one round of a dataset.
2. The inter-annotator agreement is calculated between the two annotators.
3. A confusion matrix is created, and on the basis of the discussion between the authors, the taxonomy is revised to reduce the confusion. In the revision, the annotation manual is also revised; the definitions are clarified and examples may be added. After the revision, go back to Step 1 for the next round.

We conducted five iterations in total. We used datasets B and C for the second and third rounds, respectively. We then re-used datasets A and B for the fourth and fifth rounds, respectively. We needed to re-use the data because the inter-annotator agreement did not improve as fast as we expected, and there was fear that the data would be exhausted before the inter-annotator agreement improved. Fortunately, we reached a certain level of agreement with the BU taxonomy (around 0.4 in Fleiss' κ)

and decided to stop revising the taxonomies after the fifth round. The kappa for the TD taxonomy did not improve; it was around 0.2 throughout the rounds.

As the final step of the procedure, in the sixth round, we had crowd workers annotate the dialogues in dataset D with the revised taxonomies. The aim was to ensure that the errors can be reliably annotated by anyone, not just professional annotators. Below, we describe the revised TD/BU taxonomies and show the results of their evaluation.

4.4 Revised TD Taxonomy

The major improvements in the TD taxonomy are (1) revision of confusing categories and (2) introduction of a decision flow. The decision flow gives priority to error categories, making it possible to improve inter-annotator agreement in the case of possibly overlapping categories.

As a result of the confusion analysis of the trial rounds, we found that it is difficult to distinguish errors peculiar to human-system dialogue from other categories because it depends on one's ability to understand how a system works. Therefore, we merged such categories; we merged 'non-understanding' and 'misunderstanding' into 'non-understanding' in the utterance-level and 'non-relevant topic' and 'topic switch error' into 'non-relevant topic' in the context-level. In addition, we introduced an 'Others' category for each level as a natural consequence of introducing a decision flow. Our revised TD taxonomy is shown in Table 1.

In the TD taxonomy, the main category has to be determined first. The decision flow for the main category is shown in Fig. 2. The decision proceeds from narrower scope (utterance-level) to wider scope (context-level or environment-level).

Once the main category is determined, the subcategory is decided following the decision flow for each level. In each decision flow of the subcategory, branching questions are ordered by their assumed ease of decision (Fig. 3).

4.5 Revised BU Taxonomy

The BU taxonomy also underwent (1) revision of confusing categories and (2) introduction of a decision flow.

We reduced the number of categories from 17 to 13. First, we removed 'general quality' and 'mismatch in conversation' because they were too vague and caused much confusion. We then merged 'word usage error' and 'expression error' because they focus on similar linguistic phenomena. We also merged 'ignore user question' and 'inappropriate answer' because it was found difficult to distinguish whether a question was ignored or answered inappropriately. In addition, we removed 'unclear intention' because the understanding of system intention greatly depended on the annotators.

Table 1 Revised TD taxonomy

Main category	Subcategory	Explanation
Utterance	Syntactic error	Grammatically invalid utterance
	Semantic error	Semantically invalid utterance
	Un-interpretable	Not understandable
	Others	Other utterance-level error
Response	Excess/lack of information	Utterance misses important information or contains unnecessary information
	Non-understanding	Content of utterance is false or inappropriate regarding previous user utterance
	No-relevance	Utterance does not have any relation to previous user utterance
	Unclear intention	Relation to previous user utterance is not clear
	Others	Other response-level error
Context	Excess/lack of proposition	Utterance has no content or repetition
	Contradiction	Utterance contains propositions that contradict what has been said
	Non-relevant topic	Topic of utterance is irrelevant to current context
	Unclear relation	Relation to previous dialogue context is not clear
	Others	Other context-level error
Environment	Lack of common ground	Utterance has no factual grounding
	Lack of common sense	Thoughtless utterance
	Lack of sociality	Offensive utterance
	Others	Other environment-level error

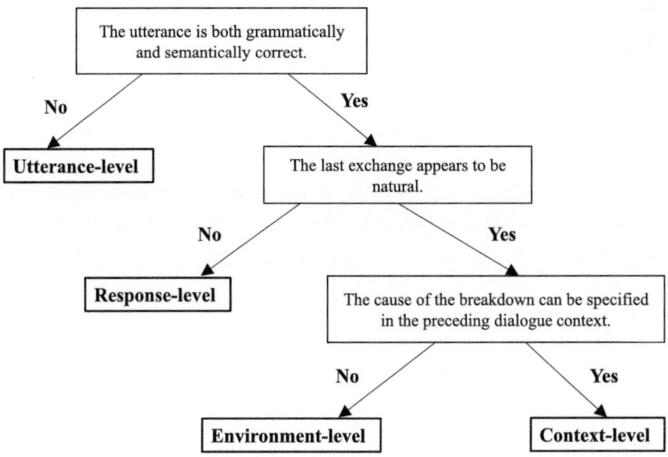

Fig. 2 Decision flow of main category for TD taxonomy

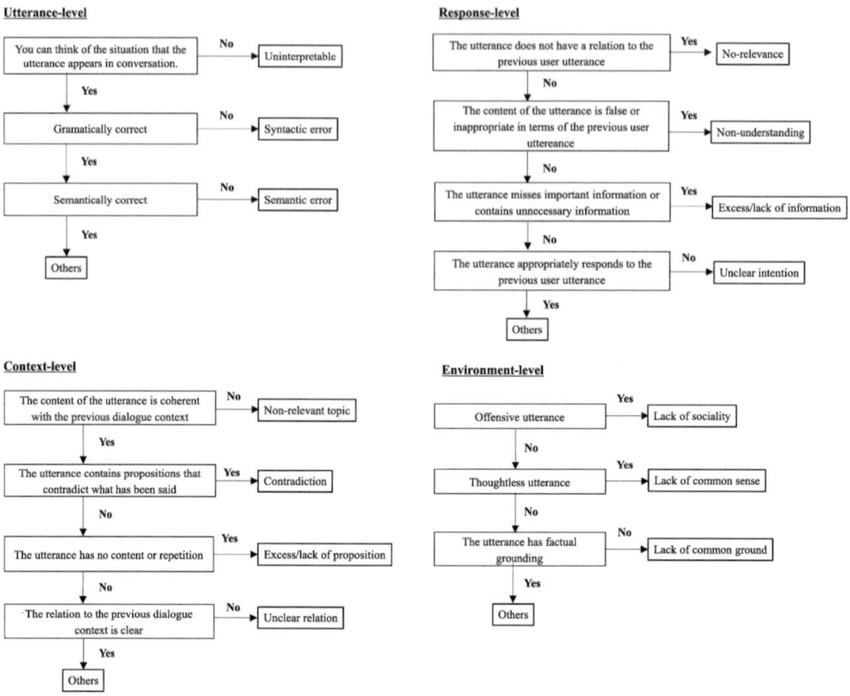

Fig. 3 Decision flow of subcategories for TD taxonomy

We initially removed 'ignore user utterance' because it was vague, but found afterwards the annotators had difficulty annotating inappropriate system responses to the user utterances that were not questions. Such errors were inevitably classified as 'Others'. This actually leveraged the inter-annotator agreement, but we had many 'Others' errors, which is not desirable. Therefore, we borrowed two concepts from the TD taxonomy that relate to inappropriate responses, i.e., 'lack of information' and 'unclear relation'. Table 2 shows the revised BU taxonomy, and Fig. 4 shows the decision flow for the BU categories. As in the TD taxonomy, the order of the branching questions is by the assumed ease of decision.

4.6 *Evaluation of Revised Taxonomies*

As mentioned above, we ran a final round (sixth round) using crowd workers.[1] We used dataset D for the annotation and evaluation. For crowd-sourcing, we recruited two groups of ten Japanese-native workers for each taxonomy. The workers were in

[1]https://www.lancers.jp/.

Table 2 Revised BU taxonomy

	Category	Explanation
1	Not understandable	Utterance is un-interpretable
2	Grammatical error	Utterance is ungrammatical
3	Word usage error	Words that do not fit context are used in utterance
4	Ignore user question	System ignores or fails to answer question
5	Topic-change error	Utterance does not reflect topic introduced by user
6	Diversion	Utterance abruptly introduces different topic
7	Contradiction	Content of utterance contradicts what has already been said
8	Repetition	Utterance is just repeated without new information
9	Lack of information	Utterance misses important information
10	Unclear relation	Relation between utterance and context is unclear
11	Social error	Utterance lacks politeness
12	Violation of common sense	Utterance lacks common sense
13	Others	Miscellaneous error

their 20–50 s, and each group consisted of four females and six males. The workers were not experts in language-annotation tasks.

Table 3 shows the results. The inter-annotator agreement of the TD taxonomy was 0.32, Cohen's κ was 0.24, and Fleiss' κ was 0.21. We can say that the inter-annotator agreement is low. On the other hand, the inter-annotator agreement of the BU taxonomy was 0.54, Cohen's κ was 0.44, and Fleiss' κ was 0.44, which is reasonably high. We additionally conducted an annotation by using the professional annotators (four annotators) and found that the inter-annotator agreement between the professionals and crowd workers was sufficiently high (0.45 in Cohen's κ).

Figure 5 shows the agreement and κ values in each round. Despite our revision, the TD taxonomy did not improve much. However, we observed a steady improvement for the BU taxonomy. In the third round, the inter-annotator agreement was rather high. This was when many errors were labeled as 'Others' because of the lack of appropriate error categories. We successfully reached the same level of agreement by appropriately modifying the taxonomy.

Figures 6 and 7 show the confusion matrices for the TD and BU taxonomies in the form of heat maps. For the TD taxonomy, it seems that it was difficult to distinguish the scope of the response from that of the context because the response can be part of the context. In addition, it seems difficult to classify errors that occur after other errors because such a case is not considered in the dialogue theories (i.e., Grice's maxims) on which the TD taxonomy is based. In fact, we compared the inter-annotator agreement of the first breakdown utterance during dialogue (error after a normal situation) and that of the rest (error after errors). The Fleiss' κ was 0.32 for the former and 0.19 for the latter, suggesting that the TD taxonomy can be effective for the analysis of a system that involves fewer breakdowns.

Fig. 4 Decision flow for BU categories

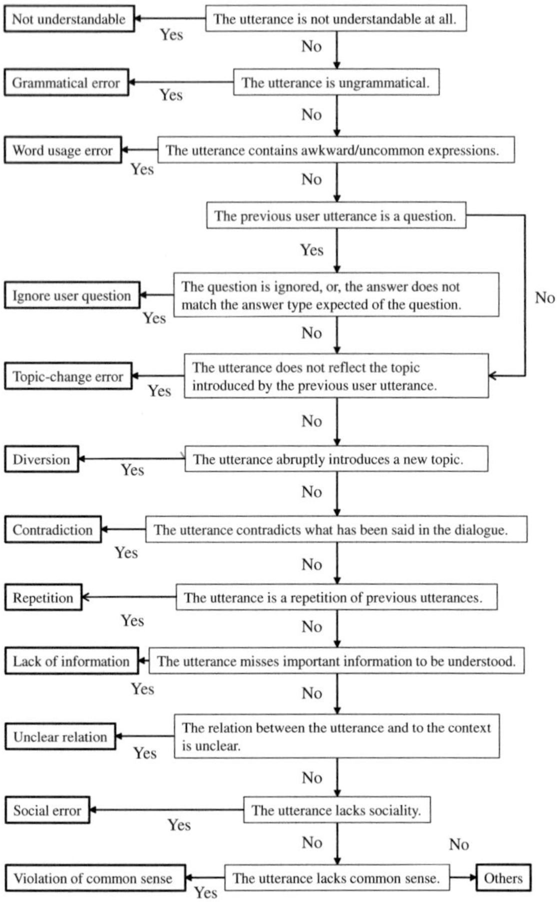

Table 3 Inter-annotator agreement, Cohen's κ, and Fleiss' κ for our revised taxonomies when crowd workers were employed as annotators. Cohen's κ was derived by averaging the κ values of all pairwise combinations of crowd workers

	Agreement	Cohen's κ	Fleiss' κ
TD	0.32	0.24	0.21
BU	0.54	0.44	0.44

For the BU taxonomy, there was a problem in distinguishing '9: lack of information' versus '10: unclear relation'; it seems difficult to distinguish 'something is missing' versus 'something is unclear'.

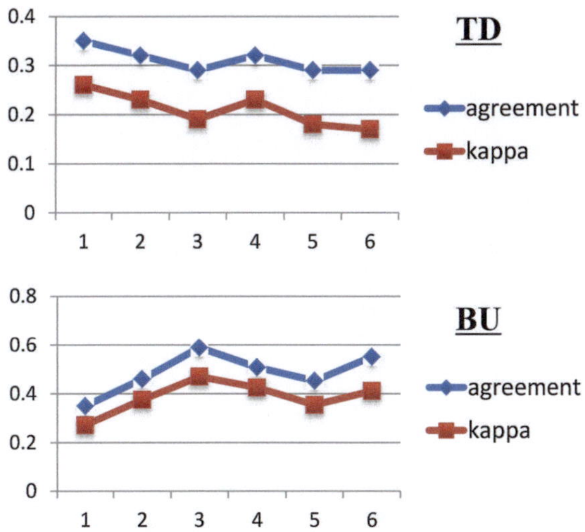

Fig. 5 Agreement and κ values in each round. We used professional annotators from rounds 1–5 and crowd workers in round 6

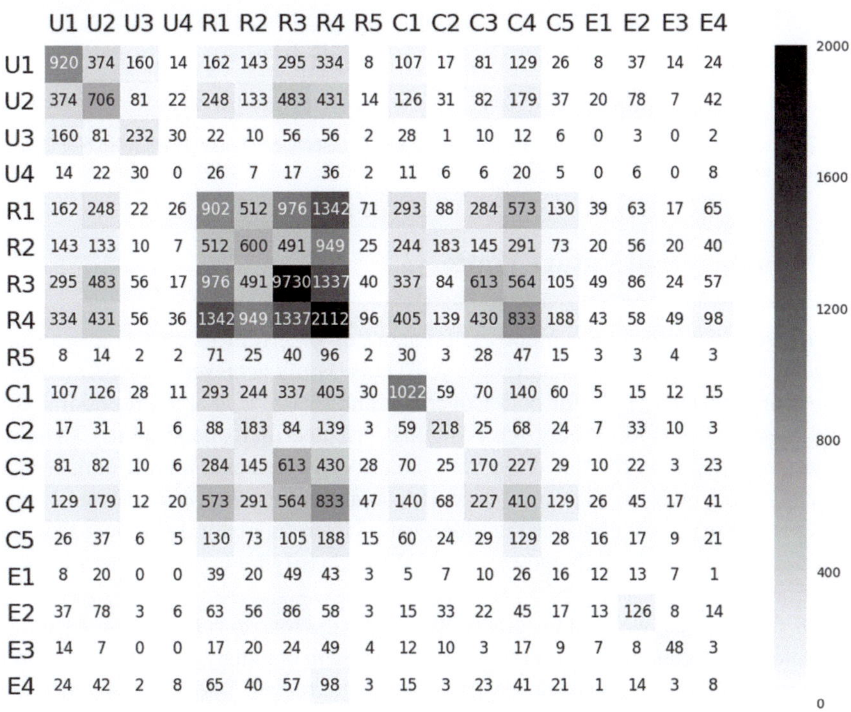

Fig. 6 Confusion matrix for revised TD taxonomy. Prefixes U, R, C, and E stand for Utterance, Response, Context, and Environment main categories. U1–E4 mean error categories from Utterance-Syntactic error to Environment-Others

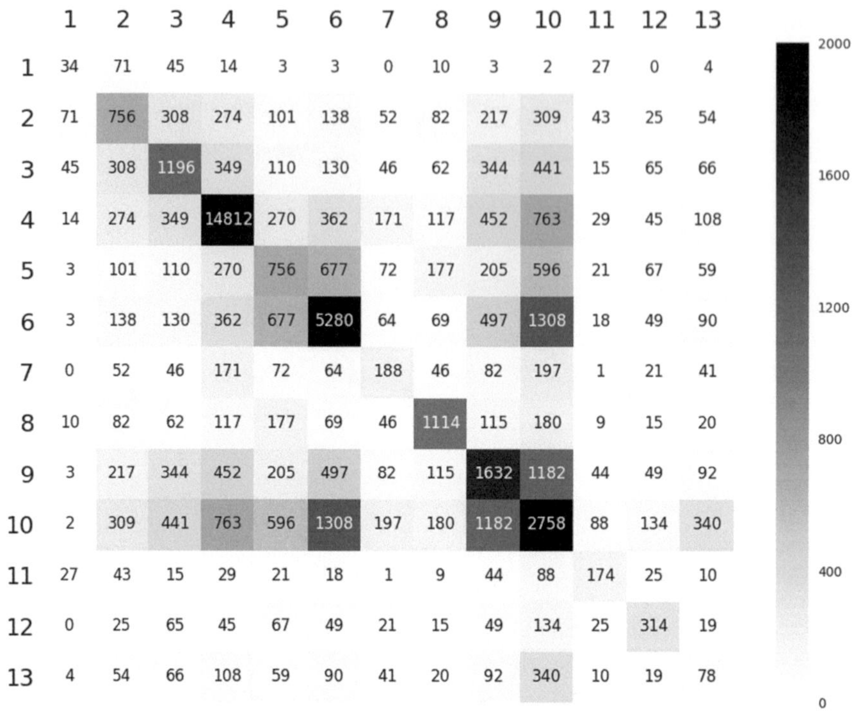

Fig. 7 Confusion matrix for revised BU taxonomy. 1–13 indicate error categories from Not understandable to Others (see Table 2)

5 Summary and Future Work

We revised two previously proposed taxonomies of errors in chat-oriented dialogue systems. The revised version of the BU taxonomy achieved reasonable inter-annotator agreement; 0.44 for both Cohen's κ and Fleiss' κ, making it possible for it to be safely used to classify errors in chat-oriented dialogue systems. These κ values were obtained by crowd workers (i.e., non-professional annotators), which is encouraging because it means that our taxonomy offers easy-to-understand conceptions of errors. We consider that the level of inter-annotator agreement that we achieved is reasonably high and can be used for classifying errors in chat-oriented dialogue systems.

For future work, we plan to further investigate the reason behind the poor inter-annotator agreement of the TD taxonomy. We noted in our analyses that it is difficult to distinguish between the response-level and context-level. One solution may be to merge these two levels. We also want to resolve cases in which one error category is the cause of the other. In our revision of the BU taxonomy, we borrowed two error categories from the TD, but it may be possible to merge BU and TD taxonomies for a better and complete taxonomy. Now that we have a reasonable taxonomy of errors, we also want to use this taxonomy for reducing errors of our chat-oriented dialogue systems.

References

1. Banchs RE, Li H (2012) IRIS: a chat-oriented dialogue system based on the vector space model. In: Proceedings of the ACL 2012 System Demonstrations, pp 37–42
2. Bernsen NO, Dybkjær H, Dybkjær L (1996) Principles for the design of cooperative spoken human-machine dialogue. In: Proceedings of the ICSLP, vol 2, pp 729–732
3. Bohus D, Rudnicky AI (2005) Sorry, i didn't catch that!–an investigation of non-understanding errors and recovery strategies. In: Proceedings of the SIGDIAL, pp 128–143
4. Clark HH (1996) Using language. Cambridge University Press, Cambridge
5. Dybkjær L, Bernsen NO, Dybkjær H (1996) Grice incorporated: cooperativity in spoken dialogue. In: Proceedings of the COLING, vol 1, pp 328–333
6. Grice HP (1975) Logic and conversation. In: Cole P, Morgan J (eds) Syntax and semantics 3: speech acts. Academic Press, New York, pp 41–58
7. Higashinaka R, Funakoshi K, Araki M, Tsukahara H, Kobayashi Y, Mizukami M (2015) Towards taxonomy of errors in chat-oriented dialogue systems. In: Proceedings of the SIGDIAL, pp 87–95
8. Higashinaka R, Funakoshi K, Kobayashi Y, Inaba M (2016) The dialogue breakdown detection challenge: Task description, datasets, and evaluation metrics. In: Proceedings of the LREC, pp 3146–3150
9. Higashinaka R, Imamura K, Meguro T, Miyazaki C, Kobayashi N, Sugiyama H, Hirano T, Makino T, Matsuo Y (2014) Towards an open-domain conversational system fully based on natural language processing. In: Proceedings of the COLING, pp 928–939
10. Higashinaka R, Mizukami M, Funakoshi K, Araki M, Tsukahara H, Kobayashi Y (2015) Fatal or not? finding errors that lead to dialogue breakdowns in chat-oriented dialogue systems. In: Proceedings of the EMNLP, pp 2243–2248
11. Martinovsky B, Traum D (2003) The error is the clue: breakdown in human-machine interaction. In: Proceedings of the ISCA Workshop on Error Handling in Spoken Dialogue Systems, pp 11–16
12. Möller S, Engelbrecht KP, Oulasvirta A (2007) Analysis of communication failures for spoken dialogue systems. In: Proceedings of the INTERSPEECH, pp 134–137
13. Onishi K, Yoshimura T (2014) Casual conversation technology achieving natural dialog with computers. NTT DOCOMO Tech J 15(4):16–21
14. Paek T (2003) Toward a taxonomy of communication errors. In: Proceedings of the ISCA Workshop on Error Handling in Spoken Dialogue Systems, pp 53–58
15. Ritter A, Cherry C, Dolan WB (2011) Data-driven response generation in social media. In: Proceedings of the EMNLP, pp 583–593. Association for Computational Linguistics
16. Tsukahara H, Uchiumi K (2015) System utterance generation by label propagation over association graph of words and utterance patterns for open-domain dialogue systems. In: Proceedings of the PACLIC, pp 323–331
17. Vinyals O, Le Q (2015) A neural conversational model. arXiv:1506.05869
18. Wallace RS (2009) The anatomy of alice. In: Parsing the turing test. Springer, Berlin, pp 181–210

Improving the Performance of Chat-Oriented Dialogue Systems via Dialogue Breakdown Detection

Michimasa Inaba and Kenichi Takahashi

Abstract Dialogue breakdown detection is a technique used for identifying inappropriate utterances in dialogue systems that has attracted increased attention, especially in chat-oriented dialogue systems. Although it is generally assumed that dialogue breakdown detection avoids generating system responses that then cause difficulties in continuing the given dialogue, this has yet to be verified experimentally or theoretically. In this paper, we apply the dialogue breakdown detection technique to generate responses for a chat-oriented dialogue system and experimentally verify that performance is improved by measuring the extent to which dialogue breakdown is avoided. Our experimental results show that dialogue breakdown detection indeed is able to improve the appropriateness of system responses; however, short, simple, and dull responses tend to increase when using this technique.

1 Introduction

Dialogue breakdown detection is a technique for identifying inappropriate utterances in dialogue systems, especially in chat-oriented dialogue systems. International competitions focused on the use of this technique are held annually, in particular the dialogue breakdown-detection challenge (DBDC) that has been held every year since 2015 [3–5].

Although it is generally assumed that dialogue breakdown detection avoids generating system responses that then cause difficulties in continuing the given dialogue, this has yet to be verified experimentally or theoretically. Further, specific application methods for using this technology remain unclear, in particular in how it can specifically improve the performance of dialogue systems.

M. Inaba (✉) · K. Takahashi
Hiroshima City University, Hiroshima, Japan
e-mail: inaba@hiroshima-cu.ac.jp

K. Takahashi
e-mail: takahashi@hiroshima-cu.ac.jp

L. F. D'Haro et al. (eds.), *9th International Workshop on Spoken Dialogue System Technology*, Lecture Notes in Electrical Engineering 579,
https://doi.org/10.1007/978-981-13-9443-0_30

Given the above, in this paper, we apply the dialogue breakdown detection technique to chat-oriented dialogue systems and experimentally verify that the performance is indeed improved by successfully avoiding breakdowns in dialogue.

2 Related Works

In [11], Sugiyama proposed a method for applying it to example-based dialogue systems. More specifically, this method searches utterances similar to input (i.e., user) utterances in a dialogue database using word vectors provided by Word2Vec [7]; next, this method re-ranks the top 20 response candidates using breakdown probabilities provided by the dialogue breakdown detection method, then uses these top-ranked utterances as its responses. In [8], Mori and Araki proposed a dialogue system using breakdown probabilities as a criterion for selecting responses from among multiple candidate utterance-generation modules; however, comparative evaluations without using dialogue breakdown detection was not conducted in these studies, and there was no evaluation of how effective the application method was in improving system performance.

3 Dialogue Systems

To properly analyze alterations in responses and the performance of systems using breakdown detection, we decided to focus on systems that generate multiple response candidates for each given input. Therefore, in this study, we do not cover typical rule-based dialogue systems, such as ELIZA [14] or ALICE [13] in which responses to inputs are fixed, because such systems cannot produce alternative candidates if the generated response is then estimated as potentially breaking down the given dialogue.

Systems that do generate multiple responses often score their candidates and rank them. In this study, we re-rank candidates by altering the given scores or ranks by applying dialogue breakdown detection, then analyze changes in performance.

For our analysis, we used the following three types of Japanese chat-oriented dialogue systems.

Example-Based System (IRS)

The example-based system that we used was an example-based dialogue system based on IR-STATUS [9] also used in DBDC2 held in 2016. This system uses Apache Lucene for its search engine and consists of 26,972 input-response pairs extracted from human dialogues as its example database. Other parameters and settings of the Apache Lucene were set to the various defaults.

To apply dialogue breakdown detection to this system, we used the top 10 matching utterances as response candidates and the similarity functionality inherent in Apache Lucene to calculate the corresponding response scores.

Neural Conversational Model (NCM)

The neural conversational model (NCM) system that we used was a dialogue system with an encoder-decoder neural network for response generation. We used KyotoNMT [1] for implementing this system. The encoder and decoder components were four-layer long short-term memory (LSTM) models, each with 1000 hidden cells. The vocabulary size was limited to 80,000 terms, and the dropout rate was set to 0.2. Further, the model was trained using 10 million tweet-reply pairs with an Adam optimizer.

In our implementation, response candidates were generated using 12 models individually trained with different initial parameters. Note that when using the encoder-decoder neural network, multiple responses can be generated using beam search; however, since our one trained model generated almost the same response candidates (e.g., as present and future tenses), we decided to use multiple models. For response scores, we used the average word probability of each response; further, if more than one model generated the same response, the highest score was used as the response score. Therefore, when the system chats with a human subject for data acquisition (as we discuss in Sect. 4), it only uses one model to reduce response times.

Neural Utterance Ranking Model (NUR)

The neural utterance ranking (NUR) system that we used was based on our neural network-based dialogue model [6]; this model ranks candidate utterances acquired from Twitter in response to the given dialogue context and uses the highest-ranked candidate as its response. The training settings and data set that we used in our present study were the same as those described in [6]; further, we used response scores for utterance ranking in our NUR model for the system response scores in our present study.

4 Dialogue Breakdown Detection

In this study, the dialogue format and various settings regarding dialogue breakdown detection are based on those of DBDC2. Here, dialogue data consists of text chats between a dialogue system and a user in Japanese; annotations are also included for each system response, where annotations are one of the following three breakdown labels:

(NB) Not a breakdown	It is easy to continue the conversation.
(PB) Possible breakdown	It is difficult to continue the conversation smoothly.
(B) Breakdown	It is difficult to continue the conversation.

For each system utterance, annotations were provided by more than 30 individuals. Next, we applied a detection method to estimate majority labels among all annotators for each utterance, as well as a distribution of breakdown labels.

4.1 Methodology

The current state-of-the-art dialogue breakdown detection method described in [12] was proposed in DBDC3; however, this method is difficult to implement because it utilizes some feature sets based on an unpublished corpus with original annotations.

In this study, we use the best performance detection method from DBDC2 [10]; this method uses typical error patterns of system responses, such as abrupt changes in the discussed topic or unnatural connections of dialogue acts as features, then estimates a distribution of labels using the extra-trees regressor [2].

As a preliminary experiment, we compared cases in which all system data were used as training data for all systems with cases in which data was divided for each system and learned individually. Our initial results indicated that the individual case was better than the all-in-one case; therefore, the breakdown detection model we used for the remainder of our study was the approach in which each system learns individually.

4.2 Dataset

As training data for the breakdown detection method, we used dialogue data from the three types of systems—i.e., IRS, NCM, and NUR—with annotations accompanying the dialogue breakdown labels. We collected dialogue data for the NCM and NUR systems in full compliance with DBDC2 data collection and annotation rules. Since the IRS system data have already been collected in DBDC2, we simply use this data as IRS data for our present study.

Resulting statistics for the above data are summarized in Table 1, indicating that the NUR system generated the fewest number of responses that caused a dialogue breakdown, and the IRS system was the opposite, based on the ratio of PB and B labels.

Table 1 Statistics of data for dialogue breakdown detection

	IRS	NCM	NUR
Number of dialogues	100	100	120
Number of user utterances	1000	1000	1200
Number of system utterances	1100	1100	1320
Number of annotators	30	30	34
NB (Not a breakdown)	31.1%	47.4%	57.7%
PB (Possible breakdown)	26.7%	32.7%	27.0%
B (Breakdown)	42.1%	17.2%	15.2%
Fleiss' κ	0.29	0.29	0.26
Fleiss' κ (PB + B)[a]	0.38	0.43	0.42

[a]Fleiss' κ when PB and B are treated as a single label

5 Experiments

Our experiments consisted of re-ranking candidate responses generated by the given systems using the results of our dialogue breakdown detection method.

To evaluate the response performance of each system after applying breakdown detection, we constructed a data set that included all context/response pairs, i.e., the chat log between two speakers as context and 10 or more response candidates generated by the given system; note that these pairs also incorporated scores and breakdown labels.

Each response candidate is then annotated using breakdown labels via the same method described in Sect. 4.2 above by at least three annotators. In evaluating response performance, we regard candidates with 50% or more annotators decided as NB as correct response and others as incorrect. Note that the data collection procedure described here is the same as that used in [6], and test data from [6] is therefore used for the NUR system in our experiments. The contexts of the data for the IRS and NCM systems were acquired from the data set used in [6] with the condition that a system must generate 10 or more response candidates. Statistics of the data are summarized in Table 2.

Table 2 Statistics for response candidate re-ranking

	IRS	NCM	NUR
Number of data points (Number of context and response pairs)	300	300	300
Number of utterances in context	1.37	2.14	2.04
Number of candidate responses per data	11.42	10.57	10.94
Number of words per candidate	18.13	9.54	10.70
Number of human annotators	4.61	5.92	3.88

5.1 Re-Ranking Method

To apply dialogue breakdown detection, we propose the three response re-ranking methods described below.

Classification-based method This method focuses on the classification results output by the breakdown detection method, lowering the ranking of response candidates classified as causing a dialogue breakdown. Here, labels with the maximum probability according to the detection model are judged as estimated results; further, we assume that only the B label is a breakdown, and therefore lump PB into the B category.

Probability-based method This method re-ranks candidates by calculating the product of the response score and the non-breakdown probability, using this product as the new score. From estimated probabilities $p(NB)$, $p(PB)$ and $p(B)$, we assume that $p(NB)$ is the non-breakdown probability and $p(NB) + p(PB)$ is a non-breakdown probability.

Regression-based method Using estimated probabilities $p(NB)$, $p(PB)$, $p(B)$ and the response score s as input features for a linear regression model, candidates are re-ranked with new score s_{new} as calculated by the model. For training the regression model, we used the mean squared error of s_{new} with the teacher score of a correct candidate as 1.0 and that of an incorrect candidate as 0.0 as the loss function. This regression-based method includes the optimization of parameters, with evaluation performed using 10-fold cross-validation.

5.2 Results

To evaluate re-ranking performance, we used the mean average precision (MAP) measure, which indicates how many correct response candidates ranked higher. Figures 1, 2 and 3 show MAP results for the top n ranked candidate utterances.

Based on this MAP measure, probability-based (NB), probability-based (NB + PB), and regression-based methods all showed performance improvements across all systems as compared to results without re-ranking. For the classification-based (B) and (PB + B) cases, the increased effectiveness differed from system to system. Individually, for the IRS results (i.e., Fig. 1), the probability-based (NB) method was the most effective, followed by the probability-based (NB + PB) method, then the regression-based method. The classification-based (B) and (PB + B) cases had very similar results, with smaller improvements versus that of other methods, but very large improvements were observed when compared to those without re-ranking. For the NCM results (i.e., Fig. 2), the observed improvement was larger for the regression-based method, then for the probability-based (NB) method, followed by the probability-based (NB + PB) method. Conversely, the classification-based (B) and (PB + B) methods showed deteriorated performance as compared to results without re-ranking. Finally, for the NUR results (i.e., Fig. 3), the regression-based

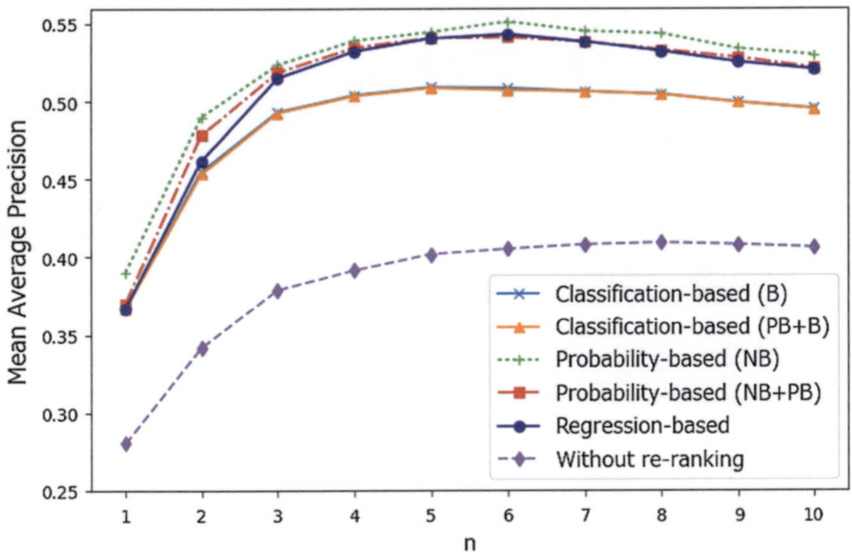

Fig. 1 MAP over top n response candidates (IRS)

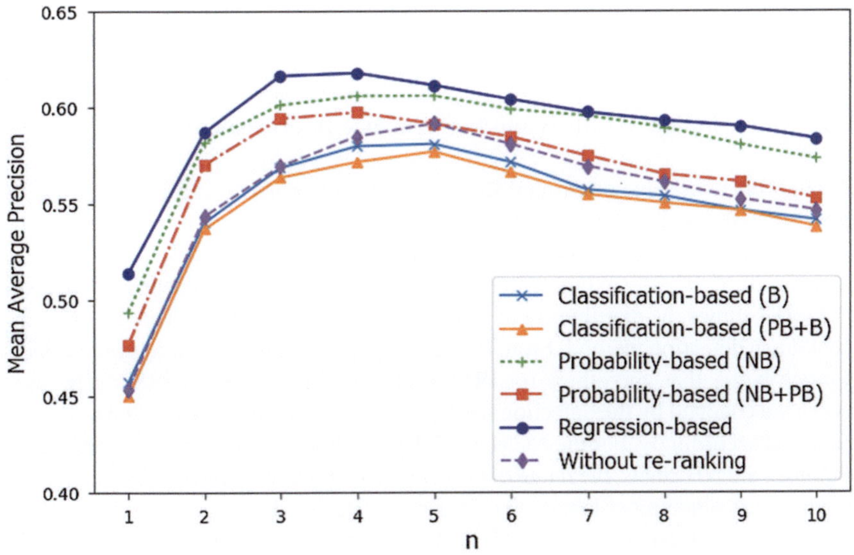

Fig. 2 MAP over top n response candidates (NCM)

method showed the best performance improvement, but improvements were smaller versus those of the other systems.

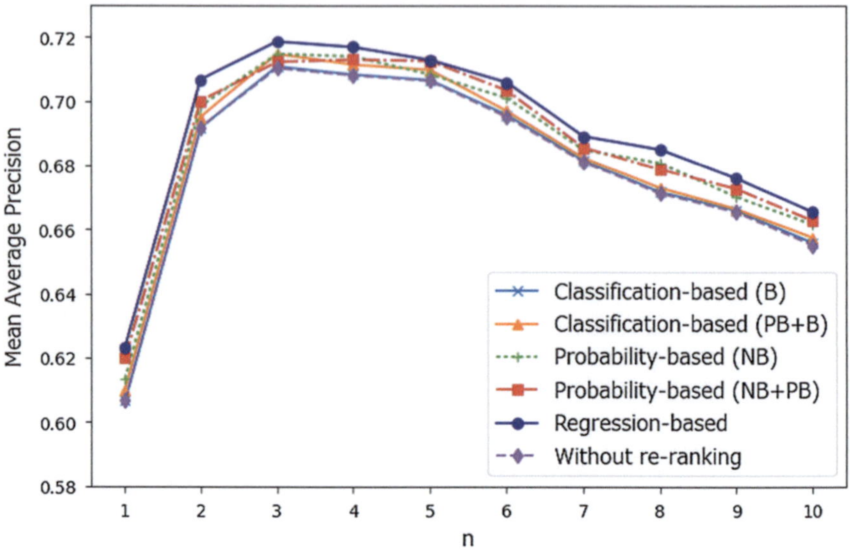

Fig. 3 MAP over top n response candidates (NUR)

5.3 Discussion

5.3.1 Distribution of Labels and Performance Improvements

Through our work, we confirmed that response performance can be improved by applying dialogue breakdown detection; however, the effective re-ranking method and degree of effectiveness here varied depending on the system. To investigate the cause, we further analyzed the output of the dialogue breakdown detection method.

Table 3 shows the distribution of labels with maximum probability output by the breakdown detection method. From the table, we observe that the ratio of the B label is much larger than the others for IRS. Therefore, in the case of IRS, since more scores are changed by the detection method, the range of performance improvement increased accordingly. For the NCM and NUR methods, since more than 80% of the responses are NB labels, the range of performance improvements became very small. In particular, since most labels were NB in NUR, the classification-based method using only the PB and B labels was less effective, whereas the probability-based and regression-based methods using NB label information were more effective.

5.3.2 Performance Deterioration in NCM

In NCM, the classification-based (B) and (PB + B) methods resulted in worse performance versus performance without re-ranking. We investigated the ratio of correct

Table 3 Distribution of labels output by the dialogue breakdown detection method. Numbers in parentheses indicate the number of labels

	NB (Not a breakdown)	PB (Possible breakdown)	B (Breakdown)
IRS	20.7% (711)	0.3% (10)	79.0% (2706)
NCM	84.9% (2862)	9.4% (317)	5.7% (193)
NUR	96.6% (3173)	2.1% (69)	1.3% (42)

responses among the response candidates with the B label, finding that IRS was 11.6% and NUR was 19.0%, whereas NCM was 39.4%, more than twice the other values. Therefore, the performance of breakdown label classification was poor for NCM, which likely caused response performance to deteriorate.

5.3.3 Dialogue Breakdown Detection and Re-Ranking Performance

We first note that it is important to improve the performance of dialogue breakdown detection to improve response performance of dialogue systems. To analyze response performance when breakdown detection performance is low, we investigated re-ranking performance by reducing the training data set size of the detection model. Figure 4 shows MAP results using the top-ranked response (i.e., MAP@1) while increasing the size of the learning data set by steps of 10. Here, we used the probability-based (NB) method for re-ranking. From the figure, we observe that MAP@1 generally improves as the size of the training data set increases, though the NUR results are only slightly improved. Therefore, we were able to confirm that it is important to improve the performance of the dialogue breakdown detection model before applying its results to dialogue systems.

5.3.4 Changes in Response

To analyze the change in response in the case of re-ranking, we calculated the number of tokens and the number of types included in the top-ranked responses when both applying and not applying our breakdown detection method. Here, we used the regression-based method for the re-ranking method.

We present our results in Table 4; as shown in the table, we found that all systems tended to decrease both the number of tokens and the number of types after applying dialogue breakdown detection. In particular, the amount of change that occurred in IRS was rather large.

Tables 5, 6 and 7 present examples of response changes after applying the regression-based method. More specifically, in Tables 5 and 6, responses generated after re-ranking tended to change into shorter and simpler responses. Unfortunately, we note here that if a system often generated such short and simple responses, while it

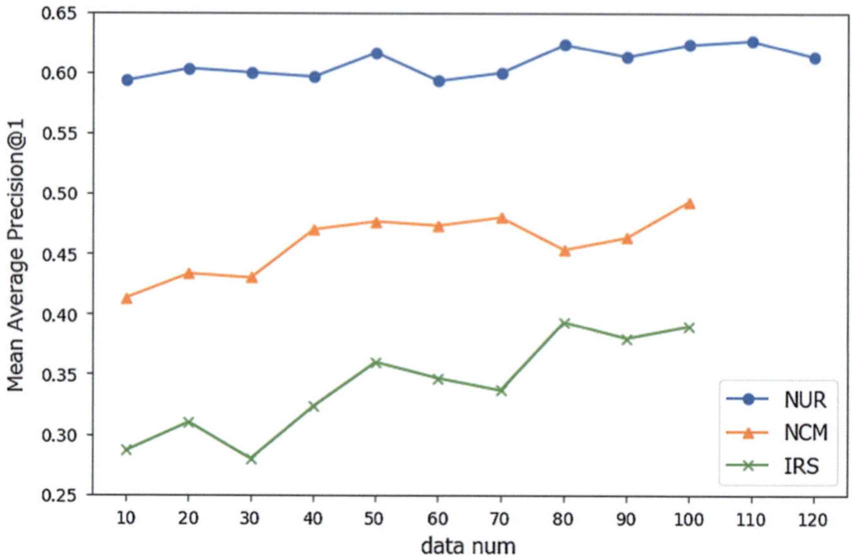

Fig. 4 Training data size and response performance

Table 4 Changes in the number of tokens and types of responses with and without re-ranking

	Number of tokens (without re-ranking)	Number of tokens (re-ranking)	Number of types (without re-ranking)	Number of types (re-ranking)
IRS	18,964	12,276	1219	761
NCM	11,143	9317	363	332
NUR	10,659	10,494	885	881

Table 5 Change in responses after applying dialogue breakdown detection (IRS)

Context	Without re-ranking	Re-ranking
A: Russian coke tastes good!	Violin! It is a waste that you practiced for 8 years but you cannot play now. Don't you begin again?	That's delicious
A: I'd like to buy a file and a card reader tomorrow, but it is hassle	It would be nice to have that. I will devote	Me too. lol
A: The timing is bad now B: Are you OK? A: No	I do not think that it will be raised so firmly	OK. Besides, it is not bad

Table 6 Change in responses after applying dialogue breakdown detection (NCM)

Context	Without re-ranking	Re-ranking
A: I did not do my homework but I came to class	That's right, you do not have any money	Really?
A: I want to take a pillow in my laboratory	You do not bring it to the laboratory?	Please make me a pillow!
A: Recently sleep time is increased B: I wish I were you A: I'm dreaming of a ghost coming out	Is it close from Kamata?	Really

Table 7 Change in responses after applying dialogue breakdown detection (NUR)

Context	Without re-ranking	Re-ranking
A: Let's calm down by saying that you can do today can be done tomorrow!	The mind and body are different	Maybe I want to live with room in my mind
A: Let's do our best ... I have to work hard ... (I go to buy Red Bull)	Red Bull and Minmin-daha are not working well for me	Red Bull is the best energy drink
A: Recent Lexus' TV commercial is funny! B: I don't like Lexus' face A: How about Mazda's?	I cannot say that Lexus is a fun car	I am really looking forward to Mazda's sports concept car

is possible to avoid dialogue breakdown, the risk increases for boring the user because the system does not provide any new or interesting information. We therefore conclude here that it is necessary to pay close attention so that only such responses will not be chosen when the dialogue breakdown detection is applied to a system in which a simple response is included in the response candidates.

6 Conclusions

In this present work, we experimentally evaluated whether response performance could be improved by applying a dialogue breakdown detection method to three different types of dialogue systems. We therefore proposed three types of response re-ranking methods for applying dialogue breakdown detection, i.e., a classification-based method that uses classification results of breakdown labels, a probability-based method that uses non-breakdown probability values, and a regression-based method that uses linear regression with the probability distribution of breakdown labels and response scores as the feature set. Our experimental results indicated that the probability-based and regression-based methods effectively improved performance across all dialogue systems. Further, the classification-based method improved the response performance of two dialogue systems, but actually caused deteriorated performance for the third system. We also found that the number of short and simple responses increased when we applied dialogue breakdown detection as compared to results without such detection.

Acknowledgements This study received a grant of JSPS Grants-in-aid for Scientific Research 16H05880.

References

1. Cromieres F (2016) Kyoto-NMT: a neural machine translation implementation in chainer. In: Proceedings of COLING 2016: system demonstrations, pp 307–311
2. Geurts P, Ernst D, Wehenkel L (2006) Extremely randomized trees. Mach Learn 63(1):3–42
3. Higashinaka R, Funakoshi K, Inaba M, Arase Y, Tsunomori Y (2016) The dialogue breakdown detection challenge 2. SIG-SLUD B5(02):64–69
4. Higashinaka R, Funakoshi K, Yuka K, Inaba M (2016) The dialogue breakdown detection challenge: task description, datasets, and evaluation metrics. In: 10th edition of the language resources and evaluation conference
5. Higashinaka R, Funakoshi K, Inaba M, Tsunomori Y, Takahashi T, Kaji N (2017) Overview of dialogue breakdown detection challenge 3. In: Proceedings of dialog system technology challenge 6 (DSTC6) workshop
6. Inaba M, Takahashi K (2016) Neural utterance ranking model for conversational dialogue systems. In: 17th annual meeting of the special interest group on discourse and dialogue, pp 393–403
7. Mikolov T, Sutskever I, Chen K, Corrado GS, Dean J (2013) Distributed representations of words and phrases and their compositionality. Advances in neural information processing systems, pp 3111–3119
8. Mori H, Araki M (2016) Selection method of an appropriate response in chat-oriented dialogue systems. In: Proceedings of the 17th annual meeting of the special interest group on discourse and dialogue, pp 228–231
9. Ritter A, Cherry C, Dolan WB (2011) Data-driven response generation in social media. In: Proceedings of the conference on empirical methods in natural language processing, pp 583–593
10. Sugiyama H (2016) Chat-oriented dialogue breakdown detection based on the analysis of error patterns in utterance generation. SIG-SLUD B5(02):81–84
11. Sugiyama H (2016) Utterance selection based on sentence similarities and dialogue breakdown detection on NTCIR-12 STC task, pp 552–553
12. Sugiyama H (2017) Dialogue breakdown detection based on estimating appropriateness of topic transition. In: Dialog system technology challenges (DSTC6)
13. Wallace R (2008) The anatomy of A.L.I.C.E. parsing the turing test, pp 181–210
14. Weizenbaum J (1966) ELIZA-a computer program for the study of natural language communication between man and machine. Commun ACM 9(1):36–45

Automated Scoring of Chatbot Responses in Conversational Dialogue

Steven Kester Yuwono, Biao Wu and Luis Fernando D'Haro

Abstract Rapid advancement in natural language processing (NLP) and machine learning has led to the recent development of many chatbot systems using various algorithms. However, in a conversational dialogue setting, creating a system to communicate with humans in a meaningful and coherent manner remains a challenging task. Furthermore, it is very difficult even for humans to evaluate the responses of a chatbot system given the context of the conversation. In this paper, we will focus on the problem of automatically evaluating and scoring the quality of chatbot responses in human-chatbot dialogue settings. We propose a novel approach of combining the word representations of human and chatbot responses, and using machine learning algorithms, such as support vector machines (SVM), random forests (RF), and neural networks (NN) to learn the quality of the chatbot responses. Our experimental results show that our proposed approach is able to perform well.

1 Introduction

Recent advancement in NLP has greatly improved the quality of chatbot systems for conversational dialogues. As a result, many chatbot systems were developed recently [3, 11, 31]. Despite the recent advances in dialogue systems, creating a system to

Luis Fernando D'Haro—Research work done while working at Institute for Infocomm Research, Singapore.

S. K. Yuwono · B. Wu
Department of Computer Science, National University of Singapore, 13 Computing Drive, Singapore 117417, Singapore
e-mail: sky@u.nus.edu

B. Wu
e-mail: e0001428@u.nus.edu

L. F. D'Haro (✉)
ETSI de Telecomunicación, Universidad Politécnica de Madrid, Madrid, Spain
e-mail: luisfernando.dharo@upm.es

Ciudad Universitaria, Avenida Complutense 30, 28040 Madrid, Spain

L. F. D'Haro et al. (eds.), *9th International Workshop on Spoken Dialogue System Technology*, Lecture Notes in Electrical Engineering 579,
https://doi.org/10.1007/978-981-13-9443-0_31

communicate with humans in a natural, coherent, and meaningful manner remains a very challenging task. The difficulty in creating a high quality dialogue system is attributed in part to the difficulties in evaluating the quality of chatbot's responses. This kind of evaluation is a challenging and subjective task, even for humans. As an evidence, the inter-annotator agreement in our dataset is relatively low between crowdsourced annotators or just fair among experts [2]. Where for the same dataset used in our experiments, [2] found that the Fishers Interclass Correlation coefficient and Cohens Kappa coefficient are 0.5454 and 0.3736 respectively that is considered fair, but that is reduced when including non-experts' annotation (0.476 for ICC and 0.317 for Kappa). In this paper, we focus on the task of automatically scoring chatbot's responses.

There are many recent studies [19, 21] which attempt to assess the quality of chatbot responses by comparing them to reference human responses. In other words, sequence to sequence comparison evaluated using BLEU and METEOR score. Our work is novel in the aspect that we are formulating our problem as a classification problem instead of a sequence to sequence problem. We believe that in a conversational dialogue, there is no single valid response but many various ways to respond that are valid in the context of the conversation. Therefore, we are not comparing the similarity of chatbot responses with their respective valid human responses but classifying chatbot responses as valid, acceptable, or invalid.

To tackle this problem, we propose an approach of using various word representations and machine learning algorithms to evaluate or score chatbot responses in a dialogue. In our SVM and random forest model, the words are represented by bag-of-words model. On the other hand, we use distributed word representations (word embeddings) with two neural network architectures, namely convolutional neural network (CNN) [18] and recurrent neural network (RNN) [12], inspired by their recent successes in multiple tasks.

RNN has proven to be successful in sequence prediction tasks such as speech recognition [13], caption generation [29], machine translation [1], and automated essay scoring [27]. On the other hand, CNN has also been very successful in image classification [6], and sequential labelling task [17].

We organize the rest of the paper as follows, we will explain the details of the chat-oriented dialogue evaluation task in the next section. In Sect. 3, we will elaborate more on our SVM and random forest approach, followed by the neural network approach in the next section. Our implementation and experimental details will be described in Sect. 5. Finally we conclude our work in the last section.

2 Chat-Oriented Dialogue Evaluation

WOCHAT,[1] a workshop on chatbots and conversational agents, releases a number of datasets containing human-chatbot conversations with 624 sessions and 20,440 turns in total. However, not all of them have annotations indicating the validity of the

[1] http://workshop.colips.org/wochat/.

Table 1 Chat snippet from TickTock dataset. `OPT`, `PES`, and `AVE` denotes optimistic, pessimistic, and averaging ground truths respectively

Instance #	Speaker	Annotated score(s)	OPT	PES	AVE	Response
9	Human	–	–	–	–	Yes, but it's more of a drama than a comedy
	Chatbot	1; 1; 0.5; 0	1	0	0.625	I really like the first Avenger movie, have you seen it before?
10	Human	–	–	–	–	Yes, I like it. did you like the batman versus superman movie?
	Chatbot	1; 0.5; 0; 0	1	0	0.375	I love his black eyes
…	…	…	…	…	…	…

chatbot responses. Only 303 sessions and 5,879 turns are annotated. A **session** consists of multiple turns. A **turn** is defined as a continuous response by either human or chatbot which can contain one or more sentences. The number of annotated turns (5,879) stated above consists of both human and chatbot turns. However, we are only interested in the annotated chatbot turns (approximately half of the number of turns above). There are 3 classes to represent the quality of a chatbot turn: **(1) Valid; (2) Acceptable; (3) Invalid**. Annotations are done by both experts and Amazon Mechanical Turk (AMT). As mentioned in the introduction, the weakness in WOCHAT datasets is that the inter-annotator agreement is very low. The Cohen's Kappa coefficient value is 0.3736 between two expert annotators [2]. Some examples of these datasets and the setup of training instances is shown in Table 1.

The task can be divided into: (1) Ground truth construction; (2) Feature extraction; (3) Learning. Since the inter-annotator agreement is very low, it is challenging to establish the ground truth. For simplicity, we assume that *valid, acceptable,* and *invalid* labels are represented by real values: `1.0`, `0.5`, and `0.0`, respectively. We will explore the following ground truth construction methods to aggregate multiple annotations for each turn: **(1) Optimistic; (2) Pessimistic; (3) Averaging**. These methods are described next.

Optimistic ground truth construction is done by obtaining the maximum score if there are multiple annotators. For example, if there are three annotations with the following scores: `1`, `0`, `0`; then the ground truth will be `1`. If the maximum score is `0.5`, we will replace it with `1`. Similarly, for **Pessimistic** ground truth, the minimum score of all annotations for each turn is used as the ground truth. If the minimum score is `0.5`, we will replace it with `0`. Therefore, we can formulate this problem as a binary classification task.

Averaging ground truth construction works by obtaining the average of the annotation scores given by multiple annotators. For example, if there are three annotations with the following scores: `1`, `0`, `0`; then the ground truth will be `0.333`. Instead of a binary classification method, we will use regression to tackle the task in averaging setup.

The idea behind optimistic and pessimistic ground truth construction is to decompose the regression (averaging ground truth) task into two simpler binary classification tasks. After the two binary classifiers are trained, the outputs can be combined (by voting/averaging) to solve the regression task.

3 SVM and Random Forest Approach

Word Representation: Bag-of-words We use a bag-of-words approach to represent words for SVM and random forest classifiers. Despite not encoding the sequential information of words, bag-of-words model is simple and effective. The vocabulary is first created using the training data. Each word in the dictionary is represented by its frequency in the current turn. Out of vocabulary words are replaced by `<unknown>` token. For each pair of human-chatbot turns, the bag-of-words representations are constructed separately and then concatenated together as the input of the model.

Support vector machine (SVM) is a supervised learning model for classification, that performs well in text classification problems [16]. To classify data points in high dimensional space, SVM constructs hyperplanes to separate training data samples by maximizing the margin between data points of different classes. By considering the bag-of-words representation as a data point in a high-dimensional space, a kernel function represents the distance between samples. There are linear and non-linear kernels (such as sigmoid and radial basis function), and the optimization can be done efficiently using iterative methods on matrix operations. SVM can be extended to solve regression problems [10], using the same concept as the original SVM but modifying the loss function with additional constraints.

Random forest (RF) is another supervised learning model which performs well for text classification [30]. It is an ensemble method of aggregating the output of multiple decision trees. Bootstrap aggregating (bagging) is used to greatly reduce the variance and hence improve the performance. Given the bag-of-words representation as the input vector, the output of the random forest is simply the majority votes (of multiple decision trees) in the binary classification case, or the average of all the real-valued scores in the regression case. The objective functions used for classification and regression are entropy and mean squared error respectively.

4 Neural Network Approach

4.1 Model Architecture

Our neural network architecture is illustrated in Fig. 1. We now describe each layer in detail:

Word Embeddings Continuous word representations where each word is represented by a word vector (or word embedding) has been proven useful for many NLP

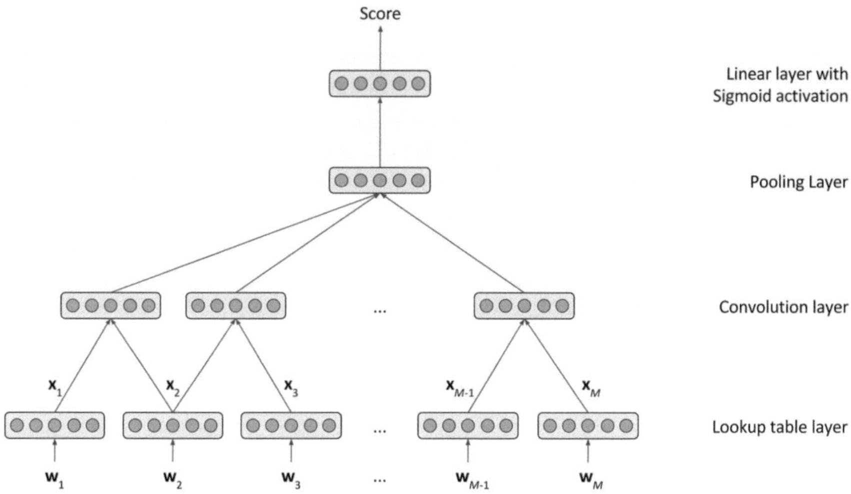

Fig. 1 The convolutional neural network (CNN) architecture (with window size 2)

tasks [4, 8, 28]. The word embeddings are usually used as additional features to the current machine learning model to improve its performance. Word embeddings can be obtained by training shallow networks such as continuous bag-of-words (CBOW) and skip-gram (SKIP) models [22], training deeper networks [5], or on aggregated global word-word co-occurrence statistics (GloVe) [24].

Lookup Table Layer: The first layer of our neural network maps each word into a vector in a d_{LT}-dimensional space. Given a sequence of words $\mathbf{W}$ represented by their *one-hot* representations $(\mathbf{w}_1, \mathbf{w}_2, \mathbf{w}_3, \ldots, \mathbf{w}_M)$, the output of the lookup table layer is calculated by Eq. 1:

$$\mathrm{LT}(\mathbf{W}) = (\mathbf{E}\mathbf{w}_1, \mathbf{E}\mathbf{w}_2, \mathbf{E}\mathbf{w}_3, \ldots, \mathbf{E}\mathbf{w}_M) = (\mathbf{x}_1, \mathbf{x}_2, \mathbf{x}_3, \ldots, \mathbf{x}_M) \tag{1}$$

where $\mathbf{E}$ is the trainable word embedding matrix and M is the number of words in a turn (by either human or chatbot).

Convolution Layer: After the dense representation of the input sequence is computed from the lookup table layer, they are then fed as an input to the convolution layer to extract *local features*. Given a window of word representations of length l, (e.g, $\mathbf{x}_1, \mathbf{x}_2, \ldots, \mathbf{x}_l$), they are first concatenated to form vector $\bar{\mathbf{x}}$, and then calculate the output vector of length d_c as shown in Eq. 2:

$$\mathrm{Conv}(\bar{\mathbf{x}}) = \mathbf{W}\bar{\mathbf{x}} + \mathbf{b} \tag{2}$$

$\mathbf{W}$ and $\mathbf{b}$ are the trainable weight and bias parameters respectively, and they are shared across all windows in a sequence.

Recurrent Layer: Alternatively, the convolution layer above can be replaced by a recurrent layer. The dense representation of the input sequence are then fed as an input to the recurrent layer to get a representation of the whole sentence(s) instead of just extracting l-gram *local features*. Following [27], we use every output of the intermediate states of the RNN and perform pooling in the next layer. There are three well-known RNN units: basic recurrent units [12], gated recurrent units (GRU) [7], and long short-term memory units (LSTM) [15]. Based on our experimental results, LSTM outperforms the other two units and hence we only report results using it.

Pooling Layer: The output of the convolution or recurrent layer $(\mathbf{h}_1, \mathbf{h}_2, \ldots, \mathbf{h}_M)$, are fed to the pooling layer. In this paper, we will be using **mean over time** (a special case of mean pooling where the window size is M) or **attention mechanism** (weighted average). Many studies have shown that attention mechanism is useful in many NLP tasks [1, 14, 25].

Mean over time is defined in Eq. 3:

$$\text{MeanOverTime}(\mathbf{H}) = \frac{1}{M}\sum_{t=1}^{M}\mathbf{h}_t \tag{3}$$

and **Attention** is defined in Eq. 4:

$$s_t = \mathbf{v} \cdot \tanh(\mathbf{W}\mathbf{h}_t), \quad s_t' = \text{softmax}(\mathbf{s})_t, \quad \text{Attention}(\mathbf{H}) = \sum_{t=1}^{M} s_t'\mathbf{h}_t \tag{4}$$

where $\mathbf{W}$ is a trainable matrix of size $d_c \times d_c$ and $\mathbf{v}$ is a trainable vector of size d_c. To learn more complex functions, $\mathbf{W}$ is introduced to increase the number of parameters and tanh is introduced to add non-linearity. $\mathbf{W}$ and $\mathbf{v}$ are *shared* across all time-steps t. To make sure that the weight for each time-step sums to 1, softmax function is performed on all of the weights $\mathbf{s} = (s_1, s_2, \ldots, s_M)$.

Linear Layer with Sigmoid Activation: The linear layer maps the pooled vector into a scalar value. This mapping is a simple linear transformation (fully connected layer). Since we need to predict a real number between 0 and 1, a sigmoid function is used to ensure the scalar value falls in the range of (0, 1). The mapping of linear layer after applying the sigmoid function is shown in Eq. 5:

$$s(\mathbf{x}) = \sigma(\mathbf{w} \cdot \mathbf{x} + b) \tag{5}$$

where $\mathbf{x}$ is output of the pooling layer, $\mathbf{w}$ is the weight vector, and b is the bias value.

4.2 Training

In our experiment, we do not utilize early stopping methods. We train the neural network for a specified number of epochs and evaluate the model on a development set in every epoch. The epoch with the best F1-score or Pearson correlation coefficient on the development set is selected as the final model. To prevent overfitting, we have adopted dropout regularization. We also clip the gradient if gradient norm is larger than a certain threshold. We use RMSProp optimization algorithm [9] to minimize loss function over the training data. To address the problem of class imbalance, we tried a loss function scaling but it did not improve our model. We adopted **binary cross-entropy** and **mean squared error (MSE)** loss function for the binary classification and regression setup respectively.

5 Experiments

5.1 Evaluation Metrics

The standard evaluation metrics for the binary classification task are recall, precision, F1-score, and accuracy. We use F1-score and accuracy as our evaluation metric for binary classification. In the optimistic case, a positive classification refers to class `1` (valid) and negative refers to class `0` (invalid). In the pessimistic case, positive refers to class `0` (invalid) and negative refers to class `1` (valid). For the regression task we use the Pearson's correlation coefficient.

5.2 Setup

There are 11 datasets (from many different chatbots) released by WOCHAT. In our experiment, we chose the three largest datasets, namely TickTock [31], IRIS [3], and Joker [11]. We use F1-score and accuracy as our evaluation metric for binary classification, and Pearson correlation coefficient for regression. Since the dataset is small, we use 10-fold cross validation. In each fold, the data is split into 80% training, 10% development, and 10% test. For experiments using SVM and RF, the development set is ignored. One instance of training is defined as a pair of 2 turns, one by human, and another one by chatbot (the reply). The ground truth is the annotated score with respect to the chatbot turn, as we are not interested in the quality of human response. The aforementioned training instances are illustrated in Table 1. `OPT`, `PES`, and `AVE` in Table 1 denotes optimistic, pessimistic, and averaging ground truths respectively. Only the annotated turns are used in our experiments because we can only evaluate our model on instances with ground truth. The number of chatbot annotated turns and the number of *valid* and *invalid* annotations (for optimistic and pessimistic ground truth) are summarized in Table 2. Figure 2 shows the histogram of the averaging ground truth we used for the regression task. We tokenized the sentences using NLTK [20], and lowercase the text.

Table 2 The number of annotated chatbot turns for each dataset

Dataset	Total	Ground truth	Valid	Invalid
TickTock	2731	Optimistic	1786	945
		Pessimistic	940	1791
IRIS	790	Optimistic	661	129
		Pessimistic	244	546
Joker	535	Optimistic	362	173
		Pessimistic	142	393

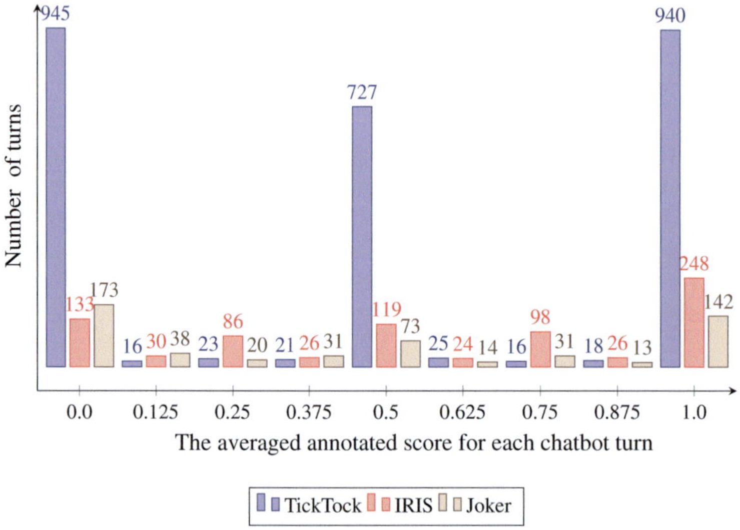

Fig. 2 The number of annotated chatbot turns for regression task

5.2.1 SVM and Random Forest

We combine each training instance consisting of two turns (one by human and one by chatbot) by concatenating their bag-of-words representations into one larger vector, and use it as the input to train the SVM and RF.

For SVM, we set the number of maximum iteration and tolerance value (for stopping criterion) to 10,000 and 0.001 respectively. Experimental results showed that more iterations did not improve the performance of the model. Different kernels such as linear, polynomial, sigmoid and radial basis function were explored. In our experiments, we set the kernel to be sigmoid as it outperforms the other kernels in this task. To address the problem of class imbalance, we tried balancing the class weights but it did not improve our model and therefore it is not reported.

For random forest, the minimum sample split is set to 2 and we do not limit the number of features in each node. We explored different number of trees and maximum tree depth. Larger number of trees can reduce variance, and limiting the depth of the trees can prevent overfitting. We limit the depth of the trees in random forest to 20. The number of trees are set to 50. The objective functions used for the classification and regression are entropy and mean squared error (MSE) respectively. Our SVM and RF models are implemented in Python using scikit-learn package [23].

5.2.2 Neural Networks

There are two turns (one by human and one by chatbot) in each training instance. Hence we designed our architecture, such that there is one NN for human turn and and another NN for the chatbot turn. The output vectors from the two NNs are concatenated, before going into the pooling layer.

We use RMSProp optimizer with decay rate (ρ) 0.9 and learning rate 0.001. Mini-batch size is 16, and we train for 30 epochs. The vocabulary is created using the most 4,000 frequent words. However, the dataset is very small such that the number of words for training is always less than 4,000 and hence all words in the training set are included in the vocabulary list. Out of vocabulary words are replaced by `<unknown>` token. The network is regularized by using dropout [26] with probability 0.8. Word embedding dimension (d_{LT}) is set to 100. The output dimension of the hidden layer for the RNN and the number of filters for the CNN (d_c) are set to 300. The convolution window size (l) is set to 3. We experimented by initializing the lookup table layer with various pre-trained word embeddings, but our best results were obtained using the publicly available word embedding for GloVe[2] (trained on Wikipedia and Gigaword 5, 6 billion tokens). Our neural network is implemented in Python using Keras[3] and the source code is publicly available at https://github.com/yulonglong/ChatbotScorer.

5.3 *Results and Discussion*

For binary classification, we adopted a simple baseline where it always predicts the majority class in the training set (Tables 3 and 4).

For regression, we perform the following steps to obtain the baseline scores. First, we count the number of *valid*, *acceptable*, and *invalid* chatbot annotations in the training data. Then we use random generator that outputs either `1`, `0.5`, or `0` with probability according to their respective distribution/count. This generator is used to predict the instances in the test set and compute the Pearson correlation coefficient against the ground truth. The generator was run 1,000 times to obtain the mean and standard deviation results of the Pearson correlation (see Table 4). Running the 1000 experiments and following the probability distribution for each category is one possible way to represent the lower bound for the correlation rather than the actual performance of a specific baseline system.

---

[2]https://nlp.stanford.edu/projects/glove/.

[3]https://github.com/fchollet/keras/.

Table 3 Averaged F1-score and accuracy with optimistic ground truth

Model	F1-score			Accuracy		
	TickTock	IRIS	Joker	TickTock	IRIS	Joker
Baseline (Majority)	0.790	0.911	0.804	0.654	0.837	0.676
SVM	0.790	0.911	0.804	0.654	0.837	0.676
Random forest	0.790	0.910	0.796	**0.660**	0.835	0.665
CNN MoT	0.784	0.904	0.788	0.658	0.828	**0.682**
LSTM attention	0.785	0.901	0.774	0.657	0.823	0.656

Table 4 Averaged F1-score and accuracy with pessimistic ground truth

Model	F1-score			Accuracy		
	TickTock	IRIS	Joker	TickTock	IRIS	Joker
Baseline (Majority)	0.792	0.817	0.846	0.656	0.691	0.735
SVM	0.792	0.817	0.846	0.656	0.691	0.735
Random forest	**0.793**	**0.884**	0.865	**0.672**	**0.820**	0.778
CNN MoT	0.783	0.870	**0.866**	0.668	0.802	**0.795**
LSTM attention	0.793	0.874	0.864	0.670	0.807	0.785

Table 5 Averaged Pearson correlation coefficient with averaging ground truth

Model	Pearson correlation coefficient		
	TickTock	IRIS	Joker
Baseline	0.0024 ± 0.061	0.0014 ± 0.104	0.0033 ± 0.138
SVM	0.225	0.457	0.333
Random forest	**0.309**	0.464	**0.465**
CNN MoT	0.277	0.481	0.455
LSTM attention	0.261	**0.505**	0.381
Voting CNN MoT	0.269	0.486	0.449

All of the results are computed by averaging the scores from 10-fold cross validation (including the baselines). For the NN experiments, we run each experiment 5 times with different seeds and average the results. The results of our best performing models are summarized in Tables 3, 4 and 5. In the cases where the systems outperform the baselines, the best score for each dataset is highlighted in bold.

Observing the results in Table 3, most of the models are unable to outperform the baseline in the optimistic case. The best model, random forest, only slightly outperforms the baseline in TickTock dataset. However, the models perform much better in the pessimistic case (Table 4) where almost every model outperforms the baseline on all datasets. Moreover, the results for regression task are also promising where almost all of the models outperform the baseline with very large margin. In addition, the combination of optimistic and pessimistic CNN performs very well.

Table 6 Examples of turns where the models perform well in the pessimistic case

Opening and closing remark	
Human	: Hi
Chatbot	: Hello…USER_NAME
Human	: Hi SYSTEM_NAME
Chatbot	: Hi
Human	: Bye bye
Chatbot	: Ok, USER_NAME…bye bye.
Human	: Bye bye SYSTEM_NAME
Chatbot	: Ok, USER_NAME…good bye…
Short response	
Human	: I will have some pizza for LUNCH
Chatbot	: Totally!
Human	: You want me to give you my number?
Chatbot	: I do.
Human	: Ok, I see, I also have my procedures
Chatbot	: Ok.

The voting is done by obtaining real-valued output of the two CNNs from both optimistic and pessimistic case, and average the two values for each test instance. Pearson correlation coefficient is then calculated using the new combined score.

Our analysis shows that the good performance of the models in the pessimistic case is because of the standard opening and closing remarks before starting a conversation. Most of the time, the words and phrases used for opening and closing conversations are highly similar, therefore the models are able to always correctly predict them as *valid*. Furthermore, the chatbot responses are most likely to be *valid* when the response is short, and the models have learnt this feature as well. Some examples are illustrated in Table 6.

Our analysis also shows that the proportion of out-of-vocabulary words in the test set is around 10% for IRIS and Joker datasets, and 3% for TickTock dataset. The models might be able to perform better using a larger vocabulary from external corpus instead of building our own vocabulary from the training set. Overall, chat-oriented dialogue evaluation is a challenging and subjective task which requires more insights and larger datasets to progress further. This claim is supported by the examples given in Table 1, where 3 out of 4 annotators gave different scores to the chatbot response. It shows that the validity of chatbot responses is very subjective and difficult for human to evaluate.

6 Conclusion

In this paper, we have addressed the task of automated scoring of chatbot responses or also known as chat-oriented dialogue evaluation. We have explored three different methods to construct the ground truth from multiple annotations for each training instance, namely optimistic, pessimistic, and averaging. We formulate the problem as simple binary classification tasks and a more complex regression task. We have explored a few machine learning models such as SVM, random forest, CNN, and RNN. Our models show good results by outperforming the baselines, especially in the regression setup. Combining the outputs of the two CNN models from the optimistic and pessimistic case also shows positive result. Therefore, it validates the idea of decomposing the regression task into two simpler binary classification tasks (optimistic and pessimistic) and combining them at a latter stage to solve the original regression task.

Acknowledgements We want to thank to Sunil Sivadas and Rafael E. Banchs from the Nanyang Technological University, A*STAR for their meaningful guidance and comments.

References

1. Bahdanau D, Cho K, Bengio Y (2015) Neural machine translation by jointly learning to align and translate. In: Proceedings of the 3rd international conference on learning representations
2. Banchs RE (2016) Expert-generated versus crowd-sourced annotations for evaluating chatting sessions at the turn level. In: WOCHAT: workshop on chatbots and conversational agent technologies
3. Banchs RE, Li H (2012) IRIS: a chat-oriented dialogue system based on the vector space model. In: Proceedings of the 50th annual meeting of the association for computational linguistics system demonstrations
4. Bansal M, Gimpel K, Livescu K (2014) Tailoring continuous word representations for dependency parsing. In: Proceedings of the 52nd annual meeting of the association for computational linguistics
5. Bengio Y, Ducharme R, Vincent P, Janvin C (2003) A neural probabilistic language model. J Mach Learn Res 3:1137–1155
6. Cao C, Liu X, Yang Y, Yu Y, Wang J, Wang Z, Huang Y, Wang L, Huang C, Xu W, Ramanan D, Huang TS (2015) Look and think twice: capturing top-down visual attention with feedback convolutional neural networks. In: Proceedings of the IEEE international conference on computer vision
7. Cho K, van Merrienboer B, Gulcehre C, Bahdanau D, Bougares F, Schwenk H, Bengio Y (2014) Learning phrase representations using RNN encoder–decoder for statistical machine translation. In: Proceedings of the 2014 conference on empirical methods in natural language processing
8. Collobert R, Weston J, Bottou L, Karlen M, Kavukcuoglu K, Kuksa P (2011) Natural language processing (almost) from scratch. J Mach Learn Res 12:2493–2537
9. Dauphin YN, de Vries H, Bengio Y (2015) Equilibrated adaptive learning rates for non-convex optimization. In: Advances in neural information processing systems, p 28
10. Drucker H, Burges CJC, Kaufman L, Smola AJ, Vapnik V (1997) Support vector regression machines. In: Advances in neural information processing systems, p 9
11. Duplessis GD, Letard V, Ligozat AL, Rosset S (2016) Purely corpus-based automatic conversation authoring. In: Proceedings of the 10th international conference on language resources and evaluation
12. Elman JL (1990) Finding structure in time. Cogn Sci 14(2):179–211
13. Graves A, Jaitly N (2014) Towards end-to-end speech recognition with recurrent neural networks. In: Proceedings of the 31st international conference on machine learning

14. Hermann KM, Kocisky T, Grefenstette E, Espeholt L, Kay W, Suleyman M, Blunsom P (2015) Teaching machines to read and comprehend. In: Advances in neural information processing systems, p 28
15. Hochreiter S, Schmidhuber J (1997) Long short-term memory. Neural Comput 9(8):1735–1780
16. Joachims T (1998) Text categorization with support vector machines: learning with many relevant features. In: Proceedings of the 10th European conference on machine learning
17. Kim S, Banchs RE, Li H (2016) Exploring convolutional and recurrent neural networks in sequential labelling for dialogue topic tracking. In: Proceedings of the 54th annual meeting of the association for computational linguistics
18. LeCun Y, Boser B, Denker JS, Henderson D, Howard RE, Hubbard W, Jackel LD (1989) Backpropagation applied to handwritten zip code recognition. Neural Comput 1(4):541–551
19. Liu CW, Lowe R, Serban I, Noseworthy M, Charlin L, Pineau J (2016) How not to evaluate your dialogue system: An empirical study of unsupervised evaluation metrics for dialogue response generation. In: Proceedings of the 2016 conference on empirical methods in natural language processing
20. Loper E, Bird S (2002) NLTK: The natural language toolkit. In: Proceedings of the 40th annual meeting of the association for computational linguistics workshop on effective tools and methodologies for teaching natural language processing and computational linguistics
21. Lowe R, Noseworthy M, Serban IV, Angelard-Gontier N, Bengio Y, Pineau J (2017) Towards an automatic turing test: learning to evaluate dialogue responses. In: Proceedings of the 55th annual meeting of the association for computational linguistics, vol 1. Long Papers
22. Mikolov T, Sutskever I, Chen K, Corrado GS, Dean J (2013) Distributed representations of words and phrases and their compositionality. In: Advances in neural information processing systems, p 26
23. Pedregosa F, Varoquaux G, Gramfort A, Michel V, Thirion B, Grisel O, Blondel M, Prettenhofer P, Weiss R, Dubourg V, Vanderplas J, Passos A, Cournapeau D, Brucher M, Perrot M, Duchesnay E (2011) Scikit-learn: machine learning in Python. J Mach Learn Res 12:2825–2830
24. Pennington J, Socher R, Manning C (2014) GloVe: global vectors for word representation. In: Proceedings of the 2014 conference on empirical methods in natural language processing
25. Rush AM, Chopra S, Weston J (2015) A neural attention model for abstractive sentence summarization. In: Proceedings of the 2015 conference on empirical methods in natural language processing
26. Srivastava N, Hinton GE, Krizhevsky A, Sutskever I, Salakhutdinov R (2014) Dropout: a simple way to prevent neural networks from overfitting. J Mach Learn Res 15:1929–1958
27. Taghipour K, Ng HT (2016) A neural approach to automated essay scoring. In: Proceedings of the 2016 conference on empirical methods in natural language processing
28. Turian, J., Ratinov, L., Bengio, Y.: Word representations: a simple and general method for semi-supervised learning. In: Proceedings of the 48th annual meeting of the association for computational linguistics (2010)
29. Turian J, Ratinov L, Bengio Y (2010) Word representations: a simple and general method for semi-supervised learning. In: Proceedings of the 48th annual meeting of the association for computational linguistics
30. Xu B, Guo X, Ye Y, Cheng J (2012) An improved random forest classifier for text categorization. J Comput 7(12):2913–2920
31. Yu Z, Papangelis A, Rudnicky A (2015) TickTock: a non-goal-oriented multimodal dialog system with engagement awareness. In: Proceedings of the association for the advancement of artificial intelligence spring symposium

Subjective Annotation and Evaluation of Three Different Chatbots WOCHAT: Shared Task Report

Naomi Kong-Vega, Mingxin Shen, Mo Wang and Luis Fernando D'Haro

Abstract This paper evaluates the performance of three different chatbots: IRIS, TickTock and Joker, that have been made available to the public online. All three retrieval-based dialogue systems are chat-oriented and designed to engage the users into all types of conversations for as long as possible. They employ different approaches to provide relevant and valid responses, and constantly utilize conversational strategies to further automatically improve its own system through machine learning. The analysis of annotations of more than 2000 responses for the three chatbots allowed us to confirm the robustness, scalability and usability of the systems, as well as to detect a few areas in which response accuracy was lacking, and propose future work to further improve the three systems and annotations scheme.

1 Dialogue Systems

As part of the WOCHAT workshop series, several chatbots are made available to the public online, where users engage in human-machine dialogues. Annotations are then manually generated to see the response accuracy delivered by the various chatbots. The collected dataset and annotations are available to the public and research community for further improvement of chatbot systems. This paper analyzes annotations of a dialogue dataset collected by 3 chatbots in English: IRIS, TickTock and Joker. All dialogs have been collected through volunteers accessing the WOCHAT

Luis Fernando D'Haro—Research work done while working at Institute for Infocomm Research, Singapore.

N. Kong-Vega · M. Shen · M. Wang
National Junior College, 37 Hillcrest Road, Singapore 288913, Singapore

L. F. D'Haro (✉)
ETSI de Telecomunicación, Universidad Politécnica de Madrid,
Avenida Complutense 30, 28040 Madrid, Spain
e-mail: luisfernando.dharo@upm.es

L. F. D'Haro et al. (eds.), *9th International Workshop on Spoken Dialogue System Technology*, Lecture Notes in Electrical Engineering 579,
https://doi.org/10.1007/978-981-13-9443-0_32

website.[1] Few instructions are given to the participants, i.e. basically to choose the chatbot they want to chat to and keep the dialogue for at least 20–30 turns maximum (this is to reduce the complexity and burden of annotating long dialogs), and in case that the chatbot provides a confusing answer, to keep the conversation flow. No indications about the topics or length of the turns are given to allow the participants to be free on exploring the capabilities of each chatbot allowing also the researchers to improve their systems.

1.1 IRIS

IRIS [1] a user-initiative dialogue system that is conversant in a large variety of topics. It is a example-based chatbot which responses are selected from a database consisting of movie quotes, but with the capability of learning new concepts from users. Candidate responses are selected by computing a vector space model representation of both the latest user input and the dialogue history so that IRIS could provide the most relevant responses in the context. Since IRIS is the chatbot that has been available in the website for longer time, it is the one having more number of files for annotation.

1.2 TickTock

TickTock [3] is a rule-based chatbot capable of conducting free-form conversations in contrast of goal-driven systems. Hence, the main objective of this chatbot is to make each conversation as long as possible. Candidate responses are selected from a database of question-answer pairs from 767 CNN interviews. Though the conversations are mainly user-initiated, when the score of the candidate response is below a threshold value, TickTock pro-actively changes the topic by raising a new open question.

1.3 Joker

Joker [2] is a user-initiative dialogue system which employs example-based dialogue modelling, using a database of indexed dialogue examples built from television drama subtitles. Besides providing the participant the most appropriated responses, it also presents social communication skills including humour.

[1] http://workshop.colips.org/wochat/main_sharedtask.html.

Table 1 Criteria for response evaluation categories and examples

VALID	This score is used when the response is semantically and pragmatically valid given the previous recent dialogue history, while allowing few or none grammatical errors e.g. **USER**: Do you like any other kind of music? **TICKTOCK**: Yes. I like classical music—VALID **USER**: What is your favourite song? **TICKTOCK**: Anything I can slowdance to—VALID
ACCEPTABLE	This score is given when the response is not necessarily semantically valid but can be acceptable, given the most recent dialogue context, from the pragmatic point of view. This could also mean that the response required slight guessing to make sense OR had severe grammatical errors e.g. **USER**: I have to get back to work **JOKER**: I'll go with you **USER**: Oh that's nice, thank you **JOKER**: No, we don't live together I mean—ACCEPTABLE **USER**: No offense, but I don't want to live with you
INVALID	Response delivered was completely irrelevant or off-topic, and caused an obvious break in the flow of conversation e.g. **USER**: Good morning **IRIS**: Gotta run. Shipment of worms coming in first thing **USER**: Are you importing worms? **IRIS**: Really?—INVALID **USER**: You tell me!

Table 2 Statistics for response evaluation for the 3 chatbots

	IRIS	TickTock	Joker
No. of. different responses annotated	1760	738	1060
(%) of VALID responses	79.7	73.5	63.3
(%) of ACCEPTABLE responses	9.1	10.8	14.1
(%) of INVALID responses	11.2	15.7	22.6
(%) of offensive or inappropriate swear language	1.4	0.1	0.5

2 Evaluation of Responses

For this paper, authors annotated a set of different dialogues from 3 chatbots using the following categories: VALID, ACCEPTABLE and INVALID, according to the guidelines in Table 1, as well as whether there were any offensive or inappropriate swear language for each turn in the dialogue.

Given that there was not full overlap of the dialogues annotated by the authors we could not calculate inter-annotator agreements, but instead all annotations were counted and averaged when yielding the statistics shown in Table 2. In the table we can see that majority of the responses for each of the 3 chatbots were VALID, therefore the response accuracy for all of them was relatively high.

Table 3 Examples of VALID responses by IRIS

Political topics	**USER**: yes, you know in trump's last rally **IRIS**: None of the papers covered the rally. Not even The Sun **USER**: but it's all over the web **IRIS**: yes, it is
General conversation	**USER**: you are scaring me **IRIS**: I'm scaring myself, because I think so much of her **USER**: is she pretty? **IRIS**: She is **USER**: this is so romantic **IRIS**: Mmm **USER**: Do you love her? **IRIS**: [USER's Name]...

Table 4 Examples of INVALID responses by IRIS

Inability to relate to contextual statements	**IRIS**: And how, and how, can these death traps be avoided? **USER**: Which death traps? **IRIS**: Which is? **USER**: I have no idea, that is why I am asking you **IRIS**: I have no idea

2.1 IRIS

Each dialogue comprised of an average of 31 dialogue turns between the user and IRIS. With the highest rate of VALID responses of 79.7% and the lowest rate of INVALID responses of 11.2%, dialogues with IRIS had the best response accuracy among the 3 chatbots evaluated. IRIS was found to respond well and carry on relatively longer conversations in political topics, and general conversation about feelings and emotions as seen in Table 3.

On the other hand, IRIS was found to be weak at relating to statements made a few turns before, and usually responded well most of the times based on the latest respond it got from the user. This also lead to abrupt changing of topics in cases that it was unable to understand the users' most recent response. IRIS was also unable to retain information (see Table 4). For example, even if the user had stated his name a few turns ago, it would address the user with another name.

2.2 TickTock

Each dialogue comprised of an average of 25 dialogue turns between the participant and TickTock. With the rate of VALID responses at 73.5% and the rate of INVALID responses at 15.7%, dialogues with TickTock had the second best response accuracy among the 3 chatbots evaluated.

Table 5 Examples of VALID responses by TickTock

Initiating conversations with topics	**TickTock**: Can you tell me something about politics? **USER**: Trump will make America great again **TickTock**: Politicians and diapers have one thing in common. They should both be changed regularly, and for the same reason **USER**: That's a good one
Vague yet relevant responses	**TickTock**: What's your idea? **TickTock**: I don't quite understand. Any thoughts?

Table 6 Examples of INVALID responses by TickTock

Vague and irrelevant responses	**USER**: I don't ski **TickTock**: I certainly try. What do you think? **USER**: How old are you? **TickTock**: Uh-huh, I do. What do you think?

TickTock, unlike other chatbots, usually initiates conversations with questions that introduce a topic it would like to talk about. This allowed the users to set the context for TickTock to respond appropriately to. TickTock was found to respond well and carry on longer conversations related to video games, music, or films, but especially general conversation about the users' life. This is because unlike the other chatbots, TickTock responds with vague and open questions, allowing the users to lead the conversations (see Table 5).

TickTock was found to be generally weaker at answering open questions asked by the user. In a few cases too, though with similar strategy of using vague responses, it produced completely irrelevant responses (see some examples in Table 6).

Like IRIS, TickTock was unable to keep up with the context of the dialogue, and could only respond based on the latest response of the user. It was also unable to retain information, as users had to repeat what the topic of discussion was in several occasions.

2.3 *Joker*

Each dialogue comprised of an average of 42 dialogue turns between the participant and Joker. With the rate of VALID responses at 63.3% and the rate of INVALID responses at 22.6%, dialogues with Joker had the lowest response accuracy among the 3 chatbots evaluated.

Joker was found to provide accurate responses to short and direct questions, and responded with brief comments that were still moderately relevant so the dialogues made sense, and could continue (see some examples in Table 7).

Unfortunately, Joker mostly responded with very short statements, and did not show initiate in the conversation. Thus, even though the dialogues kept going, there

Table 7 Examples of VALID responses by Joker

Responses to short and direct questions	**USER**: how are you? **Joker**: good **USER**: What did you do today? **Joker**: nothing
Moderately relevant brief comments	**USER**: That's great news! I'm proud of you **Joker**: mm-hmm

Table 8 Examples of extremely short & INVALID responses by Joker

Very short statements that caused frequent change of conversation topics	**USER**: I am feeling sick today **Joker**: well that's **USER**: how are you feeling today? **Joker**: okay **USER**: what's the weather like? **Joker**: good
Completely irrelevant and random responses	**USER**: so what's up today? **Joker**: they lived together **USER**: who lived together? **Joker**: where's Gordon and has he changed? **USER**: I have no idea who Gordon is

was very frequent changes in topics, and it waited for the participant to feed it information to respond to, repeating other times what was said or providing extremely short statements. Furthermore, there were many cases where Joker's responses were irrelevant and random, making dialogues confusing and short-lived (see examples in Table 8).

Like IRIS and TickTock, Joker was unable to retain and relate to previously mentioned information. Its biggest weakness was that it only gave short responses making the users to essentially come up with conversation starters very often.

3 Conclusion and Future Work

In this paper, we reported the results of annotating and subjectively evaluating the responses of three different chatbots—IRIS, TickTock and Joker. Our expectation is that by providing this analysis, we can highlight the major changes that must be done to these chatbots, but also to provide a set of important aspects that any non-oriented dialogue chatbot should have to keep engaged the users.

All chatbots were fairly proficient in giving valid responses, with all of them only having a maximum of 25% invalid responses. Unfortunately, all chatbots were found to be weak in retaining and relating to information from a few statements back, resulting in frequent changing of topics, and thus, no engaging conversations. A summary of the performance of the chatbots is given in Table 9.

Table 9 Summary of performance of chatbots

System	IRIS	TickTock	Joker
Processing of context information	Yes	No	No
Conversation length/dialogue turns	31	25	42
Retain of information	Only seems to retain information from the last user's turn	Only seems to retain information from the last user's turn	Only seems to retain information from the last user's turn
Strengths	Respond well and sustain relatively longer conversations related to politics and emotions	Initiating conversations; good responses to topics on video games, music, films and user's life	Responding to short and direct questions well; providing short comments to sustain the conversations
Weaknesses	Could not retain information. Could not respond in context	Could not retain information. Most responses to open questions were vague and irrelevant	Could not retain information. Responses are very short and do not initiate conversations

As future work, we plan to help developers to go beyond response appropriateness (i.e. a higher rate of VALID responses) by improving their chatbots' consistency in adhering to the conversation's general context, and also increasing the engagement level of the responses. As such, we plan to increase the level of annotation by providing alternative answers for the INVALID and ACCEPTABLE cases.

Besides, we are planning to provide additional annotations like the exact turn where specific information was provided. This way, if the user mentioned his name in a given turn, the annotation will point to that turn allowing training the chatbot to pay attention to the correct context in the dialogue. Finally, we expect more annotations and annotators could be used to calculate inter-annotator agreements, automatically evaluate the responses, detect different kind of non-appropriated interactions, among others.

Acknowledgements We would like to thank Dr. Rafael E. Banchs from Nanyang Technological University for providing us the chatbots data used in this project, and Mr. Kester Wong from National Junior College for providing this valuable opportunity to learn about Infocomm Technology that as students would otherwise not have been able to such depths.

References

1. Banchs RE, Li H (2012) IRIS: a chat-oriented dialogue system based on the vector space model. In: Proceedings of the ACL 2012 system demonstrations. Association for Computational Linguistics, pp 37–42
2. Duplessis D, Letard V, Ligozat AL, Rosset S (2016) Joker chatterbot re-wochat 2016-shared task chatbot description report. In: RE-WOCHAT: workshop on collecting and generating resources for chatbots and conversational agents-development and evaluation workshop programme, 28 May 2016, p 45
3. Yu Z, Papangelis A, Rudnicky A (2015) TickTock: a non-goal-oriented multimodal dialog system with engagement awareness. In: Proceedings of the AAAI spring symposium

Part VIII
Question Answering and Other Dialogue Applications

Detecticon: A Prototype Inquiry Dialog System

Takuya Hiraoka, Shota Motoura and Kunihiko Sadamasa

Abstract A prototype inquiry dialog system, dubbed Detecticon, demonstrates its ability to handle inquiry dialogs, including presenting reasonable arguments, tracking user beliefs, and refining its arguments. User interaction is supported by using a natural language interface and a graphical user interface.

1 Introduction

In the inquiry dialog framework, two participants cooperate to answer shared questions[1] [15], under the assumption that neither of them has complete domain knowledge. They make up for their lack of knowledge through mutual dialog. This characteristic distinguishes dialogs in this framework from slot-filling dialogs (e.g., [9, 14, 16]) and expert systems' dialogs (e.g., [6]), where one of the participants has complete domain knowledge.

Previous research has basically focused on theoretical aspects of inquiry dialog for autonomous agents [1, 4, 5, 7, 8, 10–12], while, to the best of our knowledge, only Sklar and Azhar [2, 3, 13] developed dialog systems that can handle inquiry dialogs with a human user. Since inquiry dialogs can occur for various situations, further analysis of human-system inquiry dialogs is needed to realize practical inquiry dialog systems for human users.

[1] A question that is common to both participants.

T. Hiraoka · S. Motoura (✉) · K. Sadamasa
NEC Central Research Laboratories, Kawasaki, Japan
e-mail: motoura@ct.jp.nec.com

T. Hiraoka
e-mail: t-hiraoka@ce.jp.nec.com

K. Sadamasa
e-mail: k-sadamasa@az.jp.nec.com

L. F. D'Haro et al. (eds.), *9th International Workshop on Spoken Dialogue System Technology*, Lecture Notes in Electrical Engineering 579,
https://doi.org/10.1007/978-981-13-9443-0_33

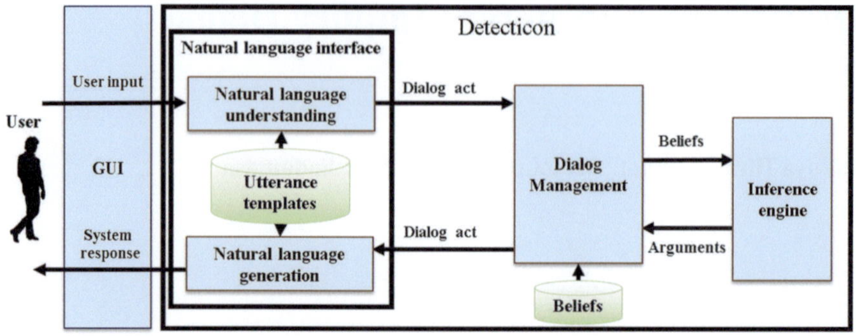

Fig. 1 Modular implementation of Detecticon features

To promote the realization of practical inquiry dialog systems for human users, we are developing a system called **Detecticon**[2] for collecting human-system inquiry dialog corpora that can be used for analysis. Detecticon can handle inquiry dialogs with a human user. We explain its abilities in Sect. 2 and then give examples of actual dialogs between Detecticon and a human user in Sect. 3.

2 Outline of Detecticon

Inquiry dialogs are a discussion in which the participants (here, Detecticon and a human user) collaborate in order to answer shared questions.[3] The participants cannot reach an answer without considering the beliefs of both participants[4] and thus must share their beliefs with the other participant. Detecticon has three features that enable it to handle inquiry dialogs with a human user: (1) it can present reasonable arguments, (2) it can track user beliefs, and (3) it can refine its arguments. It has two additional features that enable it to interact with a human user: a natural language interface and a graphical user interface (GUI). The modular implementation of these features is illustrated in Fig. 1.

The **natural language interface** translates the user's natural language input into a dialog act including beliefs and the system's dialog act including beliefs into a natural language response. The beliefs are represented by first-order logic. For example, the user's natural language input "Mr. Yoshida and Mr. Kobayashi are competitors" is translated into a dialog act including the belief Competitors($Yoshida, Kobayashi$). Translating the system's dialog act is the inverse process of translating the user's input. The "natural language understanding" module translates a user input into a

[2]Detection is meant to be a text-based inquiry dialog system to act as a detective.

[3]More precisely, inquiry dialogs are dialogs arising from an initial situation of "general ignorance" and having the main goal to achieve the "growth of knowledge and agreement". Each participant in the dialog has the goal to "find a proof or destroy one [15]".

[4]See Table 1 and upper part of Table 2 for examples.

Table 1 Example beliefs

Mr. Yoshida works for company A	Work_for_company$(A, Yoshida)$
Mr. Okamoto works for the company B	Work_for_company$(B, Okamoto)$
Mr. Tanaka works for the company C	Work_for_company$(C, Tanaka)$
An employee of the company A and an employee of the company B are competitors	$\forall x, y$ Work_for_company$(A, x) \wedge$ Work_for_company$(B, y) \rightarrow$ Competitors(x, y)

corresponding dialog act, and the "natural language generation" module translates a system dialog act into a corresponding system response. Both modules utilize utterance templates for their translations. The templates describe two types of mapping of entities, namely, (1) a natural language sentence and a dialog act type (e.g., presenting an argument); (2) a natural language sentence and a semantic content represented

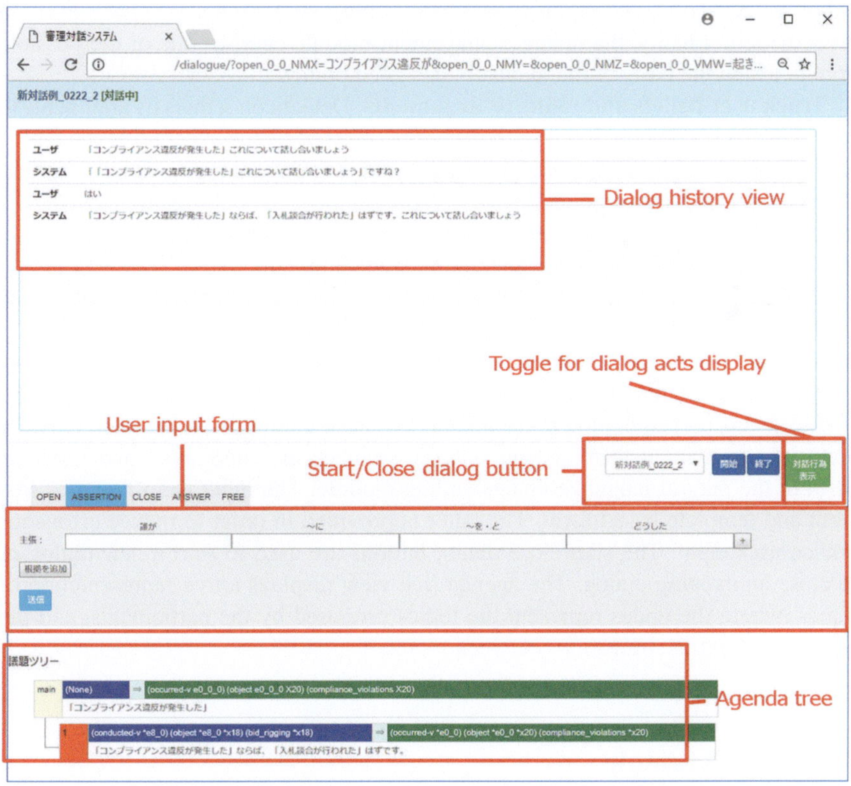

Fig. 2 Graphical user interface of inquiry dialog system

as logical formulae. The translations are performed by pattern matching of these templates.

Present reasonable arguments: Detecticon presents reasonable arguments to share what it believes with a human user. An argument consists of a claim and supporting evidence. The claim addresses the given question and be logically derived from the supporting evidence, which is a subset of the system's beliefs. For example, assume that Detecticon holds the beliefs listed in Table 1 and the question is "Who are competitors? ($\exists ?x, y$ Competitors(x, y))." Detecticon can present the following argument:

Claim: Yoshida and Okamoto are competitors. / Competitors(*Yoshida, Okamoto*)

Supporting evidence: $\left(\begin{array}{c} \text{Work\_for\_company}(A, Yoshida) \\ \text{Work\_for\_company}(B, Okamoto) \\ \forall x, y \ \text{Work\_for\_company}(A, x) \wedge \text{Work\_for\_company}(B, y) \\ \rightarrow \text{Competitors}(x, y) \end{array}\right)$

Detecticon generates an argument as follows. First, the dialog management module reads its holding beliefs (e.g., Table 1), calls the inference engine, and sends the beliefs to the engine. The engine generates possible arguments that are consistent with given beliefs. Finally, the dialog management module chooses one of the possible arguments on the basis of the manually designed policy.

Track user beliefs and refine its arguments: Detecticon refines its arguments to reflect the user's beliefs. It tracks the beliefs expressed by the user during the dialog and uses them to generate new arguments. It can thus present arguments that are not based solely on its own beliefs. Once the dialog management module receives the user's dialog act including beliefs from the natural language understanding module, it holds them and uses them for generating arguments.

Detecticon handles dialogs with the user by using the **GUI** shown in Fig. 2. The GUI provides a dialog history view, a toggle switch for displaying dialog acts, a user input form, a start/close dialog button, and an agenda tree view. The dialog history view provides the history of the dialog. The toggle switch switches the appearance of dialog acts corresponding to the natural language utterances in the dialog history view. The user input form accepts natural language user utterances and forwards them to the natural language understanding module. The input form provides free form and template based form. The latter is provided in order to reduce erroneous novice user input. The start/close dialog buttons are used to start a new dialog or to close an existing dialog. The agenda tree view displays a tree representation of topics, where the nodes represent the topics proposed by the participants, and the edges represent the transitions between topics.

3 Dialog Example

As an example within the inquiry dialog domain, we assume a situation in which the participants (Detecticon and a user) address Compliance Violation Detection and act

Table 2 Example inquiry dialog for Compliance Violation Detection. Upper table shows example participant beliefs at beginning of dialog. Lower table shows transcription of dialog translated from Japanese

User	If A and B are competitors, and A proposed an order price to B, and B accepted the proposed price, bid rigging occurred
	Kobayashi agreed that Yoshida will choose the company from which to take an order
	Kobayashi proposed an order price to Yoshida
	Takahashi accepted the order price
Detecticon	If there was bid rigging, a compliance violation occurred
	Yoshida and Kobayashi are competitors
	Yoshida proposed to Kobayashi that he (Yoshida) will choose the company from which to take an order
	Yoshida accepted Kobayashi's proposed order price

id	Speaker	Transcription
1	User	Let's talk about whether a compliance violation occurred
2	Detecticon	If there was a compliance violation occurred, bid rigging might have occurred as well. Let's discuss it
3	User	Ok
4	Detecticon	I don't have anything to tell you on the current topic (i.e., "If there was a compliance violation, bid rigging might have occurred as well."). Do you have any relevant information?
5	User	If bid rigging occurred (by suspects), (the suspect) A and (the suspect) B should be competitors, and A should propose order prices to B, and B should accept A's proposed price. Let's discuss that
6	Detecticon	Yoshida and Kobayashi are competitors
7	User	Kobayashi agreed that Yoshida will choose the company from which to take
8	Detecticon	Yoshida proposed to Kobayashi that he (Yoshida) will choose the company from which to take an order
9	User	Kobayashi proposed an order price to Yoshida
10	Detecticon	Yoshida accepted Kobayashi's proposed price
11	User	Takahashi accepted the order price
12	Detecticon	Bid rigging happened because (1) Kobayashi and Yoshida are competitors, and (2) Kobayashi proposed an order price, and (3) Yoshida accepted Kobayashi's proposed price
13	User	I have nothing else to add on the current topic (If bid rigging occurred, A and B should be competitors, and A should propose order prices to B, and B should accept A's proposal)
14	Detecticon	Me neither. Then, let's get back to the previous topic (If there was a compliance violation, bid rigging might have occurred as well). Let's talk about it
15	User	I don't have anything to tell you on the current topic (If there was a compliance violation, bid rigging might have occurred as well)
16	Detecticon	There was a compliance violation because bid rigging occurred
Dialog is closed		

as detectives. They are given different information sources (sets of beliefs) and then start a discussion aimed at answering the question, "Is there a compliance violation?" The upper part of Table 2 shows example beliefs. For this example domain, we prepare 542 beliefs (16 domain beliefs and 526 state beliefs[5]) and their natural language transcriptions. Each belief is assigned to Detecticon, its user, or neither of them, so that neither of them have complete set of beliefs.

An example dialog is shown in the lower part of Table 2. The participants set the topics for discussion at id = 2, 5. Detecticon sets the topic to discuss "bid rigging" because it believes a compliance violation can be proven if they proof that bid rigging occurred (id = 2). The user sets the topic to discuss bid rigging in more detail (id = 5). Detecticon's utterances at id = 6, 8, 10, 12, 16 are examples of **presenting reasonable arguments**. The utterances at id = 5, 9, 12, 16 are examples of **track user beliefs and refine its arguments**. Initially, Detecticon does not know (1) the definition of bid rigging and that (2) "Kobayashi proposed an order price to Yoshida". The conclusion is that there was bid rigging (id = 12) and therefore a compliance violation occurred (id = 16).

Acknowledgements We are grateful to NEC Informatec Systems and NIPPON ALGORITHM COMPANY for their cooperation in developing Detecticon. We sincerely thank Vera Götzmann for her insightful comments.

References

1. Amgoud L, Maudet N, Parsons S (2000) Modelling dialogues using argumentation. In: Fourth international conference on multiagent systems, Proceedings, 2000. IEEE, pp 31–38
2. Azhar MQ, Toward an argumentation-based dialogue framework for human-robot collaboration. PhD dissertation
3. Azhar MQ, Sklar EI (2016) Analysis of empirical results on argumentation-based dialogue to support shared decision making in a human-robot team. In: 2016 25th IEEE international symposium on robot and human interactive communication (RO-MAN). IEEE, pp 861–866
4. Black E, Hunter A (2007) A generative inquiry dialogue system. In: Proceedings of the 6th international joint conference on autonomous agents and multiagent systems. ACM, 241, pp. 1010–1017
5. Black E, Hunter A (2009) An inquiry dialogue system. Auton Agents Multi-Agent Syst 19(2):173–209
6. Buchanan BG (1984) Rule-based expert systems: the MYCIN experiments of the stanford heuristic programming project
7. Fan X, Toni F (2012) Agent strategies for ABA-based information-seeking and inquiry dialogues. In: ECAI, pp 324–329
8. Fan X, Toni F (2015) Mechanism design for argumentation-based information-seeking and inquiry. In: International conference on principles and practice of multi-agent systems. Springer, Berlin, pp 519–527
9. Levin E, Pieraccini R, Eckert W (2000) A stochastic model of human-machine interaction for learning dialog strategies. IEEE Trans Speech Audio Process 8(1):11–23

[5]The domain belief is an inference rule such that "$\forall x, y$ Work_for_company$(A, x) \wedge$ Work_for_company$(B, y) \rightarrow$ Competitors(x, y)." The state belief is a grounded atom such that "Competitors$(Yoshida, Okamoto)$".

10. McBurney P, Parsons S (2001) Representing epistemic uncertainty by means of dialectical argumentation. Ann Math Artif Intell 32(1–4):125–169
11. Parsons S, Wooldridge M, Amgoud L (2002) An analysis of formal inter-agent dialogues. In: Proceedings of the first international joint conference on autonomous agents and multiagent systems: part 1. ACM, pp 394–401
12. Parsons S, Wooldridge M, Amgoud L (2003) On the outcomes of formal inter-agent dialogues. In: Proceedings of the second international joint conference on autonomous agents and multi-agent systems. ACM, pp 616–623
13. Sklar EI, Azhar MQ (2015) Argumentation-based dialogue games for shared control in human-robot systems. J Hum Robot Interact 4(3):120–148
14. Walker MA (2000) An application of reinforcement learning to dialogue strategy selection in a spoken dialogue system for email. J Artif Intell Res 12:387–416
15. Walton D, Krabbe ECW (1995) Commitment in dialogue: basic concepts of interpersonal reasoning. SUNY Press
16. Williams JD, Young S (2007) Partially observable Markov decision processes for spoken dialog systems. Comput Speech Lang 21(2):393–422

Debate Dialog for News Question Answering System 'NetTv' -Debate Based on Claim and Reason Estimation-

Rikito Marumoto, Katsuyuki Tanaka, Tetsuya Takiguchi and Yasuo Ariki

Abstract NetTv is a question answering system for net news videos with a voice interface. When watching any news, a question answering system can answer a wide range of questions from users for deeper understanding of news. In this study, in order to help users to understand news deeply, when using NetTv, we propose a debate system which can debate with users about the watching news. The debate system estimates the user's claim (approval/disapproval/neither) and reason (presence/absence) using CNN, and uses the information to select an opinion stored in debate database and debates by presenting it to the user. In the experiment, eight subjects actually use the system to evaluate naturalness (whether the system response to the user's utterance was natural), deepness (whether the discussion was established), and user satisfaction (whether new knowledge was obtained) in 5 stages. From the result of subjective evaluation, we show that the debate system gives new knowledge and clarify the present problem of the debate system.

1 Introduction

In recent years, a system using a voice interface such as Siri developed by Apple Computer, Inc. attracts wide attention [1, 2]. NetTv developed by us is a question answering system using video indexing and voice interface in net news videos. That is, it aims at alleviating user's search burden and improving news knowledge by constructing a system that enables users to comfortably watch videos and answers the questions that occurred during watching news videos.

R. Marumoto · K. Tanaka · T. Takiguchi (✉) · Y. Ariki
Graduate School of System Informatics, Kobe University, Kobe, Japan
e-mail: takigu@kobe-u.ac.jp

K. Tanaka
e-mail: katsutanaka@puppy.kobe-u.ac.jp

Y. Ariki
e-mail: ariki@kobe-u.ac.jp

L. F. D'Haro et al. (eds.), *9th International Workshop on Spoken Dialogue System Technology*, Lecture Notes in Electrical Engineering 579,
https://doi.org/10.1007/978-981-13-9443-0_34

In this study, in order to help users to understand news deeply, we are developing a dialog system which can debate with users about any news. Several studies related to the debate have been conducted [4, 5], but in the debate system we develop, the user's claim (approval/disapproval/neither) and reason (presence/absence) are estimated through the dialog. Thereafter, our system presents its opinion against the user's claim by retrieving approval or disapproval opinions from the debate database for the discussing agenda. In the system, CNN model [3] is used for claim and reason estimation.

2 Overview of NetTv

NetTv driven by users aims to retrieve information and answer questions from users via speech recognition in a dynamically changing environment day by day. As shown in Fig. 1, the NetNews module and the QA (Question Answering) module play a key role.

In the NetNews module, text information on the website is used as meta-information for the news being broadcast on the net, and by indexing the metadata

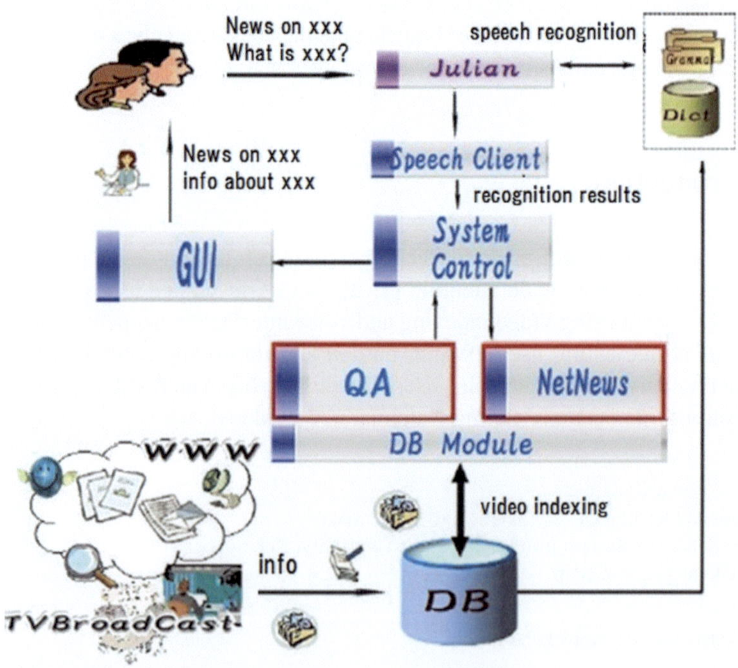

Fig. 1 NetTv architecture

to the news, voice search is enabled. In order to keep up with rapid changes of news content, the links and detailed articles are automatically collected from news sites. Then the index tables of URLs, movies, headers and words of detailed articles are created. Using these four index tables, voice search is achieved by using keywords.

In the NetTv system, to answer a wide range of questions from users, a question answering module is constructed. In the QA module, after question type is estimated to the question from the user, corresponding processing for each question type is performed and the answer is returned to the user. We prepared four question types; *factoid type*, *definition type*, *why type* and *how type*. For the *factoid type* question, answers are retrieved using Knowledge Q & A API published by NTT docomo. For the *definition type* question, the summary sentence described in Wikipedia is extracted and returned to the user as an answer. For the *why type* question, the answer is extracted based on existing study [6] and returned to the user. NetTv does not implement *how type* question answering at present.

In this paper, we extended the NetTv to have a debate function which enabled it to debate with the user on an agenda chosen by the user. In this paper, we focus on this debate system.

3 Debate System

Debate is contention in argument where opinions are in conflict with each other. Therefore, in this study, as shown in Fig. 2, the LU (Language Understanding) module estimates the user's claim (approval/disapproval/neither) and reason (presence/absence) on an agenda that the user selected before starting the debate. Thereafter, the DM (Debate Management) module decides the opinion that is against the user's claim (disapproval if the user's claim is approval) and debates with the user by presenting it. In order to present the opinion to the user, approval and disapproval opinions on any agenda are automatically collected from WWW in advance, and the debate database is constructed.

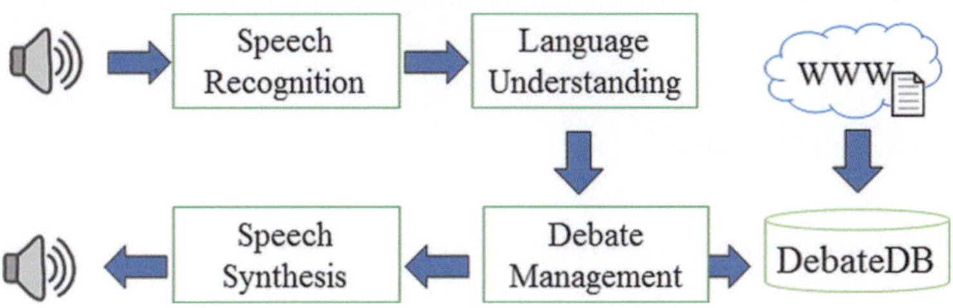

Fig. 2 Debate system architecture

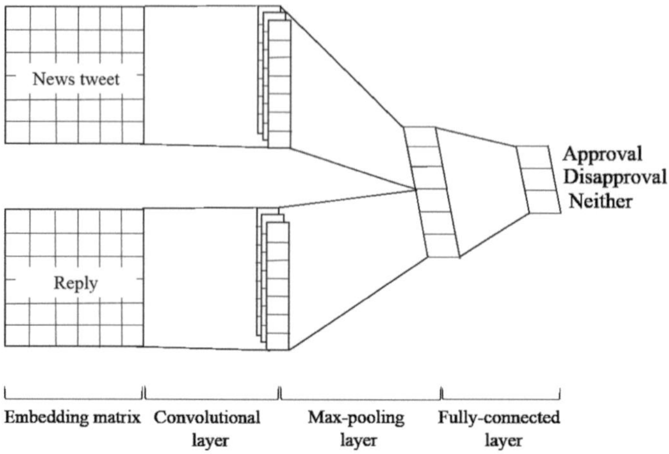

Fig. 3 CNN model architecture for claim estimation

3.1 Language Understanding

The LU module estimates the user's claim (approval/disapproval/neither) and reason (presence/absence). In this study, it is estimated based on the CNN model proposed by Shi [3]. Figure 3 shows the CNN model architecture. The input data are news Tweets of the Sankei newspaper official account and the paired Replies to them acquired by Twitter API. In the CNN model, each word in Tweet and Reply is vectorized by word2vec learned with Wikipedia. Then, the sentences of Tweet and Reply are represented as embedding matrix respectively, where the embedded word vectors are arranged in the appearing order in the sentence as shown in Fig. 3. On these two matrices, CNN is applied. CNN consists of a convolution layer, a max pooling layer, and a fully connected layer for each tweet and reply matrix, and concatenates reply and tweet information in max pooling layer. Thereafter, fully connected layer computes posterior probabilities of three claim classes (approval/disapproval/neither) or two reason classes (presence/absence) through softmax function.

3.2 Debate Database

Debate database is constructed through a two-stage search method; initial search and re-search. In the initial search, when X is an agenda, a sentence "I agree (disagree) with X" is fed to the search API and web pages containing the expression "I agree (disagree) with X" are collected. Then texts are extracted from the collected web pages, and the query sentence as well as the proceeding and succeeding sentence are grouped as an opinion candidate. Here, as a query sentence used in the initial search, a number of synonymous expressions are used in addition to the expressions mentioned in the example.

In the re-search, frequent words are extracted as related words from the set of opinion candidates obtained in the initial search. The extracted set of related words is added to the end of "X agree (disagree)" and the web pages are collected by the extended query given to the search API. The text is extracted from the collected web pages, and a passage including three or more related words is obtained as an opinion candidate. The initial search and re-search as described above are implemented for approval and disapproval respectively. In this way, opinion candidates are collected.

However, since the opinion candidates obtained in the re-search are poor in quality, by filtering it, only high-quality opinions can be selected. As such a filter, we constructed SVM classifier that classifies approval and disapproval, using high-quality opinion candidates obtained by initial search as SVM training data. Each opinion candidate obtained by initial search and re-search is fed to the constructed classifier, and the opinion candidates with score near the middle between the approval and disapproval are omitted, and only high-quality opinion candidates are stored in the database. In addition, the obtained approval and disapproval opinion candidates are individually clustered using LDA (Latent Dirichlet Allocation), and each opinion class is stored in a database.

3.3 Debate Management

The debate management module generates the system utterance. In LU module described in Sect. 3.1, when the user's claim is estimated to be "neither", or when the reason is estimated to be "absence", the system generates an utterance that asks the user about a claim.

When the system estimates the user's claim and reason, the system selects an opinion from the debate database, which is against the user's claim. For example, if the user's claim is estimated to be "approval", the system utterance is selected from the disapproval opinion stored in the debate database. As for the selection method, except for the opinions that were selected in the past, a disapproval opinion whose cos similarity is high (not low for the sake of weak disapproval) with the user's utterance is selected. In addition, by presenting the opinion in a topic (class) different from the topic (class) presented to the user in the past, it is possible to present a wide range of opinions to the user.

4 Evaluation

First, classification experiments of claim (approval/disapproval/neither) and reason (presence/absence) are described. The training data are news Tweets of the Sankei newspaper official account and the paired Replies to them acquired by Twitter API.

Table 1 Experimental result (accuracy (%)) in the claim (approval/disapproval/neither) and reason (presence/absence)

	Claim	Reason
CNN	80.0	80.0

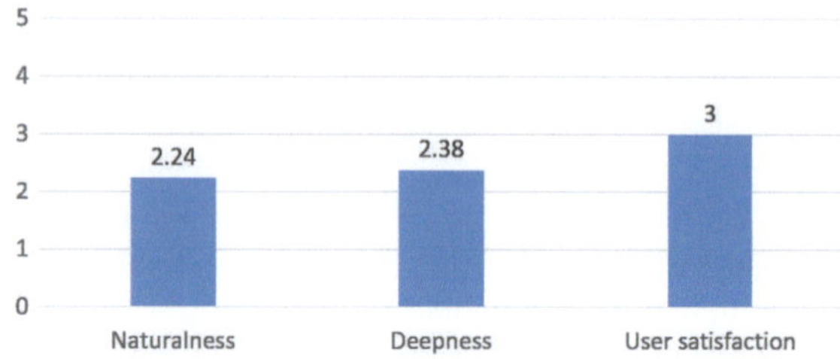

Fig. 4 Experimental result of subjective evaluation

The number of training data in the claim classification is 1,486 sentences (approval: 458, disapproval: 831, neither: 197), the number of training data in the reason classification is 1,301 sentences (absence: 1,083, presence: 218). In both classification experiments, 50 data were used for each class. The results are shown in Table 1. Accuracy was 80% in both the claim and the reason classification.

Then, a subjective evaluation experiment was conducted on the agenda items of "Casino bill", "Capital punishment" and "Resuming nuclear power plant operations". In the subjective evaluation experiment, eight subjects used the debate system and evaluated naturalness (whether the system response to the user's utterance was natural), deepness (whether the discussion was established), and user satisfaction (whether new knowledge was obtained) in 5 stages. Figure 4 shows the results of subjective evaluation experiment. The results of naturalness and deepness are 2.24 and 2.38 points, respectively. The result of the user satisfaction shows three points, which is higher than the other evaluation indices, indicating that the user gains a certain degree of knowledge. In order to further improve user satisfaction, it is necessary to increase the number of opinions of DB and improve the quality of opinion.

5 Dialog Example

Below is an example of a dialog with the system in which the user approves of a casino bill that is on the agenda (Japanese sentences have been translated into English).

System : What do you think about the casino bill?
User : I approve of the bill.
System : Please tell me why.
User : Since the bill would approve not just casinos, I think it would be good if it also approved the construction of recreational facilities, such as amusement parks.

System : Personally, I'm opposed to the casinos. I don't know what they will be built for. If we want to get more foreign currency, we should focus on sightseeing. However, if Japanese people aren't allowed to enter the casino, I think it would be all right to build them.

User : In fact, there are cases overseas in which casinos have been successful, so the construction of casinos would lead to an increase in the number of foreign tourists and also promote the economy.

System : There are already many places that provide hospitality and meet the needs of tourists from overseas. Japan is a country that prohibits and criminalizes gambling. I disapprove of the casino bill, as I've stated above.

User : Even if casinos are prohibited, I think it is necessary to increase the number of tourists and promote economic growth for a prosperous future.

System : "Atlantic Club Casino" went out of business there were too many casinos and because the number of tourists who don't gamble increased. As a result, many workers whose jobs were related to the casino were laid off, and eventually the regional economy collapsed.
(Several phrases are omitted due to the length of the sentences.)

User : Is that so ...?

System : There are also the following opinions. The government's purpose in allowing casinos is not the expected tax revenue, but the increase in the number of tourist and the resulting stimulus to the tourism industry and to regional development. Tax revenue from casinos is small, totaling 2–3 trillion yen at most, and the return rate is 97–98%, isn't it? If you take into account the tax incentives that will be given to the casinos, the tax revenue will be insignificant, I think.

As shown in the dialog example above, there may be a gap between the user's utterance and the system's response to it. One reason is that the number of opinions stored in the DB is small, and it is also necessary to devise how to collect high-quality opinion candidates. Since the opinion presented to the user is obtained from the web, the debate system has some possibility to present an opinion containing the wrong information to the user. This problem must be dealt with carefully in near future.

6 Conclusion

In this paper, we proposed a debate system in order to help users to understand news deeply in the news information retrieval and question answering system "NetTv". In the debate system, we first estimate the user's utterance claim (approval/disapproval/neither) and reason (presence/absence) using the CNN model. Thereafter, we developed a debate system which could present an opinion against the user's claim by searching from the debate database where opinions are stored by extracting from the Web for any agenda. From the result of subjective evaluation, we showed that the debate system gives new knowledge and clarify the present problem of the debate system.

References

1. Lamel L et al (2002) User evaluation of the mask kiosk. Speech Commun 131–139
2. Yoshino K, Kawahara T (2015) News navigation system based on proactive dialogue strategy. International Workshop Spoken Dialogue Systems (2015)
3. Shi H, Ushio T, Endo M, Yamagami K, Horii N (2016) A multichannel convolutional neural network for cross-language dialog state tracking. IEEE workshop on spoken language technology (2016)
4. Sato M, Yanai K, Miyoshi T, Yanase T, Iwayama M, Sun Q, Niwa Y (2015) End-to-end argument generation system in debating. ACL-IJCNLP
5. Potash P, Rumshisky A (2017) Towards debate automation: a recurrent model for predicting debate winners. Empirical methods on natural language processing (EMNLP) (2017)
6. Shibusawa U, Hayashi T, Onai R (2007) Development and evaluation of a system for extracting answers of a why type question from the web. Information processing society (in Japanese), pp 1512–1523

Question-Answer Selection in User to User Marketplace Conversations

Girish Kumar, Matthew Henderson, Shannon Chan, Hoang Nguyen and Lucas Ngoo

Abstract Sellers in user to user marketplaces can be inundated with questions from potential buyers. Answers are often already available in the product description. We collected a dataset of around 590 K such questions and answers from conversations in an online marketplace. We propose a question answering system that selects a sentence from the product description using a neural-network ranking model. We explore multiple encoding strategies, with recurrent neural networks and feed-forward attention layers yielding good results. This paper presents a demo to interactively pose buyer questions and visualize the ranking scores of product description sentences from live online listings.

1 Introduction

In online marketplaces, buyers primarily deal with sellers through chats—to ask questions, negotiate, and arrange logistics. Any improvement to these interactions can greatly increase user satisfaction and the effectiveness of the marketplace. We present a system that can automatically answer buyer questions by reading the product description, freeing the sellers' time and giving buyers quick answers.

Current question answering methods consist of generative and extractive techniques. For example, Yin et al. present a generative approach that conditions a word-level recurrent neural network (RNN) on facts retrieved from a knowledge base [13]. Others model facts using memory modules, which RNNs attend over [8, 12].

Extractive techniques extract an answer sentence or phrase from a given context. Many recent phrase-level extractive models are built and evaluated on the Stanford Question Answering Dataset (SQuAD) [5]. SQuAD provides human-generated questions along with the context and human-extracted answer spans. Much recent work has used neural networks that first encode the question and the corresponding context with word-level RNNs, followed by attention mechanisms over the context.

G. Kumar (✉) · M. Henderson · S. Chan · H. Nguyen · L. Ngoo
Carousell Pte. Ltd., 240 Tanjong Pagar Rd, Singapore, Singapore
e-mail: girishk@stanford.edu

L. F. D'Haro et al. (eds.), *9th International Workshop on Spoken Dialogue System Technology*, Lecture Notes in Electrical Engineering 579,
https://doi.org/10.1007/978-981-13-9443-0_35

Pointer networks, conditioned on the aligned representations, then point to the start and end indices of the predicted answer span, with respect to the context [6, 11].

Sentence-level extractive methods start by encoding questions and each candidate sentence in the context. Binary classifiers can be trained on the encodings to predict if a sentence answers the question, choosing the candidate sentence with the highest prediction score [1]. Masala et al. use a hinge loss to maximize the similarity between question-answer pairs and minimize otherwise [4]. We focus on sentence-level extractive techniques for their speed, to not greatly degrade chat latency.

Conversational reply methods are of interest since we deal with chats. Approaches include generative *seq2seq* models—using a RNN to encode the previous message and another RNN to generate a response word-by-word [7, 10]. Henderson et al. propose a model to rank the dot product of the previous message vector and the actual response vector favourably against other response vectors in a training batch [2]. For efficiency reasons, the ranking approach greatly appeals to us.

In this paper, we present an extension of Henderson et al.'s dot product model for answer-sentence selection. We explore adding Long Short-Term Memory (LSTM) networks and attention mechanisms to the base model. Finally, we present a demo where a user can ask questions about a product listing and view model predictions.

2 Data

2.1 Data Collection

Carousell is an online used goods marketplace. Potential buyers deal with sellers through chats. A corpus of 36 M of such chats, containing 400 M messages, was available. This corpus is encrypted and engineers can only inspect aggregated statistics across many users. We specifically collect buyer questions with seller replies repeating phrases in the product description. The description sentence with the repeated phrase is taken as the answer sentence. We also store the conversation history, i.e. buyer & seller messages till the seller's answer. 590 K examples are collected (Table 1).

The model also needs to identify whether or not the buyer's message has an answer in the product description. We collected 345 K buyer messages without an answer in the description, with the dataset containing 935 K examples in total.

2.2 Evaluation Metrics

Consider a question, q, and N candidate answer sentences, $(a_1, a_2, \ldots, a_N)$. We add a special token, a_0, to be predicted when there is no suitable answer, and write $\mathbf{A} = (a_0, a_1, \ldots, a_n)$. A question answering model must produce a distribution, P_i,

Table 1 A dataset sample, giving a conversational context, a description whose sentences are treated as candidate answers, and the correct answer. The true next message is used to identify the correct answer. **B** & **S** denote buyer & seller

Chat	Description
B: Can you do delivery? **S**: Yes, delivery is $15 **B**: Great. Is it sturdy? **S**: Yes! It's well built **B**: What colours are there?	This is one of the best cat towers we offer and your cats will love it At 185cm tall, it's a great vertical gym 8 scratch posts ensure healthy nails You've a choice of two colours We sell it in cream-white or black
Next message	Answer sentence
We have cream-white or black	We sell it in cream-white or black

over the a_i, where $P_i = P(a_i \mid \mathbf{A}, q)$ and $\sum_{i=0}^{n} P_i = 1$. Given P_i, and the labelled index of the correct answer sentence, k, we compute accuracy as follows. Note that $k = 0$ when there is no suitable answer.

$$\text{Accuracy} = \begin{cases} 1 & \text{if } \underset{i}{\operatorname{argmax}}\ P_i = k \\ 0 & \text{otherwise} \end{cases} \tag{1}$$

We compute the *Overall Accuracy* averaged over all the test samples. While our formulation requires the model to decide if q has an answer and predict an answer sentence jointly, it is possible to train separate models for each task. Therefore, we calculate, separately, the *Positive Accuracy* over test samples with $k > 0$. To evaluate how well the model predicts if there is an answer, we present a triggering accuracy:

$$\text{Trigger Accuracy} = \begin{cases} 1 & \text{if } \operatorname{sgn}(\underset{i}{\operatorname{argmax}}\ P_i) = \operatorname{sgn}(k) \\ 0 & \text{otherwise} \end{cases} \tag{2}$$

3 Model Architecture

In this section, we describe an approach to obtain $P_i = P(a_i \mid \mathbf{A}, q)$. As in [2], we estimate P_i using the dot product of two neural network functions, $\mathbf{h}(\cdot)$, $\mathbf{g}(\cdot)$:

$$P_i = P(a_i \mid \mathbf{A}, q) \approx \frac{e^{\mathbf{h}(q)^{\mathrm{T}}\mathbf{g}(a_i,\ \mathbf{A},\ q)}}{\sum_{j=0}^{N} e^{\mathbf{h}(q)^{\mathrm{T}}\mathbf{g}(a_j,\ \mathbf{A},\ q)}} \tag{3}$$

The softmax function ensures $\sum_{i=0}^{N} P_i = 1$. Separating the model into two networks allows the network to run efficiently on varying sizes of $\mathbf{A}$.

3.1 N-Gram Representation

The $\mathbf{h}(\cdot)$ and $\mathbf{g}(\cdot)$ sub-networks start by extracting n-gram features from q and $\mathbf{A}$. Embeddings are learnt for each n-gram during training. For each question and answer sentence, the embeddings of their n-grams are summed. We denote this representation as $\psi(\cdot) \in \mathbb{R}^d$. A total of 100 K unigrams and 200 K bigrams were extracted from the full conversation corpus.

3.2 Encoding Techniques

Our baseline model encodes $\psi(q)$ and $\psi(a_i)$ for $i = 1, \ldots, n$, through feed-forward neural network layers. Other encoding techniques can also be introduced before the feed-forward layers, as in Fig. 1.

LSTM Layer

The list of candidate answers $\mathbf{A}$ is an ordered sequence of sentences coming from the product description. Therefore, an LSTM could help add sequential information as context to the answer sentence representations [3]. We specifically use a bi-directional LSTM and set the initial hidden state to be $\psi(q)$. This allows the question embedding to influence the answer sentence representations.

Conversational Context

Questions in our dataset occur in the context of chats. Hence, the model could benefit from contextual information obtained from the messages before the ques-

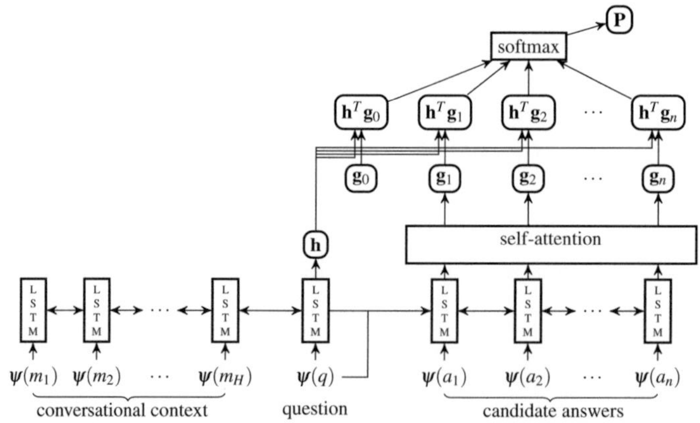

Fig. 1 Model architecture with all proposed encoding techniques. Here, $\mathbf{h} = \mathbf{h}(q)$ and $\mathbf{g}_i = \mathbf{g}(a_i, \mathbf{A}, q)$. The network is further conditioned on the conversational context

tion, $\mathbf{M} = \{m_1, m_2, \ldots, m_H\}$. We use an LSTM to model the embedded messages, taking the final LSTM hidden state as the question encoding [3].

Attention Layer

We use the feed-forward self-attention layer implemented in Tensor2Tensor [9] to further enrich the representations of the candidate answers.

3.3 *Conversational Pre-training*

The full corpus of 36 M conversations, containing 400 M messages, was used to pre-train the model on the reply suggestion task [2]. The setup is similar to the baseline feed-forward model, except that the sentences to be ranked include the actual reply and messages randomly sampled from the corpus. The reply suggestion model achieved an *Overall Accuracy* of 41.5% on the answer sentence selection task.

4 Evaluation Results

Table 2 presents results for different model variations. Recall, from Sect. 2.2, that overall accuracy is computed over all test samples while positive accuracy is computed over only samples with answers. Triggering accuracy evaluates how well the model predicts if there is an answer. All models contain 2 feed-forward or attention layers of size 500 for the baseline and pre-trained models, 128 otherwise to reduce overfitting. N-gram embeddings and LSTM layers were of size 256. For the conversational context model, the message history is at most $H = 10$ messages. Otherwise, the final 2 buyer messages are concatenated to form the question, q. The train-test split was 90–10.

Pre-training on the full conversation corpus resulted in the greatest improvement, establishing the value of general conversational information. The improvement obtained from adding LSTMs over the candidate answers suggests that contextual

Table 2 Evaluation metrics for the different models

	Overall accuracy	Positive accuracy	Trigger accuracy
Baseline	0.592	0.528	0.731
+ Pretraining	0.678	0.604	0.783
+ LSTM layer	0.688	0.618	0.786
+ Attention	0.694	0.620	0.788
+ Conv. context	0.710	0.624	0.804

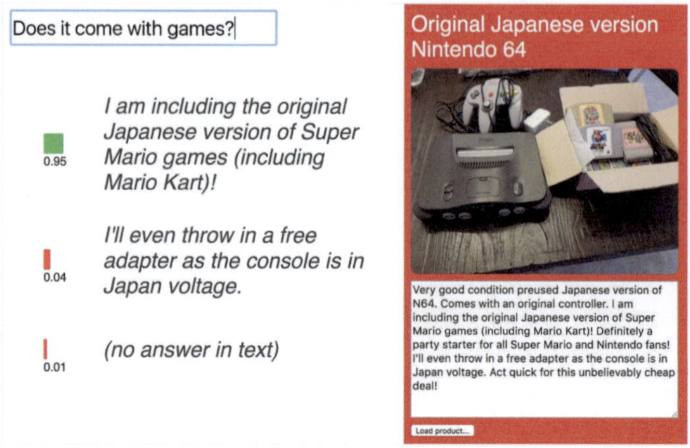

Fig. 2 Answer selection demo UI

information improves the answer sentence representations. Modelling the dialog state is also useful as evidenced by the improvement from using conversational context.

The intended use of this model is to present selected answer sentences as reply suggestions to sellers on Carousell. For research purposes, we built a demo to facilitate exploring model predictions that will be presented at the International Workshop on Spoken Dialog System Technology, 2018. The demo can import live product listings from Carousell. When given a question, the demo presents sentences from the description that the model ranks as the best answers, including the likelihood of no answer being present (see Fig. 2). The demo uses the best performing model trained on the final 2 buyer messages, i.e. Pre-training + LSTM + Attention. The conversational context model was not chosen for usability, as otherwise users would have to craft full conversations.

5 Conclusion

This paper has tackled the problem of answering buyer questions in online marketplaces using seller-crafted product descriptions. We first presented a neural-network ranking model for selecting sentences as answers. The introduction of a special no-answer token allowed the model to jointly decide whether an answer is present and to identify it if so. Multiple encoding techniques and a pre-training strategy were presented and evaluated. Finally, a demo was built to inspect model behaviour when answering questions about live products on the Carousell platform. The model performance was deemed good enough to be launched to power reply suggestions in Carousell. This means we can further fine-tune the model based on live user actions.

Future work could explore phrase-based question-answering methods to select more precise answers. It may also be interesting to study rephrasing the selected answer sentence to better fit the conversational context. Features like the listing title and product images could be introduced. We could also extend the system to answer questions without directly quotable answers in the description, e.g. yes/no questions.

References

1. Guo J, Yue B, Xu G, Yang Z, Wei JM (2017) An enhanced convolutional neural network model for answer selection. In: Proceedings of the 26th international conference on world wide web companion, pp 789–790
2. Henderson M, Al-Rfou R, Strope B, Sung YH, Lukacs L, Guo R, Kumar S, Miklos B, Kurzweil R (2017) Efficient natural language response suggestion for smart reply. arXiv:1705.00652
3. Hochreiter S, Schmidhuber J (1997) Long short-term memory. Neural Comput 9(8):1735–1780
4. Masala M, Ruseti S, Rebedea T (2017) Sentence selection with neural networks using string kernels. Procedia Comput Sci 112(C):1774–1782
5. Rajpurkar P, Zhang J, Lopyrev K, Liang P (2016) Squad: 100,000+ questions for machine comprehension of text. arXiv:1606.05250
6. Seo M, Kembhavi A, Farhadi A, Hajishirzi H (2016) Bidirectional attention flow for machine comprehension. arXiv:1611.01603
7. Shang L, Lu Z, Li H (2015) Neural responding machine for short-text conversation. arXiv:1503.02364
8. Sukhbaatar S, Weston J, Fergus R et al (2015) End-to-end memory networks. In: Advances in neural information processing systems, pp 2440–2448
9. Vaswani A, Shazeer N, Parmar N, Uszkoreit J, Jones L, Gomez AN, Kaiser L, Polosukhin I (2017) Attention is all you need, NIPS
10. Vinyals O, Le Q (2015) A neural conversational model. ICML
11. Wang S, Jiang J (2016) Machine comprehension using match-lstm and answer pointer. arXiv:1608.07905
12. Weston J, Chopra S, Bordes A (2014) Memory networks. ICLR
13. Yin J, Jiang X, Lu Z, Shang L, Li H, Li X (2016) Neural generative question answering. IJCAI

A Multimodal Dialogue Framework for Cloud-Based Companion Systems

Matthias Kraus, Marvin Schiller, Gregor Behnke, Pascal Bercher, Susanne Biundo, Birte Glimm and Wolfgang Minker

Abstract *Companion systems* are cooperative, cognitive systems aiming at assisting a user in everyday situations. Therefore, these systems require a high level of availability. One option to meet this requirement is to use a web-deployable architecture. In this demo paper, we present a multimodal cloud-based dialogue framework for the development of a distributed, web-based companion system. The proposed framework is intended to provide an efficient, easily extensible, and scalable approach for these kinds of systems and will be demonstrated in a do-it-yourself assistance scenario.

1 Introduction

Nowadays we expect technological devices to be intelligent and to adapt their functionalities to our individual needs. *Companion Technology* fulfils these requirements [1]. A *companion system* is understood as a cognitive system which models its

M. Kraus (✉) · W. Minker
Institute of Communications Engineering, Ulm University, Ulm, Germany
e-mail: matthias.kraus@uni-ulm.de

W. Minker
e-mail: wolfgang.minker@uni-ulm.de

M. Schiller · G. Behnke · P. Bercher · S. Biundo · B. Glimm
Institute of Artificial Intelligence, Ulm University, Ulm, Germany
e-mail: marvin.schiller@alumni.uni-ulm.de

G. Behnke
e-mail: gregor.behnke@uni-ulm.de

P. Bercher
e-mail: pascal.bercher@uni-ulm.de

S. Biundo
e-mail: susanne.biundo@uni-ulm.de

B. Glimm
e-mail: birte.glimm@uni-ulm.de

L. F. D'Haro et al. (eds.), *9th International Workshop on Spoken Dialogue System Technology*, Lecture Notes in Electrical Engineering 579,
https://doi.org/10.1007/978-981-13-9443-0_36

behaviour according to the individual user's needs, preferences, capabilities, and also takes into account the current situation. Therefore, it requires cognitive abilities like planning, reasoning, and the ability to conduct a dialogue. By providing a high level of cooperation capability and reliability, even in complex tasks, companion systems aim at being an everyday competent and trustworthy partner. For this purpose, those systems need to be always available and be accessible from everywhere. Hence, the exploitation of distributed, scalable, and most importantly web-based interaction design concepts has to be considered. In literature there are numerous examples for the development and implementation of such systems. For example, Fuchs et al. [2] describe a scalable architecture for web-deployment of spoken dialogue systems. The integrated dialogue service relies on grammar-based language processing and rule-based dialogue management combining the Open Source Regulus Platform [3] and the Information State approach by Larsson and Traum [4]. In order to connect user input to the respective dialogue service, a speech router relying on a session ID is used. Ramanarayanan et al. [5] present a cloud-based dialogue system relying on VoiceXML. In order to manage the dialogue, the open-source dialogue-authoring tool OpenVXML (https://github.com/OpenMethods/OpenVXML) is applied. This framework allows to specify dialogue workflows using a graphical user interface. Zhao et al. [6] present a multimodal web-based dialogue framework allowing for multi-agent conversation with real users. The proposed framework (DialPort) serves as a platform for the interaction with various remote resources, such as external end-to-end dialogue systems or knowledge bases. For dialogue management they implement a dialogue engine relying on a hierarchical policy in combination with a "Knowledge Ontology" for encoding domain knowledge.

Contrary to the approaches described above, which mostly aim at providing web-services for hosting various dialogue systems and knowledge APIs, we developed a modular and easily extensible cloud-based framework for integrating cognitive capabilities, like Artificial Intelligence (AI) planning. The design of the web-service is structured in the style of the REST (REpresentational State Transfer) paradigm, but not to a full extent, as the REST architectural constraints demand the server to be stateless. Due to the purpose of a dialogue manager to maintain dialogue state and history, our design could not comply with this limitation. In the following, the application scenario for demonstration of a first cloud-based protoype is described. Furthermore, we will provide insight into the underlying architecture and the functionality of the proposed system.

2 Application Scenario

Our current work addresses the implementation of a companion system in the do-it-yourself (DIY) home improvement domain. The system is intended to assist novice users in the performance of home improvement projects that require knowledge in the use of power tools (electric drills, saws, etc.). For this reason, the system provides support for the user while performing the task in the form of instructions as step-by-step assistance and in the form of assistance on demand, in case the user asks for specific support. By doing so, the user gains experience in the usage of individual power tools and should be encouraged to employ them in different projects.

3 Cloud-Based Architectural Concept

In order to make our *companion system* web-deployable, a client/server approach is used. On the client side the user interacts with the system through an interface. In this work, we implement a multimodal graphical interface, which is designed as a web service and applicable from every web browser. Therefore, the javascript-based framework *Vue.js* (https://vuejs.org) is used, supporting three different input modalities: speech, text, and touch input. Due to the need of handsfree communication in complex task scenarios, speech input serves as a primary modality. For Automatic Speech Recognition (ASR) we use the Google Chrome web browser's own speech-to-text service in combination with the annyang.js (https://www.talater.com/annyang) javascript library. The *annyang.js* library enables the system to allow for continuous speech detection while avoiding the processing of off-topic talk by only transmitting content to the dialogue system after a certain keyword has been spotted, similarly to the activation of current personal assistants. Dialogue output is presented depending on the user's request. As a response to planning requests, a sequence of plan slides using the *vue-slick.js* library is visualised. Explanation and background information are conveyed in the form of text and synthesized speech, also using Chrome's speech service.

The entry to the cloud-based companion system forms a HTTP API Server that handles the CRUD requests from the interface server and forwards the conveyed user input to the dialogue component. For each user of the system, there exists a resource on the HTTP API Server in the form of *example.com/companion/user/{id}*, each containing an individual instance of the dialogue manager. Hence, a user needs to authenticate himself before having access to the system. For this we make use of a token-based authentication approach (https://en.wikipedia.org/wiki/Accesstoken). This allows for handling multiple, parallel user dialogues.

For semantic analysis of the user requests a statistical-driven Natural Language Understanding (NLU) module is applied. In our architecture, Microsoft's cloud-based *Language Understanding Intelligent Service* (LUIS) [7] is used, as it provides a fast and comfortable way of generating machine-learning-powered NLU models which can easily be extended. In order to determine if the user input is addressing the planning framework, the ontology or the dialogue itself, we develop an individual NLU model for each of them. In doing so, we leverage the full power of statistical NLU that facilitates controlling the dialogue flow and creates an efficient way for interaction processing. When new user input arrives it is forwarded to all three models. After the individual LUIS models have analysed the user input, the response from the model with the highest confidence is selected and transformed into an abstract user act and forwarded to the Dialogue Manager (DM).

The task-independent dialogue management module mediates the interaction between user and the artificial intelligence components of the system and keeps track of the dialogue's state and history in order to provide context-related system output. The current implementation of the DM is based on a version of the classical Information State approach by Larsson and Traum [4]. This approach relies on the three concepts *dialogue move* (user input), *information state* (dialogue state), and *dialogue action* (system output). After each user act the dialogue state (plan- and dialogue-related context information, e.g. sequence of plan actions, previous user input) is updated and a transformed act is forwarded to the respective AI component (planning framework, ontology) of the system. Dialogue-related user acts, like affirmation of information, are directly processed in the DM.

For equipping the system with more intelligence and competence, a planning component provides procedural knowledge, i.e. it generates a valid plan, which consists of a sequence of actions a user has to take in order to reach his goal. The problem was modeled in a hybrid planning formalism [8], which allows causal reasoning with both primitive and abstract tasks. For solving these problems, the problem first gets translated into an HTN problem [9], and finally into SAT to be solved efficiently [10]. Furthermore, we make use of an ontology containing domain-specific information which hence is capable of providing declarative background knowledge to the user. It is formulated in the web ontology language OWL (https://www.w3.org/OWL). For a more detailed description of the ontology used in this work and its relation to the planning module see Schiller et al. [11]. Natural Language Generation (NLG) makes use of texts obtained from the ontology, including instructions for each plan action and descriptions and explanations for concepts in the DIY domain generated using text patterns [12].

The response of the queried module is transformed into an abstract system output and returned to the client, where the interface interprets the message and presents the content in visual and/or spoken form. In Fig. 1 the structure of the system is depicted. The interaction between user and system is based on HTTP messages using the JSON format for data exchange. The communication of the individual components of the cloud-service is handled via socket connections, while also using the JSON data format. In the following, we take a closer look on interaction design.

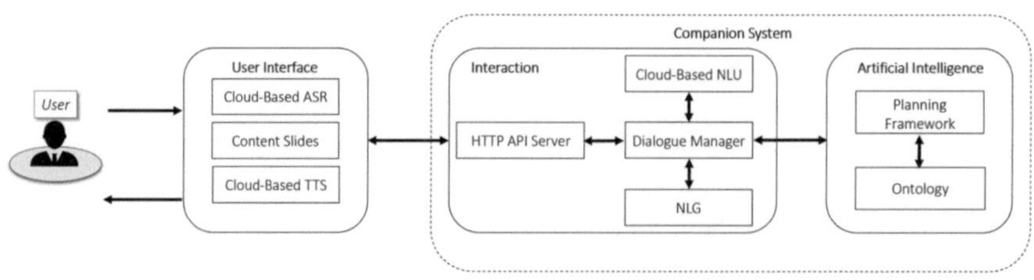

Fig. 1 Modular architecture of the cloud-based companion system. The user interacts with the system through a multimodal interface. As controlling entity of the architecture serves a dialogue manager interacting with the modules for planning and the ontology

4 Demo Interaction Design

For demonstration, we will show how the user is accompanied starting with planning a DIY project up to its completion. Each interaction starts with the user stating a planning objective, e.g. "I want to build a keyrack.". The purpose of this first dialogue is to define a task which forms a valid planning problem for the planner to solve. Once the planner has found a solution, the result is passed back to the DM. As the plan is represented as a sequence of actions with no further information besides the name of the task and the involved materials and tools, the ontology is accessed for more information on each action. This information comprises instruction texts and references to multimedia content (image, video) visualising the content. Afterwards, the enriched plan is presented to the user as a sequence of slides assisting in the performance of the task. Throughout the interaction with the system, the user may ask for background information and explanations regarding certain concepts available in the ontology. This functionality intends to provide the user with further assistance. The current implementation of the system is able to handle three different kinds of conceptual knowledge-based requests: encyclopedic requests, media request, and availability requests. In our application scenario, the user may request further information on DIY items, such as materials and tools. Encyclopedic requests concern the appearance or the purpose of a material or tool (e.g. to tell the user about the properties of a particular class of tool). For receiving an image or a video of a DIY item, a media request can be used. In order to check the availability of certain items, an availability request can be posed.

5 Conclusion

We have implemented a multimodal dialogue framework for the integration of AI components within a cloud-based architecture. An evaluation of the system was done by Behnke et al. [13]. In the next iteration of the proposed *companion system*, we aim at including proactive situation-dependent support based on user monitoring.

Acknowledgements This work is funded by the German Research Foundation (DFG) within the technology transfer project "Do it yourself, but not alone: Companion Technology for Home Improvement" of the Transregional Collaborative Research Centre SFB/TRR 62. The industrial project partner is the Corporate Research Sector of the Robert Bosch GmbH.

References

1. Biundo S, Höller D, Schattenberg B, Bercher P (2016) Companion-technology: an overview. Künstliche Intelligenz 30(1):11–20
2. Fuchs M, Tsourakis N, Rayner E (2012) A scalable architecture for web deployment of spoken dialogue systems. In: Proceedings of the 8th international conference on language resources and evaluation (LREC). ELRA, pp 1309–1314

3. Rayner M, Hockey BA, Bouillon P (2006) Putting linguistics into speech recognition: the regulus grammar compiler. Center for the Study of Language and Information
4. Larsson S, Traum DR (2000) Information state and dialogue management in the TRINDI dialogue move engine toolkit. Nat Lang Eng 6(3–4):323–340
5. Ramanarayanan V, Suendermann-Oeft D, Ivanov AV, Evanini K (2015) A distributed cloud-based dialog system for conversational application development. In: Proceedings of the 16th annual meeting of the special interest group on discourse and dialogue (SIGDIAL). Association for Computational Linguistics, pp 432–434
6. Zhao T, Lee K, Eskenazi M (2016) DialPort: connecting the spoken dialog research community to real user data. In: Spoken language technology workshop (SLT). IEEE, pp 83–90
7. Williams JD, Kamal E, Ashour M, Amr H, Miller J, Zweig G (2015) Fast and easy language understanding for dialog systems with microsoft language understanding intelligent service (LUIS). In: Proceedings of the 16th annual meeting of the special interest group on discourse and dialogue (SIGDIAL). Association for Computational Linguistics, pp 159–161
8. Biundo S, Schattenberg B (2001) From abstract crisis to concrete relief—a preliminary report on combining state abstraction and HTN planning. In: Proceedings of the 6th European conference on planning (ECP). AAAI Press, pp 157–168
9. Geier T, Bercher P (2011) On the decidability of HTN planning with task insertion. In: Proceedings of the 22nd international joint conference on AI (IJCAI). AAAI Press, pp 1955–1961
10. Behnke G, Höller D, Biundo S (2018) totSAT—Totally-ordered hierarchical planning through SAT. In: Proceedings of the 32th AAAI conference on AI (AAAI). AAAI Press
11. Schiller M, Behnke G, Schmautz M, Bercher P, Kraus M, Dorna M, Glimm B, Minker W, Dorna M, Dambier M, Biundo S (2017) A paradigm for coupling procedural and conceptual knowledge in companion systems. In: Proceedings of the 2nd international conference on companion technology (ICCT). IEEE, pp 1–6
12. Schiller M, Schiller F, Glimm B (2017) Testing the adequacy of automated explanations of el subsumptions. In: Proceedings of the 30th international workshop on description logics (DL), CEUR workshop proceedings, vol 1879
13. Behnke G, Schiller M, Kraus M, Bercher P, Schmautz M, Dorna M, Glimm B, Minker W, Dambier M, Biundo S (2019) Alice in DIY wonderland or: instructing novice users on how to use tools in DIY projects. AI Commun (Preprint), 1–27

CityTalk: Robots That Talk to Tourists and Can Switch Domains During the Dialogue

Graham Wilcock

Abstract The paper describes CityTalk, a spoken dialogue system for Nao robots that provide information for tourists about local hotels, shops and restaurants. An important feature is that the robots switch domains smoothly during the dialogue if the user asks about a new domain, for example talking about hotels using a hotel database and then talking about restaurants using a restaurant database if the user asks about restaurants. CityTalk is the first robot dialogue system to use PyDial, a research toolkit for statistical dialogue systems. A topic tracker detects from user utterances when a domain switch is required. Clarification questions appropriate for different domains are based on domain-specific ontologies. CityTalk includes new domains including Tokyo Waterfront restaurants and hotels.

1 Introduction

In many shopping centres in Japan, robots now greet customers and can provide information about local shops, restaurants and other facilities. Currently the Pepper robot from SoftBank Robotics is widely used. After being welcomed by the robot's polite *Irasshaimase!* greeting, customers select a topic from a touch-screen menu on the tablet computer that all Pepper robots carry around their neck. The robot then performs interaction for the selected topic task, for example giving information or playing a spoken quiz game such as Guess the Animal. When the topic is completed, users can select another topic from the touch-screen menu.

This paper describes CityTalk, a spoken dialogue system for Nao robots (also from SoftBank Robotics) that talk with tourists about local hotels, restaurants, and other places of interest. An important feature is that the robots can switch domains smoothly during the dialogue if the user asks about something else, without needing a touch-screen menu like the current Pepper robots.

G. Wilcock (✉)
CDM Interact, Helsinki, Finland
e-mail: graham.wilcock@cdminteract.com

L. F. D'Haro et al. (eds.), *9th International Workshop on Spoken Dialogue System Technology*, Lecture Notes in Electrical Engineering 579,
https://doi.org/10.1007/978-981-13-9443-0_37

The paper is structured as follows. After briefly reviewing related work on multi-domain systems, we describe the CityTalk robot dialogue system and the PyDial research toolkit that it uses. We then show example interactions that include domain-switching, and end with an example of an extended (10 min) interaction.

2 Related Work on Multi-domain Systems

A bottleneck for dialogue systems has been the amount of data needed to manage flexible interaction, and dialogues have usually been restricted to separate closed-topic domains such as bus timetables or pizza ordering. The required knowledge is included in the system and manually structured for the purposes of the task and interaction goals. Changing to another topic is difficult, requiring either a completely new system or an extended knowledge base that includes several topics.

Many single-domain knowledge bases have been developed for various dialogue systems. However, as they were developed by different people at different locations for different purposes, their knowledge representations and structures are different. The challenge is how to enable dialogue systems to deploy existing single-domain rules or databases and smoothly switch between different topics.

Pioneering work on this challenge was done by Komatani et al. [1] using an architecture for a multi-domain system composed of several single-domain systems plus a central module. Multiple sets of rules are kept inside the distinct single-domain systems, called *domain experts*. The central module performs speech processing as well as domain selection, and passes the speech recognition results to all the domain experts. Each expert executes its own language understanding module, updates its own dialogue states based on the given speech recognition result, and returns the information required for domain selection to the central module. Based on this information, the central module selects an expert for giving the response.

For dialogue systems to be considered *open-domain* they need to be based on interactions where the user's changing interests drive the system from one topic to another. Such systems can be based on one large knowledge base containing all the possible topics that the system can talk about, such as Wikipedia, or on a large corpus of chat logs from which suitable responses can be automatically learnt. An example of the first type is WikiTalk [2, 3] in which the robot talks fluently in several languages about an unlimited range of topics using information from Wikipedia. An example of the second type is a Japanese chat system by Higashinaka et al. [4] based on a large corpus of Twitter sentences. Recently intelligent speakers such as Google Home and Amazon Alexa have combined chatting and information access.

3 The CityTalk Robot Dialogue System

CityTalk integrates the speech processing modules of the robot with the dialogue processing modules of PyDial [5], a statistical dialogue system toolkit developed by Cambridge University Engineering Department (CUED). PyDial aims to facilitate the development of statistical dialogue systems that use machine learning and deep learning techniques. The architecture of PyDial is shown in Fig. 1.

PyDial Agent includes a dialogue pipeline with four modules: Semantic Parser performs natural language understanding to identify the user's dialogue act, Belief Tracker updates the dialogue state, Policy module decides the dialogue act for the system's response, and Semantic Output performs natural language generation. The four modules define base classes which can have domain-specific instances.

To support multi-domain dialogues, PyDial uses a set of ontologies and databases which are domain-specific, and PyDial Agent also includes a Topic Tracker module. Instead of switching between different dialogue engines as in [1], the ontologies assist the same dialogue engine to deal with different domains. When Topic Tracker detects from the user's utterances that a new domain is required, PyDial Agent loads domain-specific instances of the classes defined by the four dialogue modules.

Speech processing is not part of PyDial, which is text-based. PyDial expects ASR and TTS to be performed by an external cloud-based speech client connected to its Dialogue Server module via an HTTP/JSON interface (Fig. 1). However, CityTalk does not follow this approach as the open-vocabulary speech recognition offered by cloud speech services has not yet reached a satisfactory level for natural spoken interaction in real-world environments. Instead, following an approach that was successful in WikiTalk [2, 3], the robot's own speech components are used with multiple

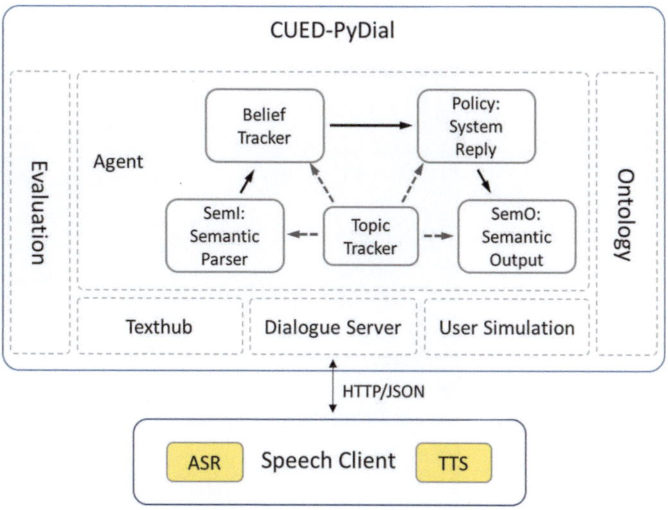

Fig. 1 The architecture of the PyDial statistical dialogue toolkit

closed speech recognition vocabularies that are switched dynamically to match the changing domains in the dialogue.

This is possible because the speech recognition module used by Nao and Pepper robots allows the recognition vocabulary to be modified dynamically, unlike most current open-vocabulary cloud speech services. When Topic Tracker detects that the user wants to switch domains, the recognition vocabulary is modified on-the-fly for the new domain. Since CityTalk detects when the user switches domains without needing to use touch-screen menus to signal the switch, we can use Nao robots which do not carry a touch-screen tablet around their neck like Pepper.

4 Example Interactions

The videos in Figs. 2 and 3 show example interactions. The first example uses the Cambridge domains that are provided with the PyDial toolkit. The user starts by asking about hotels and the robot suggests a hotel using the Cambridge hotels database. When the user asks about restaurants the robot switches domains and gives information about two restaurants using the Cambridge restaurants database.

New domains are added to the system by providing a database and an ontology for each domain. The video in Fig. 3 uses new domains about Tokyo hotels and restaurants developed by CDM Interact (www.cdminteract.com). A full transcript of the interaction in this short (2 min) video is given in Table 1. The robot gives information about restaurants in Tokyo Waterfront until the user says *What about hotels?* then it switches to the Tokyo hotels domain.

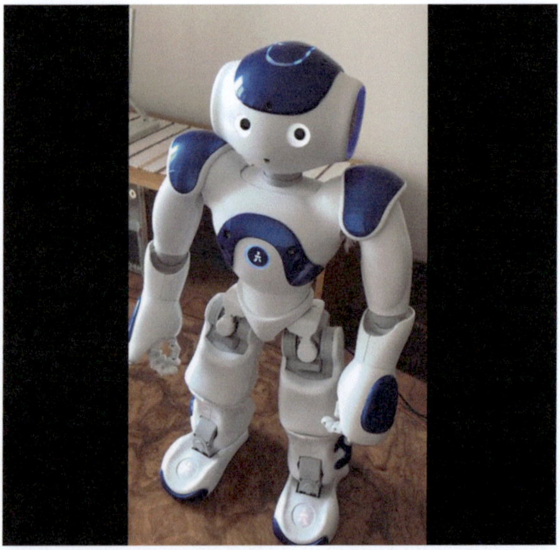

Fig. 2 An example interaction with Cambridge hotel and restaurant domains. The video can be viewed at https://drive.google.com/open?id=0B-D1kVqPMlKdZ3VyYUpjR3llaEU

Fig. 3 An example interaction with new Tokyo hotel and restaurant domains. The video can be viewed at https://drive.google.com/open?id=1qDThFaO9bOoDFO3UHfKeRlW9DaOGqGtK

The robot begins with the same question *What part of town do you have in mind?* for both the hotel and restaurant domains. Other robot clarification questions are domain-specific, for example *Would you like the place to have parking?* for the hotel domain, and *What kind of food would you like?* for the restaurant domain. The ontologies and domain-specific neural models or generation templates enable the robots to ask clarification questions that are appropriate for each domain.

The robot's utterances are often similar but usually not repetitions. For example when apologising, and listing the user's preferences to clarify why that combination is not available, the list ordering is generated independently in each case.

5 An Extended Interaction

A demo of an extended interaction with CityTalk is described by [6]. The interaction lasts 10 min and the video can be viewed at https://youtu.be/OhjIJp8XBEA.

In the first part of the video the robot gives price information about hotels in the Tokyo Waterfront area in three different price ranges. After switching domains to restaurants, the user and the robot have a prolonged interaction in the second part of the video about numerous restaurants in the Tokyo Waterfront area that serve different types of food and are in different price ranges.

This prolonged interaction includes examples of speech recognition errors and of switching domains incorrectly, but the dialogue does not break down. During the course of the dialogue the robot provides descriptions, addresses and phone numbers for all the different types of restaurants that the user is interested in.

Table 1 Transcript of the video in Fig. 3 with the new Tokyo domains

Robot:	*Hello. I can tell you about hotels and restaurants. How may I help you?*
User:	I'd like to know about restaurants
Robot:	*What part of town do you have in mind?*
User:	In Tokyo Waterfront, please
Robot:	*What kind of food would you like?*
User:	How about Italian?
Robot:	*Would you like something in the cheap, moderate, or expensive price range?*
User:	Ah, expensive, please
Robot:	*Would you like something in the cheap, moderate, or expensive price range?*
User:	Are there any expensive restaurants in Tokyo Waterfront?
Robot:	*Would you like something in the cheap, moderate, or expensive price range?*
User:	Expensive!
Robot:	*I am sorry but there is no place, serving Italian food and in the expensive price range*
User:	Oh. What about moderate?
Robot:	*I am sorry but there is no place in the expensive price range and serving Italian food*
User:	Right... Moderate? How about moderate?
Robot:	*Cafe La Boheme is a nice place. It is in the moderate price range and it serves Italian food*
User:	Oh, that's nice, OK. Erm, what about ...
Robot:	*Cafe La Boheme is a nice place. Their address is Aquacity Odaiba 4F, 1–9-1 Daiba, Minato-ku, Tokyo*
User:	Oh, thank you! Thank you. Hmm, what about ...
Robot:	*Cafe La Boheme is a nice place. It serves Italian food and it is in the moderate price range*
User:	What about hotels?
Robot:	*What part of town do you have in mind?*
User:	Again, Water..., Tokyo Waterfront
Robot:	*Would you like the place to have parking?*
User:	I don't mind, really
Robot:	*Would you like the place to have parking?*
User:	I don't mind.
Robot:	*Would you like a hotel in the cheap, moderate, or expensive price range?*
User:	Ah, I'd like an expensive hotel

Acknowledgements The PyDial statistical dialogue system toolkit was developed by Cambridge University Engineering Department. We thank the Dialogue Systems Group for making it freely available. We also thank Kristiina Jokinen of AI Research Center, AIST Tokyo Waterfront, Japan for participating in the video in Fig. 3 and for valuable discussions. CityTalk was developed by CDM Interact, Helsinki, Finland (www.cdminteract.com).

References

1. Komatani K, Kanda N, Nakano M, Nakadai K, Tsujino H, Ogata T, Okuno HG (2006) Multi-domain spoken dialogue system with extensibility and robustness against speech recognition errors. In: Proceedings of the 7th SIGdial workshop on discourse and dialogue, SigDIAL 2006, pp 9–17
2. Jokinen K, Wilcock G (2014) Multimodal open-domain conversations with the Nao robot. In: Mariani J, Rosset S, Garnier-Rizet M, Devillers L (eds) Natural interaction with robots, knowbots and smartphones: putting spoken dialogue systems into practice. Springer, Berlin, pp 213–224
3. Wilcock G (2012) WikiTalk: a spoken wikipedia-based open-domain knowledge access system. In: Proceedings of the COLING 2012 workshop on question answering for complex domains. Association for Computational Linguistics, Mumbai, India, pp 57–69
4. Higashinaka R, Kobayashi N, Hirano T, Miyazaki C, Meguro T, Makino T, Matsuo Y (2016) Syntactic filtering and content-based retrieval of Twitter sentences for the generation of system utterances in dialogue systems. In: Rudnicky A, Raux A, Lane I, Misu T (eds) Situated dialog in speech-based human-computer interaction. Springer, Berlin, pp 15–26
5. Ultes S, Rojas Barahona LM, Su PH, Vandyke D, Kim D, Casanueva I, Budzianowski P, Mrkšić N, Wen TH, Gasic M, Young S (2017) PyDial: a multi-domain statistical dialogue system toolkit. In: Proceedings of ACL 2017, system demonstrations. Association for Computational Linguistics, Vancouver, Canada, pp 73–78
6. Wilcock G (2018) Using a deep learning dialogue research toolkit in a multilingual multidomain practical application. In: Proceedings of the twenty-seventh international joint conference on artificial intelligence (IJCAI 2018). Stockholm, pp 5880–5882

Author Index

L. F. D'Haro et al. (eds.), *9th International Workshop on Spoken Dialogue System Technology*, Lecture Notes in Electrical Engineering 579,
https://doi.org/10.1007/978-981-13-9443-0

MIX
Papier aus verantwortungsvollen Quellen
Paper from responsible sources
FSC® C105338

If you have any concerns about our products,
you can contact us on
ProductSafety@springernature.com

In case Publisher is established outside the EU,
the EU authorized representative is:
Springer Nature Customer Service Center GmbH
Europaplatz 3, 69115 Heidelberg, Germany

Printed by Libri Plureos GmbH
in Hamburg, Germany